B. G. Mirkin   S. N. Rodin

Graphs and Genes

With 46 Figures

Translated from the Russian by H. Lynn Beus

Springer-Verlag
Berlin Heidelberg New York Tokyo 1984

Boris G. Mirkin
Central Economic-Math. Institute
of the USSR
Krasikov str. 32
117 418 Moscow, USSR

Sergey N. Rodin
Institute of Cytology and Genetics
630090 Novosibirsk 90, USSR

H. Lynn Beus
TMCB 232, Brigham Young University
Provo, UT 84602, USA

AMS-MOS Classification (1980): 05 C 65, 05 C 75, 92 A 10

ISBN 3-540-12657-0 Springer-Verlag Berlin Heidelberg New York Tokyo
ISBN 0-387-12657-0 Springer-Verlag New York Heidelberg Berlin Tokyo

Library of Congress Cataloging in Publication Data. Mirkin, B. G. (Boris Grigorévich). Graphs and genes. (Biomathematics; v. 11). Translation of: Grafy i geny. Includes indexes. 1. Genetics–Mathematical models. 2. Graph theory. I. Rodin, Sergeĭ Nikolaevich. II. Title. III. Series. QH438.4.M3M5713. 1984. 575.1'01'5115. 83-20339. ISBN 0-387-12657-0 (U.S.)

This work is subject to copyright. All rights are reserved, whether the whole or part of the material is concerned, specifically those of translation, reprinting, re-use of illustrations, broadcasting, reproduction by photocopying machine or similar means, and storage in data banks. Under § 54 of the German Copyright Law where copies are made for other than private use a fee is payable to "Verwertungsgesellschaft Wort", Munich.

© Springer-Verlag Berlin Heidelberg 1984
Printed in Germany

Printing: Beltz Offsetdruck, Hemsbach/Bergstr.
Bookbinding: J. Schäffer OHG, Grünstadt
2141/3140-543210

Biomathematics

Volume 11

Managing Editor
S. A. Levin

Editorial Board
M. Arbib   H. J. Bremermann   J. Cowan   W. M. Hirsch
J. Keller   K. Krickeberg   R. C. Lewontin   R. M. May
L. A. Segel

Editor's Foreword

This book is written by a mathematician and a theoretical biologist who have arrived at a good mutual understanding and a well worked-out common notation. The reader need hardly be convinced of the necessity of such a mutual understanding, not only for the two investigators, but also for the sciences they represent. Like Molière's hero, geneticists are gradually beginning to understand that, unknowingly, they have been speaking in the language of cybernetics. Mathematicians are unexpectedly discovering that many past and present problems and methods of genetics can be naturally formulated in the language of graph theory. In this way a powerful abstract mathematical theory suddenly finds a productive application. Moreover, in its turn, such an application begins to "feed" the mathematical theory by presenting it with a number of new problems. The reader may judge for himself the fruitfulness of such mutual interaction.

At the same time several important circumstances need to be mentioned. The formalization and rigorous formulation given here embraces not only the older problems, known by geneticists for many decades (the construction of genetic maps, the analysis of complementation, etc.), but also comparatively new problems: the construction of partial complementation maps, phylogenetic trees of proteins, etc. Furthermore, the formalization process, of necessity, requires the working out of a rigorous system of genetic understanding, and also the formulation and solution of a number of problems which have not attracted attention in the theory, up to this time: the problem of the uniqueness of genetic mapping, the search for a best approximation to a map, problems of invariants, etc.

It is interesting that genetics, as a field of application for graph theory, turns out to be such a rich one. Genetic methods produce information about genes, proteins and genetic systems in the form of a great variety of partial manifestations obtained through crosses. The theoretic task of reconstructing, from a great mass of data, the condensed image of a system is represented in graph theory either as the problem of representing matrices by graphs or as a problem of approximation.

It is important to emphasize that the methods of graph theory are not only useful in solving many important genetics problems, but in some cases, lead the investigator into the sphere of more fundamental problems, such as determining fine internal gene structure, studying the "semantics" of the genetic language, the genetic organization of ontogenesis, the "nonadaptive" evolution of macromolecules, etc. In this sense the attempt of the authors to unite graph theory with the theories concerning genes and genetic control systems may already be considered successful. We wish them further success in their endeavor.

Prof. V.A. Ratner

Preface to the English Edition

Mathematical genetics has traditionally been concerned with the growth of genetic systems as populations, utilizing as a basis the theory of determinate and stochastic differential equations. However, there exist a number of important areas of genetics which do not lend themselves to this approach. The genetic problems to which this book is devoted -- the analysis of basic principles of organization, functioning and evolution of genetic systems on the molecular level -- are quite adequately handled by another branch of mathematics: graph theory. This is only natural, when one considers the discreteness and quite rigid structural nature of many functions of molecular-genetic control systems.

Since the moment in 1976 when work was completed on the Russian version of the book there have been a number of changes in theories about genes, some of them being very significant. These became possible due to the rapid growth of the technology of direct biochemical "reading" of deoxyribonucleic genetic texts. And since this book is devoted to basic problems in deciphering genetic texts, one might think that its material has become significantly outdated.

But this is not so. The fact is that biochemists are deciphering only the "passive" information contained in DNA. The activation, the allocation of this information according to meaning, can only be achieved (at least, presently) by the application of independent genetic data about the interactions of different variants of some genome fragments or other. But for such an "indirect" approach the language and methods of graph theory turn out to be extremely useful. For this reason the contents of the book are practically invariant under the accumulation of new information about the structure of molecular-genetic systems.

Thus in the preparation of this edition we limited ourselves to the minimal additions and modifications necessary to sharpen positions and results in light of the latest conceptions. The single exception is the material of chapter three, which was reworked and enlarged on the basis of the appearance of recent, more subtle approaches to the analysis of organism evolution at the molecular level (section 3.1.5 was added, and section 3.2.2 was rewritten).

      Novosibirsk   B. G. Mirkin
      14 Jan. 1982   S. N. Rodin

Preface to the Russian Edition

No more surprising to us is the monstrous whale,
Than the detailed parts of a tiny worm or snail.
Great in heaven is the creator of all these things.
His greatness in earth the tiniest of all these sings.

-- M. V. Lomonosov
Epistle on the use of glass

As is well known, the theory of graphs provides a universal language for representing the structural organization of systems of arbitrary nature. Its use in describing the organization of genetic systems is of special interest, since their functioning is fundamentally determined by their structure.

In the last ten to twenty years the growth of molecular biology has elicited the fact that all basic characteristics of living systems are determined, finally, by linearly-written, discrete, hereditary information contained in the special class of macromolecules called DNA, and by a complex system for reading out this information and recoding it in the language of protein macromolecules, the activity of these being determined by their spatial structure [31]. It is quite natural to use the language of graph theory in studying both the fundamental linear genetic texts and the three-dimensional protein structures which implement DNA instructions. It is especially important in cases when direct biophysical investigations become difficult or impossible, since genetic methods based on the crossing of mutant forms have high resolving ability (at least for microorganisms). Genetic methods provide only indirect information, with graph theory serving as an instrument to draw from such information a "portrait" of the underlying genetic systems.

Recently, both in the literature of mathematics and genetics, articles along this line of thought have appeared more and more frequently. This monograph attempts (through graph theory) to systematize results obtained from the study of the structure, functions and evolution of genetic systems.

Genetic problems have led to new ideas and new problems in graph theory. In turn, mathematical methods have led to more precise descriptions of these same genetic objects and, as a result, to a more advanced understanding of how molecular-genetic control systems function. We emphasize that although independent results, both about graphs and about genes are presented in the book, the essence of our work lies in their union -- in demonstrating the fruitful applications of these mathematical methods in the investigation of specific genetic systems.

There are two stages in the application of mathematical methods to real data. The first consists in forming a mathematical model of the phenomenon and investigating its formal character. As a rule, experimental data are not represented accurately enough by the formal model, since it is impossible to cover all particular real situations. From this comes the necessity of a second stage: the construction of a mathematical theory of the transition from real data to constructed models. These two stages -- the modeling of phenomena and the processing of data -- appear in one form or another in any applied mathematical investigation.

For our purposes in this book, the first stage consists in modeling the structure of genetic objects of interest and in describing the properties of the data which may be represented by such a model. The essence of this stage lies in *representing* empirical data by ideally constructed models. The second stage consists in *approximating* experimental data, which are complex in structure, by models of more simple structure. Consequently, this book is concerned with two types of mathematical prpblems: the representation of structures, and the approximation of structures.

From this point of view we will now describe the mathematical nature of the book.

In the first chapter we investigate a model of the arrangement of mutational defects in a chromosome, i.e. a linear map. This model is simple enough to allow mathematical investigation, yet complex enough to not be trivial. The basic problem here is one of representation, i.e. the description of those data structures for which the linear map model is valid, namely, graphs and hypergraphs of intervals.

A considerable number of interesting mathematical results have accumulated in the theory of graphs and hypergraphs of intervals, and their presentation requires considerable space. At the same time, a general theory of the approximation of data by such devices does not yet exist. Consequently, in concrete examples the necessary approximation was done "by eye", since in these cases the structure of the data was either sufficiently close to the theoretical, as in Sections 1.5.1 and 1.5.3, or was so distant that it led to rejection of the model (Section 1.5.2).

The mathematical object of Chapter 2, the concept of structure, is extremely simple, and its mathematical theory is trivial. Consequently it was possible to focus on questions of approximation. The basic mathematical content of the chapter (Section 2.2) consists of the description of models and of methods of approximating data by structures

of a special form. These models constitute part of the general theory
of qualitative data analysis [22]. The complexity of the approximation
tasks examined is considerable: These are the so-called NP-complete [74a] problems
of discrete optimization, and precisely because of this their solution by
mathematical means is hardly possible. However, as our attempt shows,
the algorithms developed for the localized approximation of specific
data lead to adequate results (and not only for genetics).

In the third chapter the tasks of approximation and representation
are relatively balanced. The basic model is the notion of a weighted
tree, admitting a not-too-trivial description. A theory of approximation of data by means of trees has not yet been formulated. We examine mainly those "working" algorithms which are useful in obtaining
the genetic results of concern. In our view the concept of a weighted
tree, because of its simplicity, should attract the attention of mathematicians toward working out a corresponding theory of approximation.

It must be said that the mathematical concepts of graphs and hypergraphs of intervals, of structures, and of trees, examined in this monograph are extremely universal and applicable. Thus graphs of intervals
find application in problems of archeology and circuit design; methods
of approximating graphs by structures, in the study of structural organi-
ation and in classification problems; trees, in the organization of information systems, etc. This means the mathematical content of the book
is of use not only for genetics applications.

We will now describe the genetic content of the book. Experimental
genetic data are derived directly from interactions of nucleotide texts
in crosses (recombination analysis), or more indirectly from information about the interaction of protein products of genetic systems (complementation analysis). Most genetic analyses are based on information
of the first kind. Using such data, it is not difficult to construct
sufficiently complete and reliable portraits of genes as to be useful
in various further applications. "Functional" data about proteins are
seldom used. This is explained primarily by the indirect connection
between the interactions of proteins and the interactions of their parent nucleotide texts.

The methods described in this book are effective precisely for such
indirect information, where the usual methods of genetic mapping "do
not work." In this book, then, our primary attention is focused on the
analysis of information about the interactions of protein products.
It is the authors' hope that the methods presented here will indicate
new possibilities in complementation analysis.

In Chapter one, following presentation of the basic ideas about
genetic control systems, we show how genetic data can be used in study-

ing the organization of chromosomes. In the second chapter some highly controversial questions are discussed, dealing with the description of the functional organization of protein macromolecules. New concepts developed by V. A. Ratner and his collaborators are described, and corroborative analysis of experimental data is carried out. In the third chapter data about the structures of the proteins of a hemoglobin family are used in analyzing its evolution. Qualitative analysis of the evolutionary tree produced by this process leads to a number of important conclusions about the course of evolution.

The authors have tried to write so that the book can be read relatively independently of other sources. The necessary ideas of genetics are introduced in Section 1.1, and those of graph theory are given in an appendix. For the benefit of the reader an index of genetic terms is provided, and also an index of mathematical terms. The usual method of indices is used for internal references, including the number of the chapter and section which contain a formula or subsection.

A significant portion of the material of the book reflects the authors' own investigations, or more accurately, the investigations of the informal collective to which they belong. In this regard the authors express appreciation to all the members of the collective. We especially wish to mention the role of V. A. Ratner, not only as editor of the book, but also as leader of and active participant in all of the new genetic investigations described in it. His ideas on molecular-genetic control systems serve as pithy fundamentals to the book. Many new mathematical results were obtained in conjunction with V. L. Cooperstokh and V. A. Trophimov, who also carried out a significant portion of the machine computations. In particular, the contents of the second chapter are based to a significant degree on their work. The helpful criticisms of Yu.M. Svirežev and I. B. Muchnik significantly aided in improving the structure of the book.

Table of Contents

Chapter 1. Graphs in the Analysis of Gene Structure 1

§1. Gene systems and their maps 1
   1. Levels of the genetic language 1
   2. Mutations, recombination, complementation 12
   3. Genetic methods of investigation 16

§2. The mathematical theory of linear maps: interval graphs 20
   1. Maps and interval orders 20
   2. The description of interval graphs 25
   3. Graphs of non-covering intervals 34

§3. The mathematical theory of linear maps: interval hypergraphs 35
   1. Maps and interval hypergraphs 35
   2. Minimal hypergraphs and complons 40
   3. The construction of fictitious complons 44
   4. Non-linear hypergraphs and interval graphs 48

§4. Linear mapping algorithms 51
   1. The Fulkerson-Gross algorithm 51
   2. The uniqueness theorem 59
   3. Admissible orderings of linear matrices 61

§5. Examples of structural analysis of genetic systems ... 66
   1. The use of deletions and polar mutations 66
   2. The complementation maps of multiple mutational defects 75
   3. Complex traits and the loci which control them 81

Chapter 2. Graphs in the Analysis of Gene Semantics 92

§1. Interallelic complementation and the functioning of protein multimers 92
   1. Interallelic complementation 92
   2. The fundamental principles of organization of protein multimers 93
   3. The molecular mechanisms of interallelic complementation 97

§2. The approximation of graphs 101
  1. Approximation problems in a space of relations 101
  2. The optimal partition problem 105
  3. Detecting macrostructure 114
§3. Analyzing the spatio-functional organization of
    specific genetic systems 118
  1. Complex protein organization 118
  2. The investigation of genome spatial structure 126

Chapter 3. Graphs in the Analysis of Gene Evolution 133
§1. Trees and phylogenetic trees 133
  1. The notion of a phylogenetic tree 133
  2. The metric generated by a tree 136
  3. The construction of dendrograms 143
  4. Reconstructing the probable structure of ancestral
     successions 149
  5. Calculating the internal structure of sequences
     during tree construction 154
§2. The evolution of families of synonymous proteins 158
  1. The dendrogram of the globins and its analysis 158
  2. Analyzing the evolution of globin sequences from
     their internal structure 165

Epilogue: Cryptographic Problems in Genetics 176

Appendix: Some Notions About Graphs 180

References 188

Index of Genetics Terms 194

Index of Mathematical Terms 196

Chapter 1. Graphs in the Analysis of Gene Structure

§1. Gene systems and their maps

1. <u>Levels of the genetic language</u>. The genetic apparatus of living things stores hereditary information which ensures the reproduction of the organism. Another of its purposes, no less important, is to provide for normal functioning of each cell of the organism in the process of its individual growth (*ontogenesis*). This means that the hereditary memory of a cell (*genome*) governs all aspects of its vital activities, and the hereditary memory of the organism, in the final analysis, governs its functioning and reproduction as a whole, through the mutual interactions of cells.

The genetic program is concentrated in the chromosomes of the cell nucleus and is written down in the thread-like polynucleotide molecules of DNA -- *deoxyribonucleic acid*. The individual nucleotides, or more precisely, the nitrogenous bases contained in them, may be considered as characters of the "text" of a program. These *nitrogenous bases* (characters) are of four types: *adenine*(A), *thymine*(T), *guanine*(G) and *cytosine*(C).

Thus the primary governing information is represented as a linear record in this four-letter alphabet.

This information is used by the cell to synthesize needed proteins, and also in carrying out some other functions. It is the *proteins* which perform the basic biochemical operations in a cell, that maintain its vital activities. As a rule each protein performs one (or in some cases, a few) specific biochemical operation: catalysis of a biochemical reaction, transport of material, molecular control, etc. Therefore, in the cells of living organisms many thousands of molecules of various proteins are continuously being synthesized. A single protein molecule appears as a complex twisted *polypeptide chain* -- a sequence of hundreds of *amino acids* of twenty different kinds. In other words, a polypeptide chain is a text in a twenty-character alphabet of amino acids.

The translation of the genetic content from the language of nucleotides to the language of amino acids proceeds as follows. Each amino acid is coded by a definite ordered triplet (*codon*) of nucleotides, so that an arbitrary message in DNA is transferred by this *triplet code* into a sequence of amino acids (a protein).

Some redundancy is evident in the fact that 64 codons are used for 20 amino acids (Table 1). However, there are not 44 seemingly "meaningless" triplets, in terms of amino acids, as would be expected in a one-to-one coding, but only three. Each of a number of the amino acids is coded for by several triplets. The three "meaningless" codons play the role of punctuation marks in the genetic text. In Table 1 U is used in place of T ([31], p. 4).

The question arises, what is it that requires this representation of instructions (for each function) in two notations: the one in the language of nucleotides (for the synthesis of protein) and the other in the language of amino acids (for the direct fulfillment of this function). It seems evident that all of the necessary instructions could be written in only one language. In such a system there would be no problem of translation from one language to the other (about which, incidentally, something is said in this book).

Before answering this question we must consider several chemical properties of nucleotides and amino acids. In most cases a *molecule* of DNA consists of two coupled strands of nucleotides (the double helix of Watson and Crick). The nucleotides of one strand correspond chemically in a one-to-one way with those of the other, according to the rule: A always corresponds with T, and G with C (and the reverse). On the one hand, this makes a molecule of DNA stable under various influences, and on the other hand, it assures exact copying in each *replication*. In the replication process the helix of DNA unwinds, and on each of the strands in turn a new strand is synthesized, according to the pairing rules given above. As a result two daughter molecules are obtained, each having the same text.

At the same time, the stated characteristics of nucleotides (primarily paired interactions) are not sufficient to produce the variety of functions necessary in maintaining the vital activities of a cell.

We now consider proteins, whose spatial structure is entirely different. The initial sequence of amino acids forms the so-called *primary structure*. Due to the physico-chemical interactions of its component amino acids the primary structure sometimes acquires an extremely convoluted spatial form. The intermediate stage of this packing is called the *secondary structure*, and the final result, the *tertiary structure* of a protein. Packing does not stop with this third stage. Frequently, separate polypeptide chains unite in an integral complex called the *quaternary structure* of a protein (a *multimer*). On the tertiary and quaternary levels special sections of spatially-close amino acids known

The Genetic Code [31]

First nucleotide codon	Second nucleotide codon				Third nucleotide codon
	U	C	A	G	
U	phe	ser	tyr	cys	U
	phe	set	tyr	cys	C
	leu	set	Term.*	Term.	A
	leu	set	Term.	trp	G
C	leu	pro	his	arg	U
	leu	pro	his	arg	C
	leu	pro	gln	arg	A
	leu	pro	gln	arg	G
A	ile	thr	asn	set	U
	ile	thr	asn	ser	C
	ile	thr	lys	arg	A
	met[+]	thr	lys	arg	G
G	val	ala	asp	gly	U
	val	ala	asp	gly	C
	val	ala	glu	gly	A
	val[+]	ala	glu	gly	G

Note: Amino acids are given in their usual three-letter abbreviations (phe - Phenylalanine, leu - Leucine etc.).

+ Codon initiators (at the beginning of translation)
* Codon terminators (at the end of translation)

Table 1

as *functional centers* directly carry out the biochemical operations characteristic of a given protein. It is important to note that functional centers are conglomerations of various amino acids, so that their action is not reducible to the characteristics of the individual components.

However, amino acids do not display any kind of pronounced one-to-one pairings. Occasionally, it is true that fragments of polynucleotide chains form structures (so-called β-structures) which outwardly remind one of pair-wise-connected polynucleotides. However, this is exhibited in extremely specific localizations of amino acids, and is not characteristic of them as such [31,52]. This leads to the reverse consequence, in comparison with nucleotides. Protein molecules cannot reproduce themselves by template synthesis[†]; on the other hand, due to the associated packings of specific amino acid sequences, they are able to realize, in practice, any molecular function required by the cell.

Thus the existence of two languages means that there is a separation of responsibility: DNA is the repository of hereditary information (the hereditary memory of the cell), while proteins are the functional structures. This separation leads to high noise stability in genetic information. Having separated the hereditary memory of a cell (DNA) from participation in the indirect aspect of cell functioning, nature provided for its protection from various external influences. It "hid" *chromosomes* (structures which contain DNA) in the nucleus of the cell, while the synthesis of proteins and their succeeding reactions were removed primarily to the cytoplasm.

Such a spatial separation necessarily implies a "go-between", a messenger which carries information from DNA to the places of protein synthesis. This role is played by *messenger ribonucleic acid* (mRNA). As with DNA, it consists of four nucleotides: *adenine*(A), *guanine*(G), *cytosine*(C) and *uracil*(U). Although uracil(U) is used in place of thymine(T), specific pairings of associated nucleotides are maintained also in this case, namely A with U and G with C. In accord with this rule mRNA is synthesized on one of the strands of DNA. This process, analogous to the reading-out of information, is called *transcription*. Then the mRNA moves into the cytoplasm and joins itself to a *ribosome*, a special organelle of the cell on which protein synthesis takes place. Beforehand, amino acids are bound to molecules of RNA of a second type

---

[†]More precisely, this is not only a matter of the absence of specific paired recognition of amino acids. M. Aigen showed convincingly the basic possibility of non-template reproduction in so-called autocatalytic cycles. However these systems are not able to evolve, i.e. to reproduce changes (mutations) arising from mistakes in copying. For more details see [52].

called *transfer RNA* (tRNA), and in this form they approach the ribosomes "loaded" with mRNA. In accordance with the genetic code each amino acid is bound to a specific tRNA. A molecule of tRNA contains a particular triplet, known as an *anticodon*, which recognizes its codon in mRNA. Hence, an amino acid "riding" on tRNA is included in the synthesized polypeptide chain at the required position so that the collinearity of the chain and mRNA is ensured. This process of transferring information from mRNA to the language of polypeptides is called *translation*.

A chromosome contains information for the synthesis of not just one, but many different proteins (as many as several thousand). Even for a relatively simple organism such as the bacteriophage, a chromosome controls the synthesis of more than 50 different proteins. Portions of the text corresponding to separate proteins are called *cistrons*. Cistrons are separated from one another by special "punctuating" nucleotides: *Translation initiators* are AUG and (sometimes) GUG, which correspond respectively to the amino acids methionine and valine. *Translation terminators* are the "nonsense" codons UAA, UAG and UGA. Thus the cistron is the unit of translation. The basic action of a cistron is expressed in the functioning of a specific protein, and the phenotypic traits of an organism depend, in the final analysis, on just such protein functions. It is known that a living organism is an integral system, with each cistron influencing components of the system in one way or another. However, one can always identify one or two traits whose formation (by means of corresponding proteins) is critically dependent on a particular cistron. In this sense we say the cistron *controls* the given trait.

The term *cistron*, which arose in molecular genetics, is essentially the successor of the classical term *gene*. The historical conception of a gene arose in the specific context when individual discrete genes were associated with autonomous traits.

It should be noted that the cooperative actions of several proteins may be necessary to ensure certain functions and to form the corresponding trait. In many cases a "complex" function in a microorganism corresponds to a group of cistrons working in coordination. Typical of systems of this kind are the units of transcription, called *scriptons*. At the level of DNA, a scripton contains sequences which mark the beginning (*promotor*) and ending (*terminator*) of transcription, while the separation of this single text into cistrons corresponding to individual polypeptide chains is performed on mRNA in the translation process.

What is the physical meaning of the union of an entire group of cistrons under transcription? It turns out that the protein products of the cistrons contained in a scripton do not function independently, rather they control interconnected stages of the transformation (usually synthesis or decomposition) of materials within the cell, and frequently enter into common multimers as well.

Consider, for example, the breaking-down within the cell of some energetic raw materials such as carbohydrates or amino acids for the extraction and storing of energy. Here the associated cistrons must be switched off when the raw material is running low, and switched on when it is sufficient. The switching of a cistron is performed by a special control segment called an *operator*, situated at the beginning of the scripton. Scriptons with operators are called *operons*, and work on the principle of a control system with feedback. The operator of an operon is sensitive to a special protein, which is synthesized under control of still another special gene, a *regulator*. In the absence of needed raw material this protein (called a *repressor*) represses the action of the operon by combining with the operator and preventing transcription of the entire remaining portion (the "barrier" is closed). If there is sufficient raw material present it combines with the protein repressor and changes its spatial configuration so that it no longer recognizes the operator (the "barrier" is open). Then the operon synthesizes proteins (enzymes) which break up the raw material until its concentration falls below the level necessary for it to bind the operon repressor, and consequently the operon is again turned off.

Such control systems have been found, up to this time, only in microorganisms. Systems have now been described consisting of many operons and scriptons, and recently, in the case of the λ bacteriophage, the entire control system for individual growth has been deciphered.

In higher organisms it has been impossible to demonstrate the presence of operon-like systems (see however Section 5.3). The construction of the cells of such organisms permits a wide spectrum of possibilities for the organization of regulation, unthinkable in the lower organisms. These possibilities are connected with the spatial and temporal disconnection of the processes of transcription (the synthesis of messenger RNA in the nucleus) and translation (the synthesis of polypeptides in the cytoplasm) and to all appearances are realized through the genetic redundancy of the corresponding macromolecules of DNA and RNA. The initial transcript synthesized in the nucleus is, as a rule, much longer than the mRNA which is translated into a protein: biologists say a maturing or processing RNA takes place.

One kind of redundancy was recently deciphered. By means of direct biochemical investigation of DNA sequences in higher organisms it was revealed that a significant portion of the initial transcript of a cistron is not translated into polypeptide text. Such a cistron is divided into regions of two types: *exon* and *intron*. The exons are translated, while the introns are not. The noncoding intron fragments sometimes reach significant total length. Thus, the noncoding part of the β-cistron of mammalian hemoglobin makes up on the order ot two-thirds of its length (for more details see page 154).

The discovery of introns in 1977 gave rise to a number of serious questions concerning their roles in internal cell processes. The picture is still not clear. It can be supposed only that the intron organization of genes provides the possibility for additional means of regulating gene activity and heightens their adaptive potential. For the genes which maintain active molecular immunity in higher organisms, for example, mammals, this thesis has been demonstrated and can be given in detail.

The characteristics of the immune system's reaction to foreign substances -- *antigens*, found in the organism, are determined at the molecular level by the formation of corresponding complex protein macromolecules -- *antibodies*, or *immunoglobulins*. The diversity of the antibodies must be very great, in view of the practically unending number of different antigens. The general outline of the mechanism for generating such diversity is as follows. Initially in the genome there is a significant quantity of cistrons, coding for so-called *domains* -- spatial and functional autonomous blocks for the formation of immunoglobulin molecules. These cistrons are localized in the form of clusters, gene batteries, but separated from one another by intercistronic nucleotide sequences called *spacers*. One theory says that in response to an antigen a selection is made in the genome among all the possible combinations of these cistrons, combining them into a single DNA sequence. This selection comes about through a special recombination enzyme system, which recognized spacers and combines separate cistrons. A variety of immunoglobulins are produced by translation of the DNA sequences thus formed. The cells synthesizing an immunoglobulin necessary to inactivate a given antigen gain an advantage in propagation in order to maintain the immunity. Direct biochemical analysis (sequencing) of the genes which code for immunoglobulins, has shown that exons correspond to domains in these proteins. This permits the supposition that the introns are former spacers [66a]. On the basis of this example one can clearly see how, in principle, the intron structure can be utilized for the goals of evolution and regulation.

In wide circulation also is the idea that since without the excision of introns and some other fragments of mRNA synthesis of a single polypeptide sequence is not possible, the presence of introns may be used by the cell for the control of gene activity at the level of translation.

Thus within the genetic language there are several identifiable layers.

The "upper" layer is the language of DNA, the hereditary memory of the cell. The meanings of texts coded in the four-letter alphabet of DNA are completely unambiguous. Each word (codon) is rewritten as exactly the same word (codon) of mRNA, except that the letter T (thymine) is replaced by the letter U (uracil).

Texts in the alphabet of mRNA are also unambiguous in meaning: Each codon contains an instruction for the inclusion of a specific amino acid (in agreement with the genetic code) in the associated polypeptide chain, the "phrase" (cistron) being the instructions for synthesis of the entire chain.

Thus the third layer in the genetic language is the primary structure of proteins, i.e. linear texts in a twenty-character alphabet of amino acids. Contained in these texts are the instructions for the necessary spatial packing of polypeptides in active protein molecules. Unlike the *genetic code* (the transition from mRNA to the primary structure of proteins), these instructions (for the transition from primary to tertiary structure, i.e. for the self-organization of protein molecules) have not been completely deciphered, although recently more papers have been appearing which report progress in the solution of this complex problem.

The next layer of the genetic language is the three-dimensional structure of protein molecules. At this level, segments of proximal amino acids form *functional centers* containing instructions about "what to make" (catalytic activity) or "when and how to act" (regulatory function) or "where to go" (intracellular organization), etc., in which functions are performed as prescribed by genetic texts of the "higher" layers. Little is known about the alphabet, much less the grammar, of the language of functional centers (see page 93).

The language of the next layer is that of the functions carried out by proteins. It has had little investigation, although recently it became clear that the "dictionary" of this language is not so large as one might expect. For example, all *proteolytic enzymes*, which decompose "edible" protein molecules, act according to the same molecular scheme, differing only with respect to which adjacent amino acids are recognized as a point of cleavage.

In this way each layer of the genetic language is distinquished from the next by the alphabet and syntax of its texts, and by its *seman-*

tics, which are realized in the synthesis of texts of the next layer, and finally, in the vital activities of the cell determined by the functioning of protein molecules.

In moving through the layers of the genetic language (from DNA to the functional centers of proteins), the semantics of the genetic information becomes more and more "active." At the same time, the active form of the language of the protein layer, though structurally stable, is quite flexible. It is true that the initial polynucleotide text (in DNA) uniquely determines the molecular function of the corresponding protein. This does not mean, however, that the function of a protein requires a unique primary structure. There are key positions within the primary structure of a protein which determine its spatial packing, and consequently, its function. Thus a particular molecular function may correspond to a whole class of polypeptides, which differ only with respect to the amino acids occurring in the positions of less functional importance, the non-key positions. In other words, this spatial method of realizing the semantics of the "protein" layers allows the existence of synonyms. Actually, synonymous polypeptides do provide for some nuances of meaning within their instructions. Without the existence of such synonyms it is difficult to imagine a process of change and evolution in genetic (or any developed) systems [31,32,52]. The synonymity of the genetic language (as also the usual "human" languages) ensures the capacity for flexible response to changing environmental conditions without the creation of fundamentally new texts, by arranging (within the hereditary memory (DNA)) for only local (non-key-position) changes in the primary structure of proteins. Roughly speaking, associated with the "continuity" of change in the external surroundings there must be a "continuity" of change in the meaning of the significant constructs of the language. Thus, synonymity is characterized not only by a multiplicity of meanings but also by the blurred nature of the set of semantic values.

The transitions from layer to layer of the genetic language are used in regulating and controlling the realization of genetic information. The process of transcription (the transition from DNA to mRNA) is, evidently, central to the regulation of gene activity in such simply constructed organisms as bacteria and viruses. In higher forms the control possibilities are much broader. As an example of this, we note the system of control for the translation process (the transition from mRNA to protein) through the excision of introns. The greater the descent from the higher layers, the more restricted becomes the sphere of action of control (if elements of the control mechanism itself are not affected).

Languages of human intercourse differ significantly from the genetic language in that they are broader and richer, being much more than languages of instructions (commands) [31]. In this sense the genetic language resembles more closely the languages of socio-economic control systems, which are inherently flexible and hierarchic in nature. The comparison of these languages would probably give a deeper understanding of the general principles of organization of control in complex systems, but we know of no interesting progress in this area. It is accepted that there are essential differences between the genetic and the socio-economic control systems in the "development" of their administrative instructions. No matter how many layers are found in the genetic language, it must always be remembered that all of the necessary information is contained (in a "passive" form) in the initial polynucleotide text. Successive stages simply recode it into active "working" structures. In socio-economic control systems instructions arriving at their level of application may be noticeably changed due to conscious "creative" effects at each preceding level. However, even in genetic systems the process of recoding is not always so faultlessly exact as might be expected. In fact, every genetic control is, in one way or another, connected with a change in informational macromolecules and with corresponding modifications in their structure.

The properties of the genetic language make it similar to the languages of programming systems, which also consist of instructions, and are many-layered. Texts in a high-level programming language (the upper layer) are translated into texts of internal machine langauges (the langauge of micro commands and the language of physical processes necessary for computation), which ensure the completion of the initial prescriptions.

Considering the *collinearity* of the polynucleotide and polypeptide forms of recording genetic information (excluding introns), we will differentiate (conditionally, of course) within the genetic language its *structural* and *semantic* levels. The first level is that of the linear (collinear between them) polynucleotide (DNA, RNA) and polypeptide texts, while the second level consists of the functional centers of protein macromolecules, i.e. the level of "active" genetic information. We will keep in mind, though, that such an important stage in the realization of the semantics as the separation of one instruction from another, is effected in the polynucleotide layer (of DNA,RNA) by means of special punctuation marks [31]. It should be particularly emphasized that the differentiation within the genetic language of the levels of reproduction (DNA) and the functioning (proteins) greatly simplifies the real picture of the interactions between nucleic acids

acids and proteins. The fact is that in living cells the basic processes of recoding -- replication, transcription and translation -- can take place only with the participation of special enzymes. The hereditary memory of a cell (DNA) must, of necessity, contain instructions for the synthesis of the entire set of macromolecules (enzymes, tRNA and others) which "serve" these processes. In other words, the mechanism of self-reproduction in real cells will not work in the absence of the products it itself produces [31,52]. For this reason M. Aigen correctly remarked that even in genetic systems of relatively simple construction (viruses, bacteria) the general scheme of inter-relations among nucleic acids and proteins appears as a complex hierarchy of "closed loops" [52].

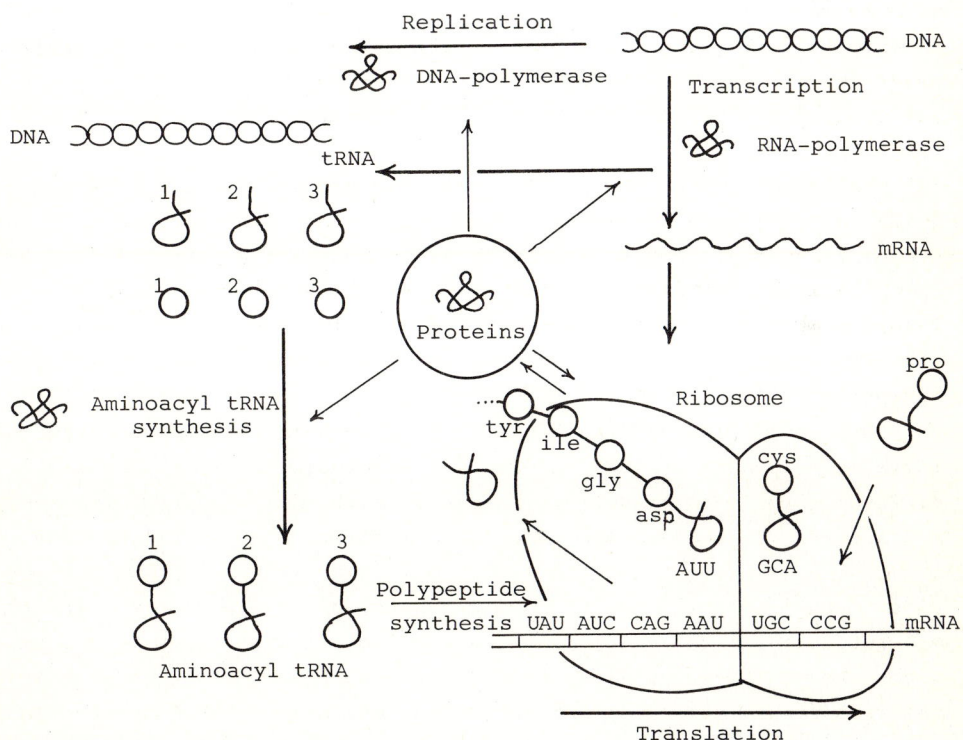

Figure 1. Inter-relations of nucleic acids and proteins in the process of their reproduction [52].

Fig. 1 shows the fundamental functional connections of a genetic system [52]:

1. DNA, on which is carried out the template synthesis of complementary chains by the enzyme DNA-polymerase.

2. mRNA being synthesized on DNA by the enzyme RNA-polymerase.

3. tRNA and aminoacyl-tRNA synthetases. These enzymes connect molecules of tRNA with certain amino acids, thus ensuring a correspondence between "words" of the polynucleotide and polypeptide texts.

4. A ribosome -- an enzymatic complex, including molecules of ribosomal RNA. A ribosome consists of two unequal subunits: In the smaller is located the acceptor segment (the center of connection with tRNA), and in the larger there is a catalytic center for forming peptide connections with neighboring amino acids. At each moment during translation there are two molecules of aminoacyl-tRNA associated with the ribosome, the anticodons of which are complementary to the corresponding codons of mRNA. In the scheme of Fig. 1 is shown one of the stages of movement of a ribosome along mRNA: The amino acid cysteine is ready to join with an already-translated polypeptide fragment (...—tyrosine—isoleucine—glycine— asparagine—).

5. The proteins appear in the center of the scheme: on the one hand, as end products of processes of recoding (thick arrows), and on the other hand, as necessary participants in these same processes (thin arrows).

In the sequel we will use the term *gene* in the broad, although inexact, classical sense, to designate the unit which controls a separate trait, keeping in mind that such a gene may consist of a single cistron or of multiple cistrons. This is important in view of the fact that in genetic experiments the investigator works, as a rule, not with molecular objects, but with the mutant variants of the trait of interest.

2. <u>Mutations, recombination, complementation</u>. Without the process of genetic reproduction described above, the basis of which is the template copying of molecules of DNA, neither growth nor reproduction of multicelled organisms would be thinkable. The process of growth is, essentially, a series of successive cell divisions, the first of which is the division of the embryonic cell, or *zygote*. The basic hereditary material of the zygote is concentrated in the nucleus in the form of a fixed number of discrete structures -- the *chromosomes*. Prior to cell division, each chromosome is doubled (by doubling DNA), and during division of the nucleus identical copies of the chromosomes are separated equally to each daughter cell. This cell division process, called *mitosis*, ensures the transfer of a full complement of chromosomes to

each daughter cell.

The basis of sexual reproduction is an essentially different process of cell division called *meiosis*. In the formation of a male or female sex cell (*gamete*) the parent cell divides twice, so that in the daughter cells there is only half the usual number of chromosomes. However, this entails no loss of genetic information. The fact is that in the cells of higher plants and animals each type of chromosome appears twice, as *homologues*. Such chromosomes are said to be *homologous*, and such cells are called *diploids*. In the products of meiosis (the *gametes*: ovum and spermatozoid) there is only a single copy of each chromosome. Such cells are called *haploids*. In the process of fertilization the male and female haploid gametes combine to form a diploid (*zygote*), re-establishing a normal diploid set of chromosomes, in which each gene again appears twice.

The genetic material of bacteria and a number of other single-celled organisms is represented by a single molecule of DNA, i.e. such organisms are typically haploids. The cell division process differs here somewhat from that described above though it retains the same purpose: to transfer a full complement of the DNA to each daughter organism.

In DNA replication occasionally there are "misfires", mistakes in reproduction, called *mutations*. The majority of mutations are the result of a simple replacement of individual nucleotides (characters). If, as a result, there is a replacement of amino acids in a protein we speak of a *missense mutation*; if the new triplet turns out to be meaningless then such a replacement is called a *nonsense mutation*. More complex mistakes are possible: the loss or insertion of separate nucleotides (such mutations lead to a displacement of the "read out frame" and the distortion of the entire subsequent text), loss of an entire DNA fragment (*deletion*), *duplication* of DNA segments, etc. The fact that all of these mistakes are reproduced in the replicating DNA means they are passed on to descendants, i.e. inherited.

Since the translation of separate genes (cistrons) is independent, mutations internal to a gene usually distort the function of only the protein it controls, and that locally.

Among all the possible nucleotide replacements a significant portion are *synonomous*. The genetic code is such that replacement of a single nucleotide in a triplet frequently produces a codon for exactly the same amino acid. There are also other, more subtle regulating properties of the genetic code and the translation apparatus which increase the noise stability of the genetic texts: Codons with similar sequences code for amino acids with similar characteristics; amino acids are incorporated more readily into protein structures, the more triplets there are that correspond to them, etc. (see Table 1 and [2,31,52]). However, the noise stability thus attained has its limits. It can be shown [2,31] that if the chromosomes of haploid descendants have a total length greater than $10^7 - 10^8$ nucleotides, then they will inevitably carry at least one *lethal* (death-causing) *mutation* (called a *lethal*).

This establishes a limit to the complexity of the organization of living systems at the haploid level (actually, existing haploid organisms have fewer than 10^7 nucleotides [2,31]). We cannot rule out the possibility that this may be the reason for the transition from a haploid to a diploid structure in the hereditary mechanism, for in the diploid system all the chromosomes are doubled, so that each gene has two representatives. Of course, in the cells of multicellular organisms there are several such pairs of homologous chromosomes (for example, the fly *Drosophila melanogaster* has 4 pairs, humans have 23, maize 10). If only one of the homologous chromosomes is mutated then, in the presence of a normal partner, its defect is not usually manifest. We will return to a more precise discussion of this question later.

Generally speaking, doubling of information may also be realized by another means: the writing down, in order, of the same text twice in a single "carrier." Some genes are repeated exactly this way. However, the doubling of hereditary information on different but homologous chromosomes not only increases its noise stability, but also leads to a fundamental increase in evolutionary possibilities. These possibilities are realized in higher organisms specifically through the process of sexual propagation described previously, which is connected with just such an information-doubling system. Indeed, in organisms propagating by sexual means, each chromosome of a homologous pair has a paternal or maternal origin, and during meiosis members of each pair are distributed to gametes randomly. As a result a random combination of chromosom pairs is produced, which in its turn leads to the appearance of mixed descendants. Thereby, the genetic diversity of that particular species of organism is significantly increased, and the possibility for the selection of new forms becomes greater.

However, an increase in genetic diversity is attained not only by the exchange of whole chromosomes, in which the genes contained in one chromosome are inherited as a group (*cohesively*). There is also a process of intrachromosomal *recombination* of genes called *crossing-over*. At one stage of meiosis homologous chromosomes draw close together collinearly (*synapse*) and exchange their homologous fragments. By this process exchange of paternal and maternal genes can occur. If, for example, homologous chromosomes differ by two mutations, only one of these being contained in each of the exchanged fragments, then as a result of crossing-over the chromosomes will be recombinant.

The vital feature of *recombinations*, as with other basic genetic processes (replication, mutation) is that purely structural rearrangements are produced by them; the semantics, the meaning of the genetic messages does not participate in the exchange process at all. The semantic aspect of genetic messages consists of the functions they perform in the genetic system, as opposed to the arrangements in which they are found.

In living organisms a significant excess concentration is characteristic for many proteins. Very frequently for a protein to carry out its function it is sufficient to have only a tiny fraction of its normal concentration. Hence, even if only one of the homologous genes of a zygote is normal, the activity of the protein coded by it is frequently sufficient for normal development. In such cases it is customary to say that the normal gene is *dominant* to the mutant form, which is said to be *recessive*. For some other proteins (for example those which play a regulatory role) the reverse is true: a mutant gene may be dominant to the normal gene, since even a small change of concentration of the normal protein may materially influence the developmental process.

We note however that *dominance* and *recessiveness* are somewhat conditional concepts. They depend on the level of phenotypic expression used in comparing normal and mutant genes. In genetics there are examples of mutations dominant at the molecular level, but recessive from higher-level points of veiw (cellular, tissue, organismal) (see, for example, [30,31]). However, if a standard level of phenotypic expression is chosen then dominance relations will have a stable meaning.

Suppose that in a diploid cell each member of some pair of homologous chromosomes carries one recessive mutation. If the mutations affect corresponding genes the phenotype of the cell will be mutant. If, however, the mutations are located in different genes, then each of the mutated chromosomes will generate an active protein product which the other is not able to generate. As a result, homologous chromosomes act in a mutually complementary way, causing a normal phenotype. Such a phenomenon is called *complementation*.

Thus, in diploids the estimation of the meaning of a genetic message depends not only on the synthesized protein, as such, but also on the confrontation with the product of the homologous gene. The result of this confrontation is manifest phenomenologically as dominance and complementation. Recessive mutations constitute part of the reserve variability of a population, which may turn out to be of vital importance under changes in the external environment.

"Thus we see that the separation of the life cycle into two phases (haploid and diploid) allows, in a remarkable way, the combining of the flexibility necessary for evolution (in sexual cells) with the stability necessary for the individual (and for control of evolutionary changes) . . . that is, the maximal reliability of transmission of hereditary information" [51].

These properties are manifest in particular in the phenomena of recombination, dominance and complementation.

3. <u>Genetic methods of investigation</u>. At present the investigation of genetic systems is conducted by mutually complementary "direct" physico-chemical means and by the "indirect" hybridization methods of genetics.

Genetic methods are based on comparative analysis of the phenotypes caused by mutant genes either singly or in combination. In genetic experiments, normal genes are never studied directly, rather only their "breakages" -- mutations. Comparing these mutations according to their phenotypic effects in various hybrids, the geneticist tries to form a picture of the genetic system being studied, by means of imcomplete, "negative" details.

To understand the essence of the genetic approach one must become somewhat acquainted with the associated terminology.

Earlier we agreed to use the term "gene" to denote a unit of heredity responsible for the manifestation of some single trait. Mutations change the structure of a gene, frequently causing a change in, or complete loss of, the function which that gene controls, thus giving rise to alternative states of that trait. The genes at a given locus associated with such alternative states (for example, eye color in *Drosophila*) are termed *allelic* or (simply) *alleles*. A diploid individual with differing alleles for a single gene is *heterozygous* for that gene, and one with identical alleles is *homozygous*. Alleles, which are situated in identical positions (loci) of homologous chromosomes, are termed *homologous* genes.

Thus, a mutation plays the role of a distinctive marker of the gene (locus) in which it arises. It is these marked fragments of the genetic apparatus which geneticists study, primarily by utilizing the appearance of structural exchange -- recombinations and functional interactions -- in the phenomena of dominance, complementation, etc. Consequently, in genetics there are two basic types of tests: for recombination and complementation.

In a *recombination test* we usually study the results of crossing

individuals which differ by two or more mutations. In the descendants,
besides the original parent forms, there may appear new ones, *recombin -
ant* with respect to the observed trait. The end product of the test is
the fraction of recombinant offspring.

The more closely two mutations are situated on a chromosome the
less chance there is of a crossing-over occurring on the segment between them, and consequently, the less frequently recombinations will
occur. In other words, the frequency of recombination reflects the
real physical distance between the positions of the mutations within
the chromosome.

It is on this basis that *genetic (recombination) maps* of chromosomes are built. The recombination test is used to detetermine the
frequency of recombination for many pairs of mutations. The collection
of all results is then used in building the genetic map. The frequencies in this collection are used to produce the recombination matrix.
From this matrix a map is formed -- a segment of the real axis with
points (corresponding to separate mutations) drawn on it in such a way
that distances between points correspond to frequencies of recombination. It is known that the more complete the initial matrix is the
more exact the map will be. However, in practice only a fraction of
the possible pairs of mutations are tested.

The construction of the recombination map is based on two properties of the organization of genetic material: its linear structure and
the monotone dependence of recombination frequency on physical distance.
For many specific situations formulas, called *mapping functions*, have
been worked out, which allow the estimation (from the frequencies) of
the physical distance between the loci carrying the mutations of interest [17,30,41]. By means of these distances it is easy to precisely
arrange the chromosome (or some part of it) along the axis, corresponding to the tested mutations. The result of such an ordering is a map
of some segment of a chromosome.

Genetic mapping serves as the basic method of clarifying the structural organization of a genome. This does not exahust its possibilities, however. If the products of separate regions of the map (corresponding to genes and gene systems) are known, then the genetic map
also helps to establish a complete representation of the functional organization of the genetic apparatus as a whole (in connection with this,
however, see Section 2.3.2). This is why the establishing of suitable
recombination maps is the most important step in solving theoretical
and practical problems in the fundamental areas of modern genetics.
Molecular genetics, classical genetics, the particular genetics of specific species, medical genetics, genetic engineering -- progress in all

of these areas depends on the procedure of genetic mapping.

The specific dependence of recombination frequency on physical distance varies greatly among species, among the chromosomes of members of a given species, and even among different sections of the same chromosome. This significantly complicates the task of constructing mapping functions to translate the measured quantities -- recombination frequencies -- into physical distances. The criterion of adequacy for such functions is evident: If the function is correct then the corresponding distances must be additive for all values of the recombination frequencies.

The theory of recombination mapping is quite well worked out for a wide class of genetic objects, and a number of articles and monographs have been written on the subject. We have not taken as our task a detailed treatment of this area, although it plays a central role in the analysis of gene structures. Besides, the "bottleneck" in such mapping tasks -- the construction of mapping functions -- has no direct connection with graph theory. Possibilities are clearly visible today for the use of graph theory in recombination analysis to clarify the topology of genetic maps without the calculation of specific distances between the mutations involved.

Evidently the first person to realize this in a sufficiently clear form was S. Benzer. In his papers [3,55] the notions of graphs and hypergraphs of intervals were actually applied as a means of doing preliminary genetic mapping using only qualitative information about the overlapping of mutational defects, without the calculation of specific recombination frequencies. The theory of graphs and hypergraphs of intervals is discussed in Sections 2-4 and its application is given in Section 5.

We point out another task of recombination analysis (where the methods of graph theory play a significant role) which arises when the chromosome locus being examined is small (in comparison with the sizes of cistrons) and the faulty segments are very near one another (they may correspond to neighboring nucleotides). In this situation the quantitative regularities do not hold, and the construction of exact mapping functions becomes a practical impossibility, since inside cistrons there are frequently observable cases of such phenomena as negative interference, polarity of recombination, allele-specificity of recombination, gene conversions, etc., each of which leads to deviations from the additivity of distance which are difficult to control (for more details see [17]). Then only the qualitative regularity remains: that the closer two defective segments are, the smaller will be the recombination frequency between the corresponding mutations. This means that the matrix of recombination frequencies has a special "linear" structure: From its diagonal the values of elements monotonically increase on "both sides", if the ordering of the mutations corresponds to their true positions in the locus (see also Section 4.3).

The second classical genetic test is the *complementation test* (also called the *cis-trans test*). In this kind of test the fundamental events occur after the formation of zygotes from gametes, each of which carries a different recessive mutation. Parents, mutant with respect to the trait of interest, are crossed. Since each is mutant for a recessive mutation, it must be homozygous at that locus: the corresponding homologous chromosomes carry identical mutant alleles. The offspring of such a cross are hybrids, i.e. heterozygotes whose homologous chromosomes carry different mutations -- one from each parent. If the heterozygote exhibits a normal phenotype then it is evident the mutations affect different genes (that is, they are non-allelic, or complementary). In the other case (non-complementation) the mutations are allelic: they affect the same gene.

The result of pair-wise complementation testing of a collection of mutations is the *complementation matrix*, the (i,j)th element of which designates whether mutations i and j are complementary or not.

If the examined mutations are *point mutations*, i.e. they cause replacement of the same amino acid in the associated polypeptide, then clearly the mutational defect does not extend beyond the bounds of a particular cistron. In that case the complementation matrix must have a particularly simple, block structure: All the mutations affecting a given cistron are pair-wise noncomplementary among themselves and, at the same time, complementary to mutations of other cistrons. Consequently, the set of mutations is arranged into groups corresponding to separate cistrons and giving rise to a block-diagonal form of the complementation matrix. Here, the question of the mutual positions of the cistrons remains open.

The real picture is much more complicated [35,43]. First, the mutations within a given cistron may be complementary. There is a whole class of cistrons in which *intracistronic*, or *interallelic* complementations have been observed. The next chapter is devoted to the possible mechanisms of interallelic complementation and to questions of its analysis by graph-theoretic methods.

In this chapter we examine a second type of deviation from a block-structured complementation matrix, connected with the existence of mutations which affect several cistrons at once. Such mutations "join" separate cistron groups (of matrix elements) and significantly complicate the structure of the complementation matrix used to construct the interallelic topography map of the genetic material.

In this case geneticists are governed by the following rules. Each mutation is characterized by a continuous interval (the defect zone) of the locus under consideration. In the complementation matrix, for each pair of mutations it can be determined whether or not their associated intervals intersect: For noncomplementarity of recessives, in point of fact, means an overlapping of the corresponding defect zones. *Complementation mapping* hinges on the construction, by means of the complementation matrix, of a representation of the given collection of mutations by a system of mutually overlapping intervals.

Although the nature of the complementation test is functional (semantic), this system of intervals provides a portrait of the structural organization of a locus. This is explained by the fact that in intercistronic complementation each cistron acts as a unit, and in that situation the complementation phenomena are not directly connected with the semantics of the cistrons, which are manifest on the protein level.

Therefore, with the degree of reduction to separate cistrons, the complementation map provides the same "structural" information as the recombination map (in other words, both these maps are collinear). This fact is widely used in the practice of genetic analysis, inasmuch as complementation testing is, as a whole, much simpler than recombination testing.

§2. The mathematical theory of linear maps: Interval graphs

1. **Maps and interval orders.** We consider a set of N objects, given by the indices $1, 2, 3, \ldots, N$. We will designate the set by $A = \{1, 2, \ldots, N\}$. With each $i \in A$ we associate an open interval of the real line I_i, with bounds at points corresponding to natural numbers, i.e. $I_i = (l_i, r_i)$ ($l_i, r_i \in \{0, 1, 2, \quad\}$), where l_i is the left boundary and r_i the right boundary of the interval I_i ($i \in A$). We designate by $\mathbf{I}$ the minimal interval containing all of the I_i. Then $\mathbf{I}$, together with the intervals I_i ($i \in A$) forms, by definition, a *map* $K = \langle \mathbf{I}, I_i \ (i \in A) \rangle$. The length of the interval $\mathbf{I}$ will be called the *length of the map* K, so that if $\mathbf{I} = (l, r)$, then the length of the map is equal to $r - l$.

The interval $\mathbf{I}$ is considered to represent a fragment of genetic material, the intervals I_i the defect zones of individual mutations within the bounds of this fragment, and the objects i the numbers (designators) of the corresponding mutations. Keeping this in mind, we will in the future call these objects mutations, not using the term in its special genetic interpretation, but only to aid in the interpretation of subsequent mathematical constructions.

The following basic postulates are associated with the notion of a map (although they frequently may not hold): the linearity of genetic material and the indissolubility of mutational defects.

As a result of genetic investigations of N mutations of some locus (as a rule, from tests of complementation) we frequently obtain an $N \times N$ Boolean matrix $r = \|r_{ij}\|$ (the *complementation graph*), which characterizes the overlapping of mutational distortions: $r_{ij} = 0$ if the ith and jth mutations are complementary (non-overlapping), and $r_{ij} = 1$, otherwise. Correspondingly we say that a map K *represents* a Boolean matrix if and only if

$$r_{ij} = 1 \leftrightarrow I_i \cap I_j \neq \phi. \tag{1}$$

As is usual with the introduction of new mathematical ideas, a number of questions arise. Which Boolean matrices (complementation graphs) are represented by maps (the existence problem)? Is a map which represents a given complementation graph uniquely determined? And if not, what relations hold among the maps representing such a graph (the uniqueness problem)? How is a map of a given Boolean matrix constructed? Etc. In this section we examine questions associated with the "internal" characteristics of matrices (graphs, relations), affecting their representation by maps, and in the next two sections we will take up the questions posed above.

For an arbitrary map $K = \langle \mathbf{I}, I_i (i \in A) \rangle$ we will be concerned with the relation "to-the-right-of" on the set A of objects, defined by

$$(i,j) \in P_K \leftrightarrow I_i > I_j, \tag{2}$$

where the relation ">" is the usual numeric > relation applied to the subset of numbers (intervals), i.e. $I_i > I_j$ means that the entire interval I_i lies to the right of the interval I_j. In other words the left end point of interval I_i is not less than the right end point of interval I_j. This relation P_K we will call the *interval order* for the map K.

The question arises, what form must a relation $P \subseteq A \times A$ take in order for there to exist a map K for which P is the interval order, i.e. $P = P_K$.

It turns out that interval orders are completely characterized by the following *condition of quasi-linearity*:

$$P\langle a \rangle \subseteq P\langle b \rangle \quad \text{or} \quad P\langle b \rangle \subseteq P\langle a \rangle. \tag{3}$$

The quasi-linearity condition means that the sets $P\langle a \rangle$ are linearly ordered by the inclusion relation. Let there be among them exactly m different ones: $P\langle a_0 \rangle, P\langle a_1 \rangle, \ldots, P\langle a_{m-1} \rangle$. Then for a suitable renumbering $F(a_0, a_1, \ldots, a_{m-1})$,

$$P<a_0> \subset P<a_1> \subset \cdots \subset P<a_{m-1}>, \tag{4}$$

where the inclusion $\subset$ is strict.

To prove the assertion we will first discuss the set-theoretic structure of quasi-linear relations.

Let $P \subset A \times A$ satisfy (3) and (4). Let

$$A_i = \{a \mid P<a> = P<a_i>\} \quad (i=0,1,2,\ldots,m-1). \tag{5}$$

The collection of these sets $\alpha = \{A_0, \ldots, A_{m-1}\}$ clearly forms a partition of the set A. Let

$$B_j = P<a_j> - P<a_{j-1}> \quad (j=1,2,\ldots,m-1). \tag{6}$$

Clearly the sets B_j ($j=1,2,\ldots,m-1$) are nonempty, since in (4) the inclusions are strict. Further, let

$$B_m = A - \bigcup_{j=1}^{m-1} B_j = A - P<a_{m-1}>.$$

If is antireflexive (and interval orders are), then $B_m \neq \phi$, since $a_{m-1} \notin P<a_{m-1}>$, i.e. $a_{m-1} \in B_m$. Consequently $\beta = (B_1, \ldots, B_m)$ is also a partition of the set A. In view of (4) both of these partitions are *strictly ordered*: the ordering of the classes is determined by the ordering of the sets $P<a>$ under inclusion.

Consider the union of all left and right parts of (6), for j going from 1 to i:

$$\bigcup_{j=1}^{i} B_j = (P<a_i> - P<a_{i-1}>) \cup (P<a_{i-1}> - P<a_{i-2}>) \cup \ldots = P<a_i> - P<a_0>.$$

The set $P<a_0>$ is empty, since if $a \in P<a_0>$ then $P<a> \quad P<a_0>$. Actually, from (3) one of the sets $P<a>$, $P<a_0>$ is contained in the other, but $a \notin P<a>$ because of the antireflexivity of P, so that $P<a_0>$ cannot be contained in $P<a>$. On the other hand the strict inclusion $P<a> \subset P<a_0>$ contradicts the minimality of $P<a_0>$ in accord with (4). Therefore $P<a_0> = \phi$ and for $a_i \in A_i$

$$P<a_i> = \bigcup_{j=1}^{i} B_j \quad (i=1,\ldots,m-1). \tag{7}$$

The equalities (7) allow us to express the relation P in terms of the partitions α and β, for from (7)

$$P = \bigcup_{a \in A} \{a\} \times P<a> = \bigcup_{i=0}^{m-1} \bigcup_{a \in A_i} (\{a\} \times P<a>) = \bigcup_{i=0}^{m-1} A_i \times P<a_i>.$$

Hence from (7) we obtain

$$P = \bigcup_{i=1}^{m-1} \left(A_i \times \bigcup_{j=1}^{i} B_j \right). \tag{8}$$

Since P was given as antireflexive the conditions

$$A_i \bigcup_{j=i+1}^{m} B_j \qquad (9)$$

hold, for otherwise there must exist $a \in A_i$ such that $a \in \bigcup_{j=1}^{i} B_j$, so that $(a,a) \in P$.

Formulas (8) and (9) completely determine antireflexive quasi-linear relations in the following sense. Suppose we have two ordered partitions of a set A into the same number of classes $R^1 = (R_0^1, \ldots, R_{m-1}^1)$ and $R^2 = (R_1^2, \ldots, R_m^2)$, which satisfy condition (9) (with the replacement of A_i by R_i^1 and B_j by R_j^2). We form the binary relation P, letting, for $a_i \in R_i^1$ ($i=0,\ldots,m-1$),

$$P\langle a_0 \rangle = \phi, \qquad P\langle a_i \rangle = \bigcup_{j=1}^{i} R_j^2.$$

This relation is clearly quasi-linear (by the definition) and antireflexive by (9). Moreover the partitions α and β derivable from it are identical, respectively, to R^1 and R^2.

We have proved the following

Theorem 1. The antireflexive relation P is quasi-linear if and only if there exist uniquely-determined ordered partitions α and β of the set A which satisfy conditions (8) and (9).

Knowing now the structure of quasi-linear relations it is not difficult to obtain the characteristics of interval orders.

Theorem 2. An antireflexive relation is an interval order if and only if it is quasi-linear.

Proof. It is clear that for any map K the interval relation $I_a > I_b$ is antireflexive. On the other hand, corresponding to the set $P_K\langle a \rangle$ is the set of all objects b whose intervals I_b are located to the left of the interval I_a. In view of the one-dimensional nature of the real line such sets are ordered by inclusion, and consequently, the interval order P_K is quasi-linear.

We will now prove the reverse statement: If an antireflexive relation P is quasi-linear then it is represented by the relation ">" on a set of open intervals of the real line. Let a be any element of A. Since α and β are partitions, we can find unique numbers i and j such that $a \in A_i$ and $a \in B_j$. From (9) it is clear that $i < j$. Therefore the correct definition of the interval $I(a)$ is $I(a) = (i,j)$.

We will show that

$$(a,b) \in P \longleftrightarrow I(a) > I(b).$$

Let $(a,b) \in P$ and $a \in A_i$. Then $b \in P\langle a \rangle = \bigcup_{j=1}^{i} B_j$ from formula (8). This means that $b \in B_k$ for $k \leq i$. The right end point k of the interval

$I(b)$ is located to the left of the left end point i of the interval $I(a)$, i.e. $I(a) > I(b)$.

Suppose now that $I(a) > I(b)$. Then the left end i of $I(a)$ is to the right of the right end k of $I(b)$: $i > k$. By definition $a \varepsilon A_i$ and $b \varepsilon B_k$. Since $k < i$, $B_k \subseteq \bigcup_{j=1}^{i} B_j = P<a>$, so that $b \varepsilon P<a>$, i.e. $(a,b) \varepsilon P$. The theorem is proved.

In the proof we constructed a map with intervals $I(a)$ $(a \varepsilon A)$ for the quasi-linear relation P, for which P was an interval order. This map gave meaning to the ordered partitions α and β connected with P. The partition α characterizes the left ends, and β the right ends, of the intervals of the graph.

This map also allows the determination of characteristics of the inverse relation P^{-1} for an antireflexive quasi-linear relation, since it corresponds to the relation "to the left of" for intervals of the map (or to the relation "to the right of" for the map when turned 180°). The relation P^{-1} is also quasi-linear and is determined by the same partitions α and β (with a reversal in their order and the order of their classes).

To conclude this section we note that there is still another characteristic of quasi-linear relations, which is distinct from (3): in "local" terms of objects (and not in "global" terms of sets $P<a>$):

Theorem 3. The relation P is quasi-linear if and only if it satisfies the condition

$$aPb \text{ and } cPd \longrightarrow aPd \text{ or } cPb. \tag{10}$$

Proof. Suppose that P is quasi-linear and aPb, cPd, i.e. $b \varepsilon P<a>$ and $d \varepsilon P<c>$. We will show that aPd or cPb. Suppose the contrary: $(a,d) \notin P$ and $(c,b) \notin P$. This means that $d \notin P<a>$ and $b \notin P<c>$. On the other hand, by assumption, $d \varepsilon P<c>$, so that $P<c> \not\subseteq P<a>$, and $b \varepsilon P<a>$, so that $P<a> \not\subseteq P<c>$, but this contradicts the quasi-linear condition for $P<a>$ and $P<c>$.

Suppose now that for arbitrary a,b,c,d (10) is satisfied. We will show that P is quasi-linear. Suppose that $P<a> \not\subseteq P<c>$, where $P<a> \neq \phi$, $P<c> \neq \phi$, i.e. we can find $b \varepsilon P<a>$ such that $b \notin P<c>$, i.e. $(c,b) \notin P$. Then for any $d \varepsilon P<c>$ from (10) the relation aPd holds, i.e. $d \varepsilon P<a>$, so that $P<c> \subseteq P<a>$, as was to be shown.

We note that for algorithmic testing of the quasi-linearity of a relation condition (3) is more useful than (10) since it requires the fewest examinations of objects and their relationships. The properties of interval orders in terms of (10) were given by P. Fishburn [60] independently of the work of B. G. Mirkin [19], from whom the contents of this section were borrowed.

2. **The description of interval graphs.** We will examine, for an arbitrary map $K = \langle \mathbf{I}; I_i(i\varepsilon A)\rangle$, the intersection relation I_K: $(i,j) \varepsilon I_K \leftrightarrow I_i \cap I_j \neq \phi$, defined on the set A of mutations. Clearly I_K is connected with P_K in the following way: $I_K = \overline{P_K \cup P_K^{-1}}$. In other words those and only those intervals intersect, no one of which lies "to the left of" another. The relation I_K is called the *interval equivalence* relation for the map K. It is clear that I_K is reflexive and symmetric. However, it is not necessarily transitive: two intervals do not necessarily intersect though they both intersect a third interval lying "between" them.

What characteristics must an arbitrary reflexive, symmetric relation $I \subseteq A \times A$ have in order for it to be the interval equivalence for some map K? The answer to this question is the more interesting, since the initial complementation matrix gives precisely the intersection relation for intervals of the desired map. In other words, complementation mapping consists in constructing, for a given $I \subseteq A \times A$, a map K for which $I = I_K$.

It is customary to call an ordinary graph corresponding to an interval equivalence an *interval graph*. Following our practice of identifying corresponding notions in the terminology of relations, graphs and matrices (see the appendix), we will consider the notions of an interval equivalence and an interval graph synonomous.

The matrices of interval graphs possess a regular structure. If we renumber the objects (mutations) $i\varepsilon A$ corresponding to the ordering of the left ends of their associated intervals, then in the intersection matrix the ones in each row will follow in succession, beginning at the main diagonal:

$$r_{ij} \geq r_{i,j+1} \qquad (j=i, i+1, \ldots, N-1), \qquad (11)$$

for if I_i intersects I_{j+1} for $j \geq i$ then because of the renumbering I_i must also intersect I_j, the left end of which lies between those of I_i and I_{j+1}.

Matrices satisfying condition (11) are called *quasi-diagonal*.

Theorem 4. For a complementation matrix to admit a mapping it is necessary and sufficient that it be quasi-diagonalizable under a suitable renumbering of its objects (i.e. simultaneous permutation of rows and columns).

Proof. If a map exists then, as shown above, renumbering the objects to correspond with the ordering of left ends of intervals leads

```
      1  2  3  4  5  6  7  8  9 10 11 12
   1  1  1  1  1  1  1  1  1  1
   2  1  1  1  1  1  1  1  1
   3  1  1  1  1  1  1  1
   4  1  1  1  1
   5  1  1  1     1  1  1  1  1  1  1
   6  1  1  1     1  1  1  1  1  1
   7  1  1  1     1  1  1  1  1
   8  1  1        1  1  1  1  1  1  1
   9  1  1        1  1  1  1  1  1
  10  1           1  1  1  1  1  1  1
  11              1  1        1  1  1  1
  12              1           1     1  1  1
```

a)

```
                1
                    2
          3
                          5
                             6
                       7
                                8
                                   9
                                     10
                                        11
   4                                       12
   ←————————————— Complons —————————————→
```

b)

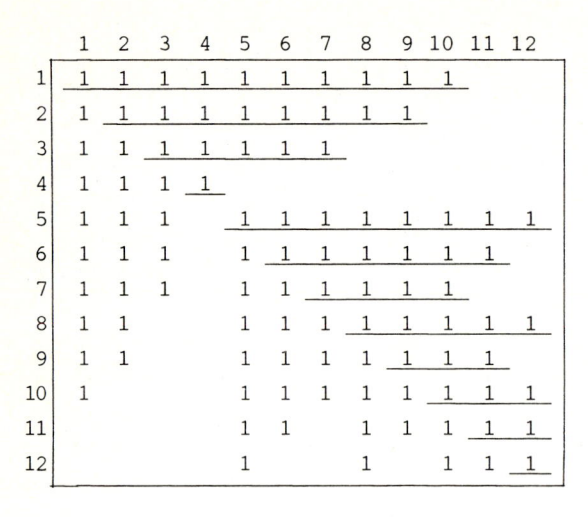

Figure 2. The matrix (a) and minimal map (b) of complementation at the locus pan-2 in the fungus *Neurospora crassa* [43]; the matrix is given in quasi-diagonal form.

to a quasi-diagonal matrix.

The converse remains to be proved: If a matrix can be brought to quasi-diagonal form, then there exists a map. To prove this we will show that the quasi-diagonal form of the matrix of (11) itself defines a map. As the intervals of this map we take the segments defined by $I(i) = (i, n_i)$, where n_i is the number (index) of the first zero element to the right of the diagonal in row i, so that $r_{i,n_i-1} = 1$, but $r_{i,n_i} = 0$ ($n_i > i$). Now the statement that $I(i) \cap I(j) \neq \phi$ means that there exists k such that $i < k < n_i$, $j < k < n_j$. Suppose for example, that $i < j$. Then clearly $i < j < k < n_i$ so that $r_{ij} = 1$ (the mutations i, j are non-complementary).

If, on the other hand, for $i < j$ $r_{ij} = 1$, then $i < j < n_i$ so that (i, n_i) and (j, n_j) intersect. This completes the proof.

The proof of theorem 4 shows that the quasi-diagonal form of a complementation matrix can be easily carried over to a map: It is sufficient, for example, to simply underline, in each row, the set of ones beginning at the diagonal and going to the right (Fig. 2,a). The segments thus defined correspond to intervals.

Hence the mapping problem

is directly connected with the problem of bringing a complementation matrix into quasi-diagonal form, the solution of either problem leading immediately to the solution of the other. However in terms of the complementation matrix itself there is little one can say about the construction of a "quasi-diagonal" ordering of the mutations. In order to describe the processes of local permutations the more "local" language of graphs and relations is required, to which we now turn.

From the fact that $I_K = \overline{P_K \cup P_K^{-1}}$ it is clear that I is an interval equivalence when and only when one can find an interval order (quasi-linear antireflexive relation) P such that $I = \overline{P \cup P^{-1}}$ (see Section 3). Therefore, to investigate I we must move to the complementary relation $\overline{I}$. If I is an interval equivalence, then we can find an interval order P such that $I = \overline{P \cup P^{-1}}$. The graph of P is obtained from the complementary graph $\overline{I}$ by orienting all of its edges. If for a given ordinary graph I there exists such an orientation of its complement $\overline{I}$ satisfying the quasi-linear condition, then I is an interval graph. We will reformulate the condition for the existence of a quasi-linear orientation in the local terminology of condition (10).

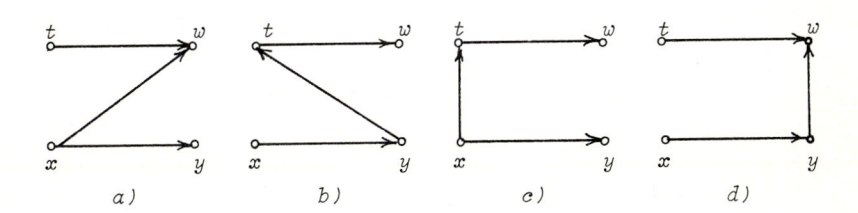

Figure 3

In (10) the objects a, b, c and d are arbitrary and some of them may coincide (though of course $a \neq b$ and $c \neq d$, since P is antireflexive). If, of these four, only two actually differ then clearly condition (10) places no additional restrictions on P. If three of the objects are different it is easy to show by direct substitution that condition (10) becomes simply the transitivity condition for P. For example, if $b=c$, then we have aPb and $bPd \to aPd$ or bPb, since bPb is impossible aPb and $bPd \to aPd$, which is transitivity. Finally, if all four of the objects are distinct, condition (10) (together with the transitivity of P) asserts that if in $\overline{I}$ there are two non-adjacent edges on the set $\{a,b,c,d\}$, then there must be a third, adjacent to both of them (i.e. "parallel" edges are always connected by a third edge). If (10) holds, then the presence of a connection between parallel edges in $\overline{I}$ follows trivially from it, if the arcs of P are deoriented. Conversely, if two edges in

$\bar{I}$ are connected, then under orientation of $\bar{I}$ we obtain one of the four configurations of Fig. 3, where the "parallel" edges are shown horizontally, and cases a) - d) reflect all possible placements of the connecting arc, obtained from an original edge by orientation.

In case a) condition (10) is fulfilled automatically; in case c) for vertices x,t,w, in case d) for vertices x,y,w, but in case b) for x,y,t and then x,t,w the relation P must contain, by transitivity, an arc (x,w), leading to fulfillment of (10) for all four objects. Analyzing this argument, we conclude that the existence of a quasi-linear P is equivalent to the existence of a transitive P and the presence of connections between parallel edges in $\bar{I}$ (or, what amounts to the same thing, fulfillment of (10) for any four distinct objects a,b,c,d).

The second condition can be easily reformulated as follows. We will say that an edge u is a *triangulator* of adjacent edges xy and yz if $u=xz$, i.e. the vertices x,y and z form a "triangle." A *triangulator of a cycle* is a triangulator of any two of its adjacent edges.

The existence of a connection between every pair of non-adjacent edges in $\bar{I}$ is equivalent to the existence of a triangulator for every "quadrilateral" (cycle on four vertices) of the initial graph I. The fact that the edges xy, yz, zw and wx form a cycle without a triangulator in I means that in $\bar{I}$ there are two non-adjacent edges xz and yw without any connector, and this proves the assertion. We sum up our discussion in

<u>Theorem 5</u>. An ordinary graph I is an interval graph if and only if it satisfies one of the equivalent conditions:

a) The complementary graph $\bar{I}$ is transitively oriented, with every quadrilateral in I containing a triangulator;

b) The complementary graph $\bar{I}$ is transitvely oriented, with each of its transitive orientations P satisfying the quasi-linear conditions (10) or (3).

Condition b) needs some comment. The initial assertion of the theorem requires the existence of at least one quasi-linear orientation. Why then did b) refer to the quasi-linearity of every result of a transitive orientation of $\bar{I}$? The fact is that the triangularity condition for quadrilaterals in I is in no way connected with the orientation of $\bar{I}$, and therefore, in connection with the discussion of Fig. 3 we must conclude the quasi-linearity of any transitive P for which $P \cup P^{-1} = \bar{I}$.

This type of result: from one fact formulated in terms of the initial relation I, are deduced consequences true for an arbitrary orientation of $\bar{I}$, will be encountered repeatedly in the following analysis.

We remark that condition a) was first proved by Gilmore and Hoffman [67] in a direct investigation of interval graphs (but not by reducing it to a case of quasi-linearity).

Theorem 5,a) shows that the checking of the interval equivalence of I leads to the investigation of all quadrilaterals in I, and to the elucidation of whether $\bar{I}$ is transitively orientable.

As will become clear in what follows, the most simple transitive orientability is verified by direct construction of the corresponding transitive P. In this case it is not necessary to consider all possible quadruplets of objects: In accord with condition b) of the theorem it is sufficient to verify the quasi-linearity of this P. This is much easier than examining quadrilaterals, since it only requires a single examination of P, with the goal of constructing all sets of the form $P<a>$, and the subsequent verification of whether or not they are completely ordered by inclusion. The construction of the sets $P<a>$ also directly provides a map, representing I, as was described in the proof of theorem 2.

However, the formulation a) of theorem 5 represents supplementary interest as a step toward characterizing interval graphs using only interval graph terminology, without bringing to bear "additional" operations such as going to complementary graphs and complete or partial orderings of vertices. For such an "internal" description in agreement with this formulation it remains to characterize those graphs which admit transitive orientations. This question has independent significance of its own, but the extent of the present work precludes its examination. Existence criteria and algorithms for transitive orientation are considered in [7]. We will use the *Gilmore-Hoffman criterion*, which states that an ordinary graph admits a transitive orientation if and only if every nonrepeating cyclic route of odd length in it has at least one triangulator [67,7].

At present we know that I is an interval graph if and only if all quadrilaterals in I and all cyclic routes of odd length in $\bar{I}$ admit triangulators. The remaining step is to move from conditions on $\bar{I}$ to conditions on I.

The conditions for the transitive orientability of $\bar{I}$ mean that for any sequence of vertices $a_0, a_1, \ldots, a_{2n}$ such that among the pairs (a_i, a_{i+1}) $(i=0,\ldots,2n)$ none are identical, the condition

$(a_0, a_1) \in \bar{I}$ and $(a_1, a_2) \in \bar{I}$ and $\ldots$ and $(a_{2n}, a_0) \in \bar{I} \longrightarrow$

$$\exists i \leq 2n [(a_i, a_{i+2}) \in \bar{I}]$$

is fulfilled. For indices k greater than $2n$ we will find it necessary to use instead indices obtained as remainders on division of $k-1$ by $2n$.

For example, 0 in place of $2n+1$, 1 in place of $2n+2$, etc.

In terms of I this condition, having the form "$u \to v$", is reformulated in the inverted form of "not $v \to$ not u":

$$\left(\forall i \; [a_i I a_{i+2}]\right) \to \exists j \; [a_j I a_{j+1}].$$

We renumber the elements of the sequence $a_0, \ldots, a_{2n}$ as follows: At the beginning we place elements with even indices, and following them, those with odd indices (in order by indices): $a_0, a_2, a_4, \ldots, a_{2n}, a_1, \ldots, a_{2n-1}$. Then the condition of transitive orientability of $\bar{I}$ in the terminology of I means that if the sequence $a_0, a_2, \ldots, a_{2n}, a_1, \ldots, a_{2n-1}$ is such that among the pairs (a_i, a_{i+1}) there are no repetitions, and it forms a cyclic route, then in I we can find an edge which divides it "into two parts" (the edge $a_j a_{j+1}$ joins an "even" vertex with an "odd" vertex and divides the set of vertices of the route into two parts with n and $n+1$ vertices).

It is easy to conclude that each of the parts thus obtained also gives a cyclic route, and one of them contains an odd number of vertices, so that there must again exist an edge dividing it in two, and so forth, to the point when one of the parts contains three vertices, so that the dividing edge (which led to this part) is also a triangulator of the initial route. We thus arrive at the result that fulfillment of the Gilmore-Hoffman criterion for $\bar{I}$ leads to its fulfillment for I also, so that I and $\bar{I}$ simultaneously admit (or do not admit) a transitive orientation.

However this is not so. Examine the graph of Fig. 4 a), which represents the map of Fig. 4 b).

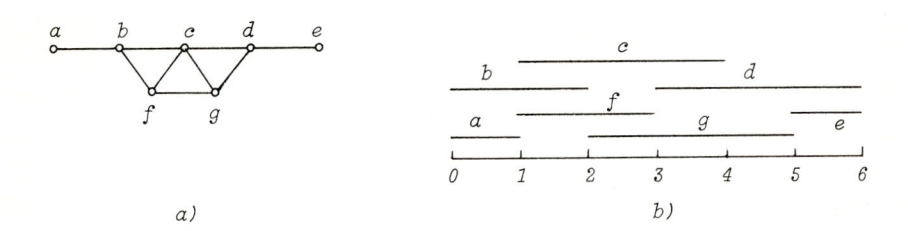

Figure 4

In this graph the cyclic route $abcdedgfba$ has no triangulator, although its complementary graph unquestionably admits a transitive orientation (the ordering $abfcgde$ leads to quasi-diagonal form for the matrix of the graph).

The mistake in our earlier argument lies in the assertion that the process of successive divisions necessarily leads to a triangle. This is true only in the case when each dividing edge is not contained in the route, as for example, when the route is properly a cycle. If some edges are encountered twice (in a nonrepeating route this is the maximum number of times an edge can be encountered), then the dividing edge may belong to the route and yet not be a triangulator. In our example either of the edges fc and cg are dividers for the route $abcdedgfba$. Selecting fc, we obtain $abcfba$ as a route of odd length, for which the edge bf is already a divider. Yet bf belongs to the route and is not a triangulator.

A similar mistake appeared in [83] and was corrected by Fisher [61]. Thus, in general, the conclusion we reached does not hold. However, it is true for cycles of odd length: if the Gilmore-Hoffman condition is satisfied for $\overline{I}$, then in I each cycle of odd length has a triangulator. In fact there is a stronger property of graphs of intervals: each cycle in I has a triangulator.

To prove this assertion we suppose there exists in I a cycle of odd length without a triangulator (it is also possible to prove this by direct consideration of the map representing I). Consider the one (cycle) among them having minimal length. This cycle clearly has no chords. For, from its minimality, each of its chords must divide it into two "subcycles" of odd length, and these cycles, from the foregoing, must have triangulators. These triangulators are either triangulators of the initial cycle, which is not possible by definition, or they are chords of the initial cycle, which divide it into subcycles of even length, which is impossible due to its minimality.

Thus, in $\overline{I}$ all the vertices of a given cycle, except neighboring ones, are connected with one another by edges. We consider the first five vertices $1,2,3,4,5$ (the situation of only four vertices is not possible by theorem 5,a). From the above, there is in $\overline{I}$ a cyclic route of length 5: 352415 which has no triangulators: any of its pairs which are connected by edges have, as triangulators, edges of the original cycle in I. This contradicts the Gilmore-Hoffman condition.

Thus, in an interval graph all cycles must have triangulators. A graph in which all cycles possess triangulators is said to be *triangulated* We have shown the triangulatedness of interval graphs.

Is the triangulatedness condition for a graph not only necessary but also sufficient to characterize interval graphs?

To answer this question we have to determine whether, in the complement $\overline{I}$ of a triangulated graph I, the cyclic routes of odd length must be triangulated or not.

An edge of a cyclic route in $\overline{I}$ either is encountered twice, or belongs to a cycle within the route, and consequently, if the route has no cycles of odd length, the overall length of the route must be even. Therefore a cyclic route of odd length must contain a cycle of odd length. This cycle clearly must be trangulated, since in the complementary graph $\bar{\bar{I}} = I$ all cycles of odd length are triangulated, and this, as we saw, leads to the existence of triangulators also in the cycles of the "original" graph $\bar{I}$.

Are the triangulators of this cycle also trinagulators of the route? Not if the triangulated neighboring edges are not encountered in the route one after the other, but "separately." For this to be possible it is necessary to have edges in the route which are incident on the cycle at vertices joining these edges, as illustrated in Fig. 5.

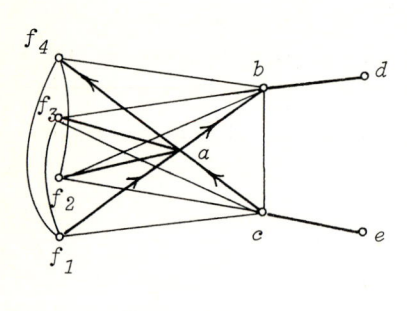

Figure 5

Here ca and ab are successive edges of the cycle under consideration, with triangulator bc. The route "enters" the cycle along edge f_1a, continues in the "upper" part of the cycle ab... (possibly with exits from the cycle such as $af_4af_3af_2af_1$ at the point a, because of the necessity of the absence of triangulators), and returns to f_1 by the route $caf_4af_3af_2af_1$. To make Fig. 5 more readable edges belonging to the route are drawn darkly.

In the simplest case a cycle traversable by a route is a triangle abc such that for all three vertices of the cycle there are "branches" of the type illustrated for vertex a of Fig. 5 (naturally, the number of edges leaving these vertices need not be four) in order to have the non-triangulatedness of the route.

It is easy to understand that in the language of the original graph I, such a route in $\bar{I}$ is non-triangulated if and only if the vertices a,b,c form in I a so-called *asteroidal triplet*. A collection of three pair-wise nonadjacent vertices a_1, a_2, a_3 is called an asteroidal triplet of a given graph, if for each pair a_i, a_j ($i \neq j$) there exists a chain joining them, such that none of its vertices is adjacent to the third vertex a_k ($k \neq i,j$). For example, for the graph $\bar{I}$ of Fig. 5 the vertices b and c are joined in I by the chain $bf_1f_2f_3f_4c$ not adjacent to a.

Thus, for I to be an interval graph it is necessary for it not only to be triangulated but also that it have no asteroidal triplets: This ensures the absence in the complementary graph of non-triangulated routes of odd length encountered in three-vertex cycles. In fact, a more general statement is true: The absence of asteroidal triplets in a triangulated I ensures the absence of non-triangulated routes of odd length in I, i.e. the representability of I by a map. This means that the following criteria, formulated in terms of I, hold.

A graph is said to be *asteroidal* if it contains asteroidal triplets.

Theorem 6. An ordinary graph is an interval graph if and only if it is triangulated and is not asteroidal.

This theorem was first established by Boland. Its proof, based on a direct inductive (on the number of vertices) construction of a map representing a given triangulated non-asteroidal graph, was published in a paper by Lekkerkerker and Boland [80]. A main notion in their construction was that of a simplicial point, corresponding to a complon of the map (see below, §3).

We leave the proof of Theorem 6 from Theorem 5 as an (quite trivial) exercise for the reader (see also [96]).

In concluding this section we note that Theorem 5 allows one to obtain very elegant criteria (although little suited to practical constructions) for the representability of a graph in terms of the absence in it of graphs of a special form (that reminds one of the well-known planarity criterion for graphs [78]).

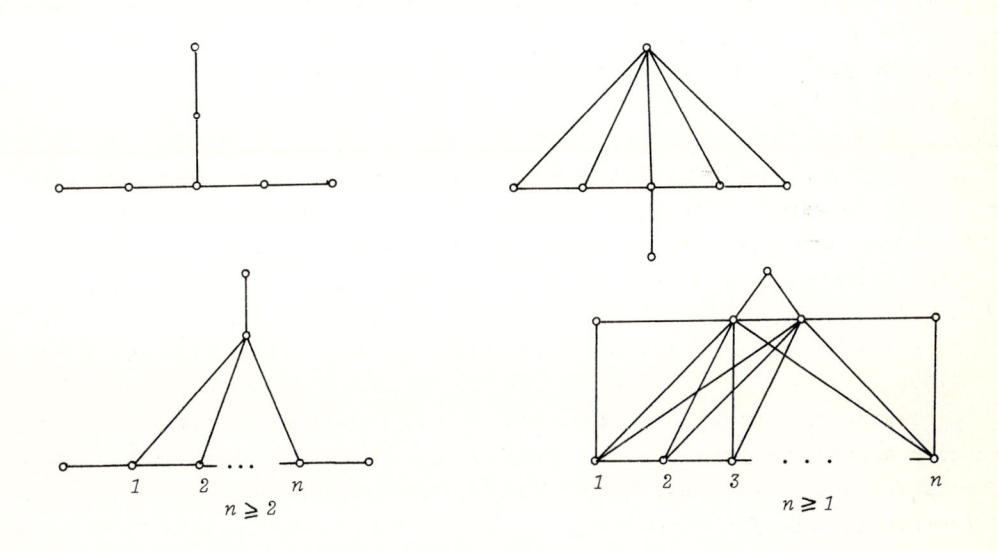

Figure 6

In [80] it was established that triangulated non-asteroidal graphs containing no proper asteroidal subgraphs, cannot contain subgraphs of any of the four types illustrated in Fig. 6. This means that the following criterion holds:

Theorem 7. A graph is an interval graph if and only if it contains no subgraphs of the types illustrated in Figs. 6 and 7.

Fig. 7 illustrates the minimal non-triangulated graphs.

3. **Graphs of non-covering intervals.** We will consider those interval graphs which are represented by maps with intervals I_i ($i \in A$), no one of which is contained in another, i.e. $I_i \subseteq I_j \rightarrow I_i = I_j$. We call these *graphs of non-covering intervals*.

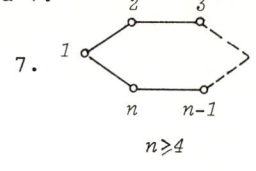

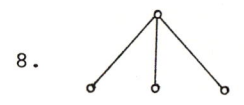

Figures 7 and 8

For such graphs it is not possible, in particular, to have a situation in which one interval intersects three other mutually intersecting intervals, i.e. inside one mutational defect there are not three other mutually complementary distortions. This fact characterizes graphs of non-covering intervals.

Theorem 8. An interval graph I is a graph of non-covering intervals if and only if it contains no subgraphs of the form shown in Fig. 8.

This theorem was proved by Roberts [89] and at the same time in undergraduate work done under the direction of one of the authors by V. A. Kogan (Novosibirsk State University, 1970). It will not be proved here. We note only that the proof by V. A. Kogan was a simple translation to the language of interval equivalences of the following result about interval orders (analogous to obtaining Theorem 5,a from Theorem 4).

Theorem 9. An antireflexive quasi-linear relation $P \subseteq A \times A$ is a relation of an interval order for a map of non-covering intervals if and only if it satisfies one of the following equivalent criteria:

$$(a,b) \ \varepsilon \ P \text{ and } (b,c) \ \varepsilon \ P \rightarrow (a,d) \ \varepsilon \ P \text{ or } (d,c) \ \varepsilon \ P, \tag{11}$$

$$P\langle a \rangle \subseteq P\langle b \rangle \text{ or } P^{-1}\langle a \rangle \subseteq P^{-1}\langle b \rangle. \tag{12}$$

Relations of an interval order for a map of noo-covering intervals were first considered in the theory of psychological measurements by R. Luce [82], and by P. Suppes and D. Scott [91], who called them semi-orders. The characterization of semi-orders in the "local" terms of (10) and (11) was done by the authors. Conditions (3) and (12) were obtained in B. G. Mirkin [19].

The necessity of condition (12) for non-covering intervals is determined by its geometric meaning: For any intervals a and b the right or left end of interval a is to the right of or to the left of the corresponding end of interval b. The sufficiency follows easily from the structure of Theorem 3: If $I(a) = (i,j)$ and $I(b) = (i',j')$ with $i < i'$,

$j > j'$, then from (8) $P^{-1}\langle a \rangle \subset P^{-1}\langle b \rangle$ and $P\langle a \rangle \subset P\langle b \rangle$, where the inclusion is strict, since for b and a condition (12) is not satisfied.

This shows that on the map obtained in Theorem 3 the covered interval and the covering interval must coincide at their left or right ends. Such a map can be changed into a map for non-covering intervals without violating the intersection relation of the intervals. If, for example, the left ends i, i' of the intervals $I(a)$ and $I(b)$ coincide, where $I(a) \subset I(b)$, then each interval whose left end is not greater than $i=i'$, "behaves" identically with respect to $I(a)$ and $I(b)$: simultaneously intersects or does not intersect them both. Therefore we can introduce at the point $i=i'$ a unit interval, including it in all intervals of the map which contain $i=i'$ as well as in $I(a)$. Now $I(a) \not\subset I(b)$ and the intersection relationship is unchanged.

By carrying out suitable insertions for all cases of inclusion of intervals we obtain a map of the desired form, as was to be shown. We leave the proof of the equivalence of (11) and (12) to the reader.

We will also give, without proof, a characterization of the matrices of non-covering intervals.

A Boolean matrix will be called *row-linear* (or *column-linear*) if in each of its rows (or columns) the units (ones) are arranged consecutively. A matrix is said to be *linear* if it is row-linear or column-linear. For symmetric square matrices the three notions coincide.

Theorem 10. A graph is a graph of non-covering intervals if and only if its matrix can be brought into linear form by simultaneously permuting rows and columns.

Clearly linear matrices are quasi-diagonal. However, the construction of an ordering of vertices (i.e. of rows and columns) which will produce a matrix of linear form is easier than ordering the vertices of an arbitrary interval graph. An algorithm for such a construction is described in detail in Section 4.1.

§3. The mathematical theory of linear maps: interval hypergraphs

1. **Maps and interval hypergraphs**. The results described in the preceding section give the characteristics of those complementation data representable by maps. They may also be used to construct such maps. These results are not sufficient, however, for the analysis of real complementation data.

This is primarily because of the many possible representations (in general) of a complementation matrix by graphs: A single initial data set may correspond to many maps on which the same mutations are representable by many intervals of various lengths. To obtain an encompassing genetic interpretation one must try to identify the invariants of the maps, determined only by the original complementation matrix.

Of no less interest is the question of what happens if a complementation graph is not an interval graph (possible reasons for this are discussed in Section 5.2). Real complementation matrices, involving 20 or more mutations cannot usually be brought into quasi-diagonal form. Can the techniques of interval graphs really help in the investigation of this situation?

The difficulties we described (non-uniqueness and impossibility of an interval representation of complementation data) are not so heterogeneous as may appear at first glance. The fact is that the invariants of maps must somehow or other reflect the influence of the elementary complementation units, i.e. *"complons"*, and consequently admit characterization in terms of "complons." The non-representability of complementation graphs is naturally interpreted in these terms as the impossibility of a non-contradictory "linear arrangement" of complons. This naturally leads to consideration of the possibility of a "nonlinear" arrangement of complons for real complementation matrices.

To pursue this path we must study the maps themselves and not just the intersection graphs of their intervals.

The set-theoretic object corresponding to a map is an interval hypergraph. We will call a system of subsets S_1, S_2, ..., S_N of a set $X = \{x_1, x_2, ..., x_n\}$ an *interval hypergraph*, $\Gamma = (X, \{S_i\}_{i \in A})$ if there exists an ordering of the set X of vertices for which the sets S_1, S_2, ..., S_N are intervals. As usual a set S is called an *interval of the ordering* P, if for any two objects x, y contained in S, all objects in P lying between x and y are also contained in S.

There is a one-to-one correspondence between maps and interval hypergraphs. The set of vertices of an interval hypergraph Γ is formed by all the unit map intervals x_i of the form $(i, i+1)$ $(i=0, 1, ...)$, and a set S_j consists of those unit intervals x_i for which $x_i = (i, i+1) \subset I_j$. Conversely, every hypergraph of intervals gives a map whose intervals I_j correspond to the sets S_j $(j=1, 2, ..., N)$.

What features are peculiar to the matrices of interval hypergraphs? Clearly for that ordering of the rows (and of the set X) for which all the S_j become intervals of X, in each column of the matrix the ones are situated consecutively. In the inverse case a set S_j corresponding to a column in which this condition is violated is not an interval for the given ordering of X. This means that a rectangular Boolean matrix is the matrix of an interval hypergraph if and only if it can be brought to column-linear form by some permutation of its rows. This permutation of the rows completely characterizes the map of the interval hypergraph

Thus an interval hypergraph is still not a map: The set X of vertices of the hypergraph differs from the base interval $\mathbf{I}$ of a map in that X is not ordered as $\mathbf{I}$ is. X is an unordered collection of unit intervals.

There are still other properties of interval hypergraphs. We consider the relations γ_X and γ_A associated with the hypergraph $\Gamma = (X, \{S_i\}_{i \in A})$:

$$(S_k, S_l) \varepsilon \gamma_A \leftrightarrow S_k \cap S_l \neq \phi,$$

$$(x_i, x_j) \varepsilon \gamma_X \leftrightarrow \exists k \varepsilon A [x_i, x_j \varepsilon S_k].$$

Clearly Γ is an interval hypergraph if and only if γ_A is an interval graph. Moreover, any map representing γ_A yields an interval hypergraph Γ' with $\gamma_A' = \gamma_A$.

At the same time, the relation γ_X does not generally provide information about the characteristics of the interval hypergraph. By the definition of γ_X

$$\gamma_X = \bigcup_{k \in A} S_k \times S_k.$$

Thus if $S_l \subseteq S_k$, knowledge of γ_X does not provide any information about S_l, since $S_l \times S_l \subseteq S_k \times S_k$. The relation γ_X is defined by only those "edges" S_k ($k \varepsilon A$) which are not proper subsets of other edges. We will designate the collection of all indices of such maximal edges of the hypergraph by A'. Then

$$\gamma_X = \bigcup_{k \in A'} S_k \times S_k$$

Moreover, from γ_X it is easy to reconstruct all the maximal edges: They are simply the maximal cliques of the graph of γ_X. Non-maximal edges are subsets of the maximal cliques, but information only about γ_X does not allow the determination of which subsets are edges and which are not.

An ordering of the set X for which all the maximal cliques of γ_X turn out to be intervals, provides a partial interval hypergraph only for the maximal edges. The non-maximal edges will not necessarily be intervals of this ordering.

Ordering the elements k of the set A' according to the order assigned the left ends of the intervals S_k in X, we find that in the matrix of the partial interval hypergraph $\Gamma' = (X, \{S_k\}_{k \in A'})$ the ones in each row are consecutive, since the S_k do not cover one another, and consequently the same thing holds for an ordering of the right ends of S_k ($k \in A'$). This means that Γ' is row-linear, and consequently its transpose matrix Γ'^T also corresponds to an interval hypergraph

$$\Gamma'^T = (A', \{T_i\}_{i \in X}),$$

where

$$T_i = \{k | x_i \in S_k\}.$$

Thus the graph γ_X is a graph of non-covering intervals if and only if the partial hypergraph defined by its maximal edges is an interval hypergraph.

In particular, if all the edges of the original hypergraph are maximal (i.e. the hypergraph corresponds to a graph of non-covering intervals) then the graphs γ_X and γ_A are simultaneously interval graphs or not interval graphs.

The matrices G of the relations γ_X and γ_A may be considered as Boolean products of the $n \times N$ matrix of the hypergraph Γ and the transposed $N \times n$ matrix Γ^T.

$$G_X = \Gamma \times \Gamma^T \quad \text{and} \quad G_A = \Gamma^T \times \Gamma.$$

Boolean matrix multiplication is the same as the usual matrix multiplication except that in place of (scalar) multiplication the operation of taking the minimum (the *conjunction* or ANDing) is used, and in place of addition the operation of taking the maximum (the *disjunction* or ORing) is used. In fact, the (i,j)th element of the Boolean matrix $\Gamma \times \Gamma^T$ is equal to 1 if and only if there exists a column of the matrix Γ in which ones appear at positions i and j, i.e. there exists an edge S_k containing both x_i and x_j. Similarly, the (k,l)th element of the Boolean matrix $\Gamma^T \times \Gamma$ is equal to 1 if and only if there exists a row in which ones appear at positions k and l, i.e. there exists an x_i contained both in S_k and S_l, so that $S_k \cap S_l \neq \phi$.

Curiously, in an analogous way an interval hypergraph Γ can be characterized in terms of the usual matrix products $\Gamma\Gamma^T$ and $\Gamma^T\Gamma$. For an arbitrary hypergraph Γ the (i,j)th element of the $n \times n$ matrix $\Gamma\Gamma^T$ specifies the number of edges of the hypergraph containing both x_i and x_j, and similarly the (k,l)th element of the $N \times N$ matrix $\Gamma^T\Gamma$ specifies the cardinality $|S_k \cap S_l|$.

Clearly if Γ is an interval hypergraph, and X is ordered according to an interval representation of Γ on a map, then the symmetric matrix

$\Gamma\Gamma^T$ has a simple organization: In each row the maximal element appears on the main diagonal, and the values of elements decrease (more accurately, do not increase) monotonically in both directions from this diagonal element. This property is an extension of the linearity property of Boolean matrices. We will therefore call a square Boolean matrix $B = \|b_{ij}\|$ *row-linear* (*column-linear*) if in each of its rows (columns) the elements are monotonically non-increasing in both directions from the diagonal element, i.e.

$$b_{ij} \geq b_{ik} \text{ for } i \leq j \leq k \text{ or } i \geq j \geq k$$

for rows, and

$$b_{ji} \geq b_{ki} \text{ for } i \leq j \leq k \text{ or } i \geq j \geq k$$

for columns.

A *linear* matrix is one that is both row-linear and column-linear. For symmetric matrices all three conditions are, of course, equivalent.

We give the linearity property of the matrix $\Gamma\Gamma^T$ for interval hypergraphs in the form of the following theorem [75]:

Theorem 1. If the Boolean matrix Γ can be brought into column-linear form by some permutation of its rows, then the same permutation (of rows and columns, simultaneously) will bring the matrix $\Gamma\Gamma^T$ into linear form.

The analogous property does not generally hold for the matrix $\Gamma^T\Gamma$ because of the possibility of the covering of intervals. For example, the intervals of the ten-element set $X = \{1,...,10\}$, $S_1 = (1,2,3,4,5)$, $S_2 = (2,3,4)$, $S_3 = (4,5,6,7,8)$, $S_4 = (5,6)$, $S_5 = (6,7,8,9,10)$ form the submatrix of $\Gamma^T\Gamma$:

$$\begin{Vmatrix} 5 & 3 & 2 & 1 & 0 \\ 3 & 3 & 1 & 0 & 0 \\ 2 & 1 & 5 & 2 & 3 \\ 1 & 0 & 2 & 2 & 1 \\ 0 & 0 & 3 & 1 & 5 \end{Vmatrix},$$

which cannot be brought to linear form (by the theory of linear matrices, see §4).

It can be proved, nevertheless, that the matrix $\Gamma^T\Gamma$ carries quite complete information about the hypergraph Γ [66]:

Theorem 2. Suppose, for two rectangular Boolean matrices Γ and Δ, the condition $\Gamma^T\Gamma = \Delta^T\Delta$ is satisfied. Then these two matrices simultaneously either are or are not matrices of interval hypergraphs. If Γ is an interval hypergraph and Δ has the same number of rows as Γ, then Δ is a matrix of the same interval hypergraph as Γ.

The proof follows from the results of §4 (page 59).

A description of interval hypergraphs without regard to the ordering of objects, may be obtained, by using Theorem 2.7, on the basis of the analysis of hypergraphs whose γ_A-graphs are illustrated in Figs. 6 and 7.

A. Tucker proved the following statement [96]:

Theorem 3. A rectangular matrix is the matrix of an interval hypergraph if and only if it contains no submatrices of the forms I_n, II_n, III_n ($n \geq 1$) and IV, V:

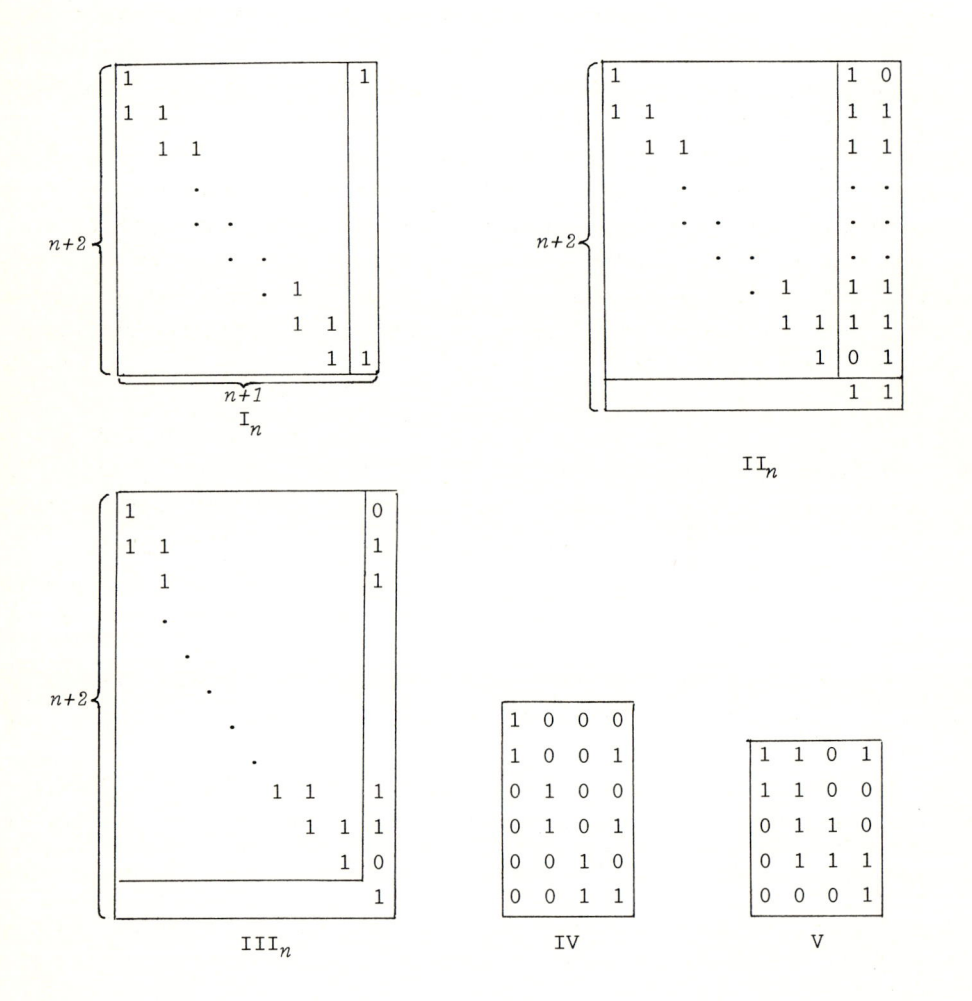

$$I_n \quad II_n \quad III_n \quad IV \quad V$$

$$IV = \begin{pmatrix} 1 & 0 & 0 & 0 \\ 1 & 0 & 0 & 1 \\ 0 & 1 & 0 & 0 \\ 0 & 1 & 0 & 1 \\ 0 & 0 & 1 & 0 \\ 0 & 0 & 1 & 1 \end{pmatrix} \quad V = \begin{pmatrix} 1 & 1 & 0 & 1 \\ 1 & 1 & 0 & 0 \\ 0 & 1 & 1 & 0 \\ 0 & 1 & 1 & 1 \\ 0 & 0 & 0 & 1 \end{pmatrix}$$

2. **Minimal hypergraphs and complons.** The invariants of the initial information are not reflected in every representative map, but only in the minimal one. This and the following subsections of this section are

devoted to sharpening this thesis.

For a given interval graph I, we will say that a representative map is *minimal* if it has the shortest length among all the representative maps of I. The set of vertices of the associated hypergraph has minimal cardinality with respect to all Γ's for which $\gamma_A = I$. This interval hypergraph will also be called *minimal*.

It may seem that to obtain the minimal map we need only remove some objects from A, leaving only those whose *neighborhoods* $I<a>$ differ; and similarly: some of the vertices of X, leaving only those to which correspond differing rows of the representative hypergraph. But this is not so.

In fact, the construction of the minimal map is given by Theorem 3 §2: for any map, in terms of the associated interval order the left ends of those intervals with identical inverse images $P^{-1}<a>$ are identified (as one), and similarly, with the right ends of intervals whose images $P<a>$ are identical. By this technique we obtain a map of length m, and this length cannot be decreased without violating the intersection relation. The fact is that, by the construction of this map, for every "internal" natural number k ($0<k<m$) there exist intervals for which k is the left end and intervals for which k is the right end (since the classes A_k and B_k are non-empty).

However, it does not follow that the map thus constructed is minimal: The size of m may, generally speaking, vary from map to map (the maps being taken as initial maps). The minimality of such a map follows later (see Corollary 2), and therefore we will not give a direct proof here.

For the time being we simply note the mentioned necessary condition for minimality of a map:

Lemma 1. A map $<I, I_i (i \varepsilon A)>$ is non-minimal if there exists an internal natural number $k \varepsilon I$ which is not the right (or left) end of any of the intervals I_i $(i \varepsilon A)$.

In fact, if k is not the right end of any interval, the intersection of the intervals is still maintained if we eliminate the unit interval $(k-1, k]$, i.e. "shorten" the map so that the points $k-1$ and k fall together (without changing any remaining parts of the map).

Actually, the exclusion of a unit interval can only diminish the number of intersecting pairs of intervals. For intervals I_i and I_j become non-intersecting if the left end of one of them is at $k-1$ and the right end of the other is at k. But this is not possible by definition. Q.E.D.

In reality the reverse assertion is true: A map is minimal if and only if each internal natural k is simultaneously both a left and a right end of some intervals of the map (see Corollary 2).

We will call the unit intervals of a minimal map *complons* [23,40]. They correspond to the vertices of the associated hypergraph.

The genetic equivalent of a complon is an elementary portion of a genetic locus, distinguishable by complementation testing of a given set of mutants. It is known that if the number of mutants is varied the number of complons may vary also.

An object $i \in A$ is said to be *complonic* if the associated interval of the minimal map is a complon (later we will show (Corollary 1) that this definition is justified: A complonic object in any minimal map is represented by a complon).

Let $I \subseteq A \times A$ be a reflexive symmetric relation, characterized by a complementation matrix. Consider the following condition:

$$I<i> \subseteq \bigcap_{j \in I<i>} I<j> \tag{1}$$

To check this condition only the initial matrix is used, not requiring even that it be reducible to quasi-diagonal form.

Clearly, vertices satisfying (1) are characterized by the condition that the collection $I<i>$ be a complete subgraph-clique. This clique is clearly maximal. Such vertices were called *simplicial vertices* by Boland and Lekkerkerker [80]. They used the properties of such vertices in proving their interval criteria for graphs.

Independently in [50] it was noted that a *mutation* $i \in A$, for which (1) is not satisfied, cannot be last in the quasi-diagonal form of the complementation matrix. This fact was of fundamental use in [50], where V. V. Shkurba developed an algorithm to bring a matrix to quasi-diagonal form. In accordance with this algorithm, at each step of the construction of the desired ordering the simplicial points (among those points not yet ordered) are examined. Each of these provides its own "branch" for construction of the desired ordering.

The authors observed in [23] that condition (1) actually characterizes complonic mutations.

Theorem 4. An object $i \in A$ is complonic if and only if it satisfies condition (1).

Proof. If an object $i \in A$ is complonic then (1) clearly holds, since every interval of the map intersecting $I_i = (k, k+1)$ contains I_i.

Conversely, let $i \in A$ satisfy condition (1). Suppose it is not complonic. This means that for some k I_i includes the interval $(k, k+2)$ of length 2. By the minimality of the map, and Lemma 1, there exists an interval I_s for which the point $k+1$ is the right end, and an interval I_t for which $k+1$ is the left end.

These intervals do not intersect: $I_s \; I_t = \phi$; yet clearly $s, t \varepsilon I \langle i \rangle$ and by (1) s and t must be adjacent. This contradiction proves the theorem.

Condition (1) is formulated in terms of the initial matrix of complementations and is not dependent on a specific map. Therefore the following corollary flows directly from Theorem 4:

Corollary 1. Every minimal map which represents a given interval graph has the same set of objects as its complons.

We have already noted that the real prototypes of complons are sections of genes, which are able to mutate both separately (complonic mutations) and in various combinations (non-complonic mutations). Therefore in the general case Corollary 1 shows the possibility of dependably revealing such elementary fragments of the genetic system.

It is necessary, however, to make a significant stipulation. In intracistronic complementation phenotypic effects are connected with the quaternary structure of the proteins, or more precisely, with the constitution of the functional centers of protein molecules (see Section 2.1.1), each of which appears under complementation testing as a unit. Consequently it would seem, for sufficiently complete and detailed initial data, including a large number of mutations affecting only one center, condition (1) would allow the unique identification of all the special centers of a protein molecule. From this one understands the importance of Corollary 1, according to which, such real complons are independent of the specific map invariants.

However, as we shall see below (Section 2.1.1), this very rule of complementation mapping is found to be in serious conflict with mechanisms of mutational complementation at the level of proteins with quaternary structure. Therefore in §2.1 we give a more adequate procedure for representing intracistronic complementation by graphs of another type, which reflect the real picture of the interconnections of the functional centers in complex proteins.

Complons of a minimal map are *real* if they correspond to complonic objects (i.e. the corresponding unit intervals are map intervals). Other (non-real) complons are *fictitious*.

We first examine the case when all the complons are real. In this case the set of complonic objects determines the set of vertices of the corresponding hypergraph; with the remaining objects associated with subsets of complonic objects, adjacent to them. Thus, condition (1) in this case permits the construction of a minimal interval hypergraph, with its set of vertices formed by the set of objects i, satisfying (1), and having as its edges subsets of complonic objects, adjacent to each of the remaining objects.

In this case the construction of a minimal map leads to an ordering of the hypergraph vertices for which all edges are intervals. The ordering of an interval hypergraph is simpler than the ordering of an interval graph. We will examine the first process in §4.

However, the question arises whether all complonic objects have been given. In case they haven't it would be useful to try to artificially add new "fictitious" objects corresponding to fictitious complons. The addition of new complons is of interest not only for constructing minimal interval hypergraphs, but also as a means of analyzing situations in which the initial graph is not an interval graph. In this last case knowledge of all the complons permits their relatively easy arrangement in a non-linear structure, by the rule "place together those complons which correspond to a single mutation", whereas in terms of the initial matrix non-linear structures are unidentifiable.

3. **The construction of fictitious complons.** In this section we will give a simple constructive solution to the problem. An algorithm [23] will be given for constructing new complonal objects corresponding to fictitious complons. This algorithm terminates when all such objects have been obtained, thus permitting, in particular, the determination of whether all the complons were manifest at first by condition (1).

We will describe the algorithm in successive stages, illustrating its use on the graphs G_1 and G_2 of Fig. 9.

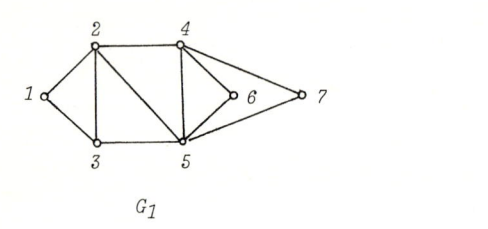

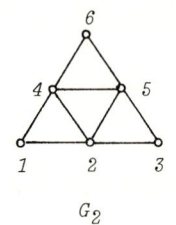

Figure 9

1. By means of (1) find all the complonic objects (simplicial vertices) in the initial set A.

For graph G_1 vertices *1,6* and *7* are simplicial. For graph G_2 the complons are *1,6* and *3*.

2. Examine the subgraph I' on the set of vertices A' obtained from A by removing the vertices found in step 1.

3. Determine the set of simplicial vertices of I', and call it L.

For each $i \varepsilon L$ fix the set S_i of its complonic objects (of the initial graph) which are adjacent to i.

For G_1' with $A' = \{2,3,4,5\}$ the set $L = \{3,4\}$, with $S_3 = \{1\}$ and $S_4 = \{6,7\}$. The graph G_2' forms a triangle such that L coincides with A', $L = \{2,4,5\}$, with $S_2 = \{1,3\}$, $S_4 = \{1,6\}$ and $S_5 = \{6,3\}$.

4. Exclude from further consideration those $i \varepsilon L$ for which

$$S_i \cap I<j> \neq \Phi \text{ for all } j \varepsilon I'<i> \qquad (2)$$

Excluding the corresponding vertices from A', return to step 3. If A' is empty the algorithm terminates.

Relation (2) means there are no non-complonic objects j, adjacent to i and such that the interval I_j does not contain the complons from S_i. In this case the interval I_i, though i is complonic in I', does not contain information for the construction of new objects.

For the graph G_1 there are no objects satisfying (2). For G_2 all the objects of L satisfy (2) so the process ends, having produced no fictitious objects.

5. For each of the remaining objects $i \varepsilon L$ introduce a new object $\bar{i} \notin A$, making it adjacent with the vertices of $I<i> \cap I<j>$, where $j \varepsilon I'<i>$ is selected so that $S_i \cap I<j> = \Phi$ (otherwise its selection is arbitrary). The object $\bar{i}$ corresponds to a unique fictitious complon covered in the minimal map by the interval I_i.

If, for two fictitious objects $\bar{i}$ and $\bar{j}$, $I<\bar{i}> = I<\bar{j}>$ then they correspond to the same complon. Keep only one of them.

For the graph G_1 we designate $\bar{3}=a$, $\bar{4}=b$. Then $I<a> = \{a,2,3,5\}$, $I<b> = \{b,2,4,5\}$.

6. Call this new augmented graph I, the new set of vertices A, and return to step 2. Executing step 1 again would be superfluous, as all the complonic mutants are known at this stage: They are the initial ones plus the new ones.

In actual use of the algorithm it is not necessary to build the augmented graph, since all of the same complonic objects are eliminated in A'. The introduction of fictitious objects is reflected in the changes of initial information used in steps 3 and 4.

For the graph G_1 now $S_3 \cap I<j> \neq \Phi$ for any $j \varepsilon I'<3> = \{1,2,3,5\}$, since $a \varepsilon S_3$ and $a \varepsilon I'<j>$, and similarly for S_4. Hence 3 and 4 must now be excluded from consideration.

This leaves $A' = \{2,5\}$, 2 and 5 being adjacent, so that $L = \{2,5\}$, $S_2 = \{1,a,b\}$, $S_5 = \{6,7,a,b\}$, with S_2 and S_5 intersecting $I<2>$ and $I<5>$ owing to the addition of the new objects a and b. After the exclusion of the objects 2 and 5 the process terminates.

The hypergraphs obtained for G_1 and G_2 are these:

$$\begin{array}{c|cccc} & 2 & 3 & 4 & 5 \\ \hline 1 & 1 & 1 & 0 & 0 \\ 6 & 0 & 0 & 1 & 1 \\ 7 & 0 & 0 & 1 & 1 \\ a & 1 & 1 & 0 & 1 \\ b & 1 & 0 & 1 & 1 \end{array} \qquad \begin{array}{c|ccc} & 2 & 4 & 5 \\ \hline 1 & 1 & 1 & 0 \\ 6 & 0 & 1 & 1 \\ 3 & 1 & 0 & 1 \end{array}$$

Here rows correspond to complons, and columns to non-complonic mutations. The maps associated with graphs are illustrated in Figs. 10 and 11. We leave it to the reader to verify that application of the algorithm to the matrix of Fig. 2,a gives the map of Fig 2,b.

The graph G_2 is not representable by a linear map. In Section 4 we will return to the analysis of the nonlinear case, but now we prove the correctness of this algorithm.

Theorem 5. If there exists a map representing I, then the algorithm of steps 1-6 will find all its complons.

Proof. By definition S_i is a set of real complons (i.e. complons corresponding to mutations already selected in step 1), which are covered in the minimal map by the interval I_i. Clearly $i \varepsilon L$ if and only if I covers no more than one fictitious complon (which does not correspond to any $j \varepsilon S_i$), and does not cover any interval of the map with length greater than 1. Otherwise, since the removal of real complons does not

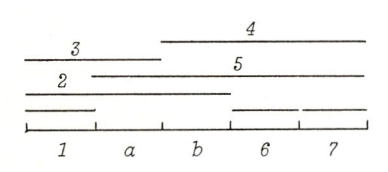

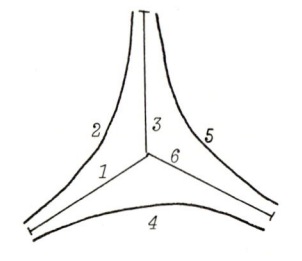

Figure 10

Figure 11

diminish the number of fictitious ones, repetition of the discussion in the proof of Theorem 1 would lead to the contradiction of (1), satisfied for $i \varepsilon L$ with respect to I'.

Condition (2) isolated those $i \varepsilon L$ for which I_i does not in general cover any fictitious complon, or else the covered fictitious complon is located within I_i, so that real complons are situated on both sides of it in I_i.

Indeed, if (2) is not satisfied, then there exists an interval of the map I_j ($j \varepsilon A$) such that $I_i \cap I_j \neq \Phi$, but I_j does not cover any real complon encountered in I_i. Clearly, this is possible only in the situation when the single fictitious complon is located within I_i at one end, i.e. its left (or right) end is the left (respectively, right) end of I_i.

If all the complons are real, i.e. correspond to initial objects or objects added at earlier steps, condition (2) is satisfied for all $i \varepsilon L$, since there are no fictitious objects. After removal of these i a new again-excluded set L of mutants is obtained, with intervals which cover complons and map intervals which have already been removed. This continues until all $i \varepsilon A$ are excluded, when the algorithm stops.

We now consider the case when there are fictitious complons. There then exists $i \varepsilon L$ for which (2) does not hold.

From Lemma 1, in the minimal map every natural number $k = 0, 1, \ldots \ldots, m-1$ (m is the map length) is the left end of some interval I_i. In particular there is a real complon $(m-1, m)$. Let $(k, k+1)$ be the fictitious complon with maximal k, so that all complons located to its right, i.e. $(k+1, k+2), \ldots, (m-1, m)$ are real. Consider the map interval $I_i = (k, n_k)$, where n_k is as small as possible. Clearly $n_k > k+1$, since $(k, k+1)$ is a fictitious complon.

The object i belongs to L, since I_i covers only one fictitious complon and does not cover any interval from the map of length greater than 1: All such intervals cover only real complons and are eliminated at step four of the algorithm.

Now consider an arbitrary interval I_j with right end $k+1$. Clearly $I_i \cap I_j = (k, k+1) \neq \Phi$, but $S_i \cap S_j = \Phi$, since the objects from S_i correspond to the complons $(k+1, k+2), \ldots, (n_k-1, n_k)$ lying to the right of I_j, and intersecting with it. Thus for i with $I_i = (k, n_k)$ i is an element of L and does not satisfy (2). Q.E.D.

The formal object $\bar{i}$ introduced for this i is such that $I_{\bar{i}}$ intersects with I_i but not with the complons $(k+1, k+2), \ldots, (n_k-1, n_k)$. This means the right end of $I_{\bar{i}}$ is equal to $k+1$. But the left is equal to k, since I_i does not intersect with intervals lying to the left of k. Thus $I_{\bar{i}} = (k, k+1)$ and the object $\bar{i}$ is a complon.

We have shown that the algorithm finds fictitious complons, if they exist, and stops when all complonic objects have been constructed. The theorem is proved.

From Theorem 5 it follows that the total number of complons is determined according to algorithm 1-6 only by means of the interval graph, without taking into account its possible representations, and consequently, it is fixed. From Theorem 1 the complonic objects for all minimal maps are the same. This means that no map which represents an interval graph can have fewer than m complons.

Thus the procedure described in Theorem 2.3 for identifying left and right ends of intervals of maps must lead to a minimal map. In other words the following statement holds, since both the theorem and the algorithm made use only of Lemma 1.

Corollary 2. A map is minimal if and only if each of its internal natural points is the left (and right) end for some intervals of the map.

In addition, since the length of an interval of a minimal map is equal to the number of complons covered by it, and this number is determined only by the initial complementation matrix, the following holds:

Corollary 3. A given object $i \varepsilon A$ corresponds in every minimal map to intervals of the same length.

The final solution of the uniqueness problem will be given in Section 4.2. Here, we note in conclusion that the selection of complons performed by algorithm 1-6 may be considered as characteristic of interval graphs.

Corollary 4. A graph is an interval graph if and only if the complon selection process of the algorithm, applied to the graph, produces a minimal interval hypergraph.

4. <u>Non-linear hypergraphs and interval graphs</u>. A map $K = <\mathbf{I}, I_i (i\ A)>$, where $\mathbf{I} = (0,m)$ may be considered as an undirected chain joining vertices $0, 1, 2, \ldots, m$, with the intervals I_i ($i \varepsilon A$) corresponding to subchains, sets of consecutively arranged edges represented by map complons.

This sets the stage for the following general definitions. Let G be an ordinary graph with the set X of edges. The set of edges $S \subseteq X$ of a chain of the graph G is a G-<i>interval</i>. The collection of G-intervals form a hypergraph on the set X which is called the G-<i>interval hypergraph</i>. The corresponding graph of the intersections of G-intervals (and graphs isomorphic to it) is the G-<i>interval graph</i>. These definitions lead to the usual "linear" graphs and hypergraphs of intervals in the particular case when G is represented by a chain.

The complementation mapping task (in the absence of linear representations) is to construct a graph G, such that the complementation graph is a graph of G-intervals. It is evident that every graph is a graph of G-intervals for a complete graph G, having sufficient vertices. Our interest is only in those graphs G which have a minimal number of edges among all G for which the given graph is a graph of G-intervals. Such a minimal graph G together with the corresponding G-intervals

($i \varepsilon A$) is naturally called a minimal map $K = <G, I_i (i \varepsilon A)>$ of the given graph, and the edges of G, complons. The edges of G are vertices of the associated G-interval hypergraph, which in this case has a minimal number of vertices.

For the notions here introduced there are a number of analogues of the statements proved in preceding sections. To formulate these analogues we will use the names and numbers of their "prototypes", adding a prime to numbers.

Lemma 1'. In a minimal map $K = <G, I_i (i \varepsilon A)>$ for every complon x there exist two intervals $I_i, I_j (i, j \varepsilon A)$ such that $I_i \cap I_j = \{x\}$.

Assuming the opposite, it is easy to see that the complon x can be removed from G by "contraction" of its intervals (by disjunctively uniting corresponding rows (and columns) of the matrix of G) without changing the intersection graph of the sets I_i (since nonempty intersections are still nonempty after the removal of x). Moreover every $I_i (i \varepsilon A)$ is a chain in the derived graph G', i.e. a G'-interval. The contradiction to the minimality of G thus derived proves the lemma.

From Lemma 1 follows

Theorem 4'. A mutant $i \varepsilon A$ is complonic if and only if it satisfies condition (1).

For non-complonic i, I_i consists of at least two neighboring complons x and y. From Lemma 1' there exist I_k, I_l, I_m, I_n such that $I_k \cap I_l = \{x\}$ and $I_m \cap I_n = \{y\}$. Hence, in particular, it follows that $I_k \cap I_l \cap I_m \cap I_n = \phi$, though $k, l, m, n \varepsilon I<i>$, which contradicts (1) and proves the theorem.

Corollary 1'. Every minimal map for a given graph has the same set of complons.

The selection of components in algorithm 1-6 is valid in this case also, since it is based solely on relation (1) (see the sample graph G_2 in Section 3). However, the algorithm frequently fails to select a complon. For example, on the map of Fig. 12 where the graph G is a cycle with vertices $1, 2, 3, 4$ and intervals $I_1 = \{12, 23\}$, $I_2 = \{23, 34\}$, $I_3 = \{34, 41\}$, $I_4 = \{41, 12\}$, none of the complons is real, so that algorithm 1-6 simply has nothing to catch hold on. It is necessary to make a significant modification of the algorithm so that it is applicable in this situation as well. The authors have a number of working procedures, but a strict proof of their completeness is lacking. Therefore we can formulate only the following statement,

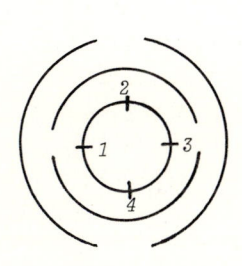

Figure 12

in analogy to Theorem 5.

An ordinary graph is a *tree interval graph* if it is a G-interval graph, where G is a tree.

Theorem 5'. If a graph is a tree interval graph then algorithm 1-6 will construct all complons.

It is sufficient to note that the proof of theorem 5 works for any minimal map of G for which, at each step of the algorithm, there is a sequence of real complons, associated with a fictitious complon x, which together with x constitute one of those intervals I_i, I_j for which $I_i \cap I_j = \{x\}$. A tree is clearly related to such maps because of the presence of pendant complons, trivially real by Lemma 1' (analogous to the complon $(m-1,m)$ in the proof of Theorem 5).

Corollary 4'. If algorithm 1-6 finds all complons then they determine a minimal G-interval hypergraph representing the initial G-interval graph.

Corollary 4' can be modified, considering in place of algorithm 1-6 one of its modifications that permits the analysis of situations of the type shown in Fig. 12.

Theorem 6 clearly remains true, with the replacement of the notion of an interval hypergraph by that of a G-interval hypergraph.

With respect to the characteristics of nonlinear interval graphs, results have so far been obtained only for *graphs of arcs* of circles, i.e. G-interval graphs for G defined by cycles [97,42]. Arc hypergraphs are easily characterized as follows [42].

A hypergraph G is an arc hypergraph if and only if its matrix can be brought, by permutations of its rows, to a form in which in each column the zeros or the ones are located consecutively.

We note that such a permutation corresponds to the sequential numbering of neighboring edges of a cycle of G, so that the existence of a column $k \varepsilon A$ with consecutively-placed ones means that I_k is an interval of this permutation, and a column k with consecutively-places zeros means that the complement $\overline{I}_k$ is an interval of the permutation, so that the set I_k itself is also an interval of the cycle (complementary to $\overline{I}_k$). From this remark still another criterion follows. For an arbitrary matrix Γ we construct a matrix Γ' of the same dimensions, with a column s'_k of Γ' identical with s_k of Γ if it begins with a zero. If, on the other hand, the kth column of Γ begins with a one, then s'_k is obtained from it by Boolean complementation.

Clearly Γ is an arc hypergraph if and only if Γ' is an interval hypergraph. The construction of Γ' actually means that for certain of the complons of Γ we replaced all the arcs passing through them by complementary arcs. From this it is clear that we obtain a linear map.

since a given complon does not intersect any intervals, and may be removed from the cycle G.

Graphs and hypergraphs of arcs are of definite genetic interest, since the chromosomes of a number of microorganisms (bacteria, viruses, bacteriophages, etc.) have circular form. In the first place, with their help one can establish the "closed" topological character of the corresponding recombination map. Second, by analyzing indirect "functional" data, one can arrive at the circular form of a chromosome. For example, if a circular chromosome is sufficiently well "covered" by mutations of the deletion type then, as one can easily understand, the complementation matrix corresponding to it is representable by a cyclic map. Moreover, under several reconstructions of a linear chromosome (inversions of its fragments) the deletion of several genes, which cover the inverted section, may also give the typical picture of intergenic "cyclical" complementation (see Section 5.2).

It should be said that our definition of G-interval graphs differs from precise definitions used in the literature, since for us a chain is a collection of edges and not vertices of a graph. Therefore, in particular, known properties of tree interval graphs such as triangulatedness [65a] do not hold in our case.

§4. Linear mapping algorithms

1. **The Fulkerson-Gross algorithm.** In this section we present algorithms for solving the linear mapping problem by ordering the objects involved.

If we have available for analysis the initial Boolean matrix $r = \|r_{ij}\|$ of the results of complementation testing ($r_{ij}=0$ if the ith and jth mutational defects do not overlap, and $r_{ij}=1$ otherwise), then we can take two routes.

In the first, we try to bring the matrix r to quasi-diagonal form and to obtain the map, as described in Theorem 1 (see Fig. 2). To bring the matrix to quasi-diagonal form we can use the idea of V. V. Shkurba, that the last complon in a map must be real, or we can use the algorithm for the orientation of the complementary graph $\bar{r}$ in order to satisfy Theorem 2.4. As a rule this approach does not work, because the initial matrix cannot be brought to quasi-diagonal form (the complementation graph is not an interval graph). Therefore we will focus on the second mapping method. First it is necessary to construct the hypergraph by the algorithm of Section 3.3 (the matrix of "complon-mutations"), and then try to order its vertices. This approach is more suitable, since the construction of fictitious complons can be carried out even if the complementation graph is not an interval graph. In this section we present the Fulkerson-Gross algorithm for the second part of the process: ordering the vertices of an interval hypergraph, i.e. the construction of a representative map (see page 37).

This algorithm may also be interpreted as an algorithm for ordering the rows of a Boolean matrix so that in each of its columns the set of ones is consecutively arranged (i.e. the matrix becomes column-linear). Terms, associated with this algorithm, finally allow us to solve the uniqueness problem for the representation of interval graphs by means of maps.

Thus, let an arbitrary hypergraph $\Gamma = (X, \{S_k\}_{k \in A})$ be given, where $X = \{x_1, \ldots, x_n\}$, $A = \{1, \ldots, N\}$. The corresponding $n \times N$ matrix of Γ has the property that its columns S_k ($k \in A$) do not coincide with one another: $S_k \neq S_l$ for $k \neq l$, since all the edges of the hypergraph are different. We are required to find an ordering of the set X (rows of the matrix Γ) such that all edges S_k become intervals (the one appear consecutively in all columns).

The first thing that comes to mind is to successively look over the sets S_k, collecting "together" those vertices which do not violate the interval character of the ones already examined. This is precisely the idea realized in the Fulkerson-Gross algorithm, but as we shall see, its execution is associated with a number of substantial auxilliary actions.

We first select a system of subsets S_k ($k \in A$), for which a relatively independent ordering of vertices is possible.

Consider the graph $G(\Gamma)$ with vertex set A, and edges defined by pairs kl for which $S_k \cap S_l \neq \phi$ and neither of which is a proper subset of the other (in which case we say that S_k and S_l overlap). The graph $G(\Gamma)$ differs from the intersection graph γ_A of hyperedges in one significant aspect: edges which connect covering and covered sets S_k are absent from it.

The components of the graph $G(\Gamma)$ are designated by $A_1, \ldots, A_p$, ($\bigcup_1^p A_s = A$, $A_s \cap A_t = \phi$ ($s \neq t$)). We examine the oriented graph $D(\Gamma)$ whose vertices are the components $A_1, \ldots, A_p$, with an arc (A_s, A_t) belonging to the graph if for some $i \in A_s$ and $j \in A_t$ $S_j \subset S_i$.

Let us characterize the arcs (A_s, A_t) of the graph $D(\Gamma)$ in more detail. Consider $i \in A_s$ such that the set B of those indices $j \in A_t$ for which $S_j \subset S_i$

is nonempty. We will show that $B=A_t$. If $B \neq A_t$ then, since A_t is connected in $G(\Gamma)$, we can find an object $k \epsilon A_t - B$ such that $S_k \cap S_j \neq \phi$ for some $j \epsilon B$. Then clearly $S_k \cap S_i \neq \phi$ also. Since k and i belong to different components, A_t and A_s, one of the sets S_k and S_i must be a subset of the other. But $S_i \subset S_k$ is impossible since $S_j \subset S_i$, and $S_j \subset S_k$ contradicts the fact that j and k belong to the same component A_t. So $S_k \subset S_i$, i.e. $k \epsilon B$! This contradiction proves the assertion.

We consider now the relationship of the sets $S_k (k \epsilon A_t)$ with each of the remaining sets $S_l (l \epsilon A_s)$. Since S_k and S_l cannot overlap each other the only remaining possibilities are that $S_k \cap S_l = \phi$ for $k \epsilon A_t$, or that $S_k \subset S_l$ for all $k \epsilon A_t$.

We have proved the following

<u>Lemma 1</u>. An arc (A_s, A_t) belongs to the graph $D(\Gamma)$ if and only if for any $l \epsilon A_s$ one of the following holds: a) all the sets $S_k (k \epsilon A_t)$ are contained in S_l; b) none of the sets $S_k (k \epsilon A_t)$ intersects S_l, with the objects l for which a) holds necessarily existing.

From Lemma 1 it follows directly that the graph $D(\Gamma)$ is antisymmetric. Moreover $D(\Gamma)$ is transitive:

$$(A_s, A_t) \epsilon D(\Gamma) \text{ and } (A_t, A_k) \epsilon D(\Gamma) \longrightarrow (A_s, A_k) \epsilon D(\Gamma),$$

since every $S_l (l \epsilon A_s)$ which contains all $S_i (i \epsilon A_t)$ also contains all $S_m (m \epsilon A_k)$. At the same time every $S_l (l \epsilon A)$ which does not intersect any $S_i (i \epsilon A_t)$ also fails to intersect the sets $S_m (m \epsilon A_k)$, since each of them is contained in at least one of the $S_i (i \epsilon A_t)$.

Thus the graph $D(\Gamma)$ is a graph of a partial order, which characterizes the mutual disposition of the systems of subsets corresponding to different components of the overlap graph $G(\Gamma)$: Each of these systems is wholly contained in the intersection of some subset of the sets of the preceding system.

Each of the components $A_s (s=1,\ldots,p)$ of the graph $G(\Gamma)$ corresponds to a partial hypergraph $\Gamma_s = (X, \{S_k\}_{k \epsilon A_s})$.

<u>Theorem 1</u>. A hypergraph Γ is an interval hypergraph if and only if each $\Gamma_s (s=1,\ldots,p)$ is an interval hypergraph.

<u>Proof</u>. In one direction the statement is trivial: If Γ is an interval hypergraph then for some ordering of the set X all the subsets $S_k (k \epsilon A)$ are intervals and thus for each $s=1,\ldots,p$ the subsets $S_k (k \epsilon A_s)$ are intervals since $A_s \subseteq A$.

Now let all the Γ_s be interval hypergraphs. We are required to show that Γ is also an interval hypergraph, i.e. that the intersection of nonempty sets of the admissible orderings of the hypergraphs Γ_s for each $s=1,2,\ldots,p$, is nonempty (it consists of admissible orderings of the hypergraph Γ). This follows from the fact that the sets S_k for $k \varepsilon A_s$ affect only a local segment of the set X, the ordering of which does not depend on the ordering of the components of the graph $D(\Gamma)$ which precede or are unconnected with A_s.

We will give a strict proof.

For clearness the argument will be carried out in terms of the matrices Γ_s corresponding to the hypergraphs $\Gamma_s(s=1,\ldots,p)$. Then the matrix of the hypergraph Γ can be represented as a sequence of submatrices $\Gamma_s(s=1,\ldots,p)$: $\Gamma = (\Gamma_1, \Gamma_2, \ldots, \Gamma_p)$ ordered one after another in accord with the graph $D(\Gamma)$. Thus Γ_p must correspond to a minimal component of $D(\Gamma)$ (i.e. a vertex from which there are no outgoing arcs).

We first note that in an admissible ordering identical rows may be placed together. The exclusion of rows (i.e. vertices of X) cannot make an interval graph into one that is not. Therefore we can exclude one of the identical rows, find an admissible ordering of the resulting matrix, and then insert the excluded row next to its twin, since this does not violate the linearity of the matrix. Repeating this operation we obtain an admissible ordering of the rows in which identical rows are situated together.

Now consider the submatrices $\Gamma^s = (\Gamma_1, \ldots, \Gamma_s)$ formed by the first s components of the graph $D(\Gamma)$ ($s=1,\ldots,p$). Clearly the matrix Γ_s corresponds to the minimal A_s in the sense of the partial order $D(\Gamma^s)$.

The proof will proceed by induction on s. For $s=1$ $\Gamma^1 = (\Gamma_1)$, but Γ_1 is column-linearizable by agreement. Suppose Γ^{s-1} is column-linearizable in such a way that identical rows appear consecutively. We will show that Γ^s is also column-linearizable.

We arrange the rows of Γ^s so that Γ^{s-1} is column-linear. It will have the form $\Delta^s = (\Delta_1, \ldots, \Delta_{s-1}, \Delta_s)$, where $\Delta^{s-1} = (\Delta_1, \ldots, \Delta_{s-1})$ is column-linear. In the submatrix Δ_s, let i_1 be the index of the first row (from the top) in which a one appears, and let i_2 be the last row in which a one appears, so that all rows above i_1 and below i_2 contain only zeros. We will show that in the matrix Δ^{s-1} all the intermediate rows i, with $i_1 \leqslant i \leqslant i_2$, are identical.

In $D(\Gamma)$ A_s is minimal among $A_1, \ldots, A_s$, so that every A_t ($1 \leqslant t < s$) either precedes A_s or is not connected with it. If A_t is not connected with A_s then in Δ_t the rows i_1 and i_2 are zero (since they are zero in Δ_s) and are consequently identical. If, on the other hand, A_t precedes A_s, by Lemma 2 the rows i_1 and i_2 in Δ_t are also identical. This means that in Δ^{s-1} the rows i_1 and i_2 coincide, and consequently, by the induction hypothesis, so do all intermediate rows.

Thus every permutation of the rows of Δ_s lying between i_1 and i_2 maintains the linearity of the matrix Δ^{s-1}, but only such permutations (by the definition of i_1 and i_2) are necessary to bring Δ_s to column-linear form. Selecting one of these permutations we obtain a column-linear form for the matrix Γ^s, which proves the theorem.

The foregoing proof shows that to order Γ it is sufficient to order each Γ_s individually, in the sequence which corresponds to the graph of components $D(\Gamma)$.

It remains to give the algorithm for ordering the rows of a matrix Γ_s corresponding to a component A_s.

We first examine the case in which the component $G(\Gamma_s)$ has only three vertices, i.e. the matrix has three columns s_1, s_2, s_3.

Consider now the scalar products of these columns. Clearly $s_i \cdot s_i$ is the number of ones in column s_i, $s_i \cdot s_j = s_i \leftrightarrow S_i \subseteq S_j$, $s_i \cdot s_j = 0 \leftrightarrow S_i \cap S_j = \phi$ ($i, j = 1, 2, 3$).

Since the columns s_i are contained in one component, at least two pairs of corresponding edges overlap. Let these be, say, $S_1 \cap S_2 \neq \phi$ and $S_2 \cap S_3 \neq \phi$. In this case we will call the ordered triplet s_1, s_2, s_3 an *overlapping triplet*. Of course $s_1 \cdot s_2 \neq 0$ and $s_2 \cdot s_3 \neq 0$.

Two cases are possible

$$s_1 \cdot s_3 \quad \min(s_1 \cdot s_2, s_2 \cdot s_3) \tag{3}$$

or

$$s_1 \cdot s_3 \quad \min(s_1 \cdot s_2, s_2 \cdot s_3) \tag{4}$$

It is easy to show that in the linear form of a hypergraph matrix each of these cases corresponds to a specific mutual arrangement of the subcolumns of ones (i.e. intervals S_1, S_2, S_3) following one another. Case (3) corresponds to the ladder-like arrangement of Fig. 13, and case (4) to the "inverted hump" of Fig. 14. Here the sections of consecutive ones of corresponding columns are shown as vertical segments, with zeros not shown.

The situation of Fig. 13 is obtained by placing $s_1 \cdot s_1$ ones together in the left column, and then $s_2 \cdot s_2$ ones in the second column with an intersection of $s_1 \cdot s_2$ ones with the first column, and then $s_3 \cdot s_3$ ones in the third with an intersection of $s_2 \cdot s_3$ ones with the second column. The analogous situation obtains for Fig. 14 with respect to (4).

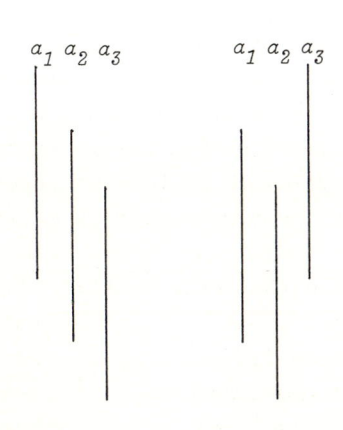

Figure 13

Figure 14

We emphasize that there are two possibilities for the placement of the ones in the second column: lower than those in the first, as in Figs. 13 and 14, or higher. Once this selection is made the position of the ones in the third column is uniquely determined from (3) and (4), since the information about the scalar products determines the relative positions of the segments up to reversals.

Having fixed the relative positions of the ones, it remains to determine the number of common ones in the first and third columns. If this quantity is not equal to $s_1 \cdot s_3$ then as a result of the uniqueness of the construction we can say that Γ_s is not linearizable. However, if equality holds, we can conclude that any ordering connected with S_1, S_2, S_3 as described, is admissible.

In the case when contains more than three columns, the construction is easily extended to all possible overlapping triplets.

For this it is sufficient to construct the spanning tree of the component A_s of the overlap graph. As is shown in the appendix (page 187) the spanning tree of a component (which, itself, may not be specified in advance) is determined by the following procedure. Beginning with an arbitrary vertex, examine all those adjacent to it, and for each of these new vertices in succession, those adjacent to it, and so forth, until no new vertices can be found (which marks the exhaustion of that component). It is important that edges of the spanning tree be formed only between a given vertex (at any stage) and new adjacent vertices.

Once the spanning tree is formed, the overlapping triplets are those triplets whose members are successively connected by edges of this tree. These are enough, since in any other overlapping triplet the first two columns are contained in some overlapping spanning-tree triplet, so that their relative disposition is uniquely determined by the spanning tree.

The process begins with the ordering of an arbitrary overlapping spanning-tree triplet of columns. In the general stage (of the algorithm) we consider a new overlapping triplet, whose first two columns are already ordered, so that positioning of the ones in its third column is uniquely determined by the associated scalar products. After the ones of the third column are positioned in accord with Fig. 13 or Fig. 14, it is necessary to verify that there is no contradiction in the configuration obtained so far. To do this the third column (of this latest triplet) is compared with each of the previously ordered columns, with respect to their scalar products and their common ones, more precisely, the number of vertices in the intersection of the corresponding hyperedges.

As a result of the uniqueness (for a given skeleton) of the configuration obtained so far, inequality in any of the comparisons means the matrix Γ_s is not column-linearizable. If equality is found in all the comparisons the admissibility of the ordering so far obtained is supported, and the next overlapping triplet (with one new column) is selected for examination. The process ends when all columns have been examined.

We obtain as a result a mutual disposition of the intervals S_k $(k \varepsilon A)$ which easily allows the ordering of the vertices, or in general, the transition to a minimal map by observation of the interval order as was done in Section 2.1.

Thus the Fulkerson-Gross algorithm for ordering the vertices of a hypergraph consists of the following steps:

1. Build the overlap graph $G(\Gamma)$.

2. Through the construction of spanning trees, separate out the components A_s of the graph $G(\Gamma)$ $(s=1,2,\ldots,p)$. The spanning tree of each component A_s is, at once, the spanning tree of the graph $G(\Gamma_s)$ for the partial hypergraph Γ_s associated with A_s.

3. Build the graph $D(\Gamma)$ of the partial order of the components A_s and renumber the A_s so that $(A_s, A_t) \, \varepsilon \, D(\Gamma) \longrightarrow s<t$.

4. Set s to 1.

5. Calculate the matrix of scalar products of all pairs of columns of the Γ_s matrix.

6. Using the spanning tree of the component A_s order the ones in the columns of Γ_s, by comparing their scalar products (page 55). If these comparisons lead to a contradiction then the desired ordering does not exist.

7. Increment s by 1. If $s \neq p$ go to step 5. Otherwise, stop. The ordering of ones in the columns of the hypergraph matrix Γ is consistent. The corresponding ordering of the rows of Γ is the one desired.

For example, let the matrix Γ_s have the following form:

	1	2	3	4	5	6
1	1	1	0	0	1	0
2	1	1	1	1	0	0
3	1	0	0	0	0	0
4	0	0	1	0	0	1
5	0	1	1	1	0	1
6	1	1	0	1	1	0

The component of the overlap graph of columns is shown in Fig. 15,a. Beginning with vertex 5 we obtain the spanning tree of the graph shown in Fig 15,b.

a)   b)

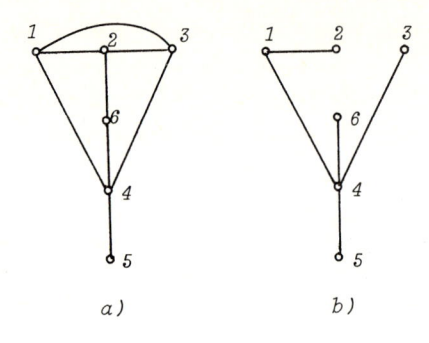

Figure 15

The matrix of scalar products is:

	1	2	3	4	5	6
1	4	3	1	2	2	0
2		4	2	3	2	1
3			3	2	0	2
4				3	1	1
5					2	0
6						2

We will successively examine the overlapping triplets of the spanning tree, beginning with 214. Condition (4) is satisfied for this triplet, so that we obtain the "inverted hump" configuration.

	2	1	4
	1		1
	1	1	1
	1	1	1
	1		1
			1

The number of common ones in columns 2 and 4 is three, and $s_2 \cdot s_4 = 3$ so that we can continue the process.

Column 3 is added (from the triplet 143, satisfying condition (3)):

	2	1	4	3
				1
	1		1	1
	1	1	1	1
	1	1	1	
	1	1		
				1

Comparing $s_2 \cdot s_3 = 2$ and $s_1 \cdot s_3 = 1$ with the numbers of ones in the associated pairs of columns shows no contradictions in the configuration obtained so far.

Adding column 6 from the triplet 146 satisfying condition (3):

	2	1	4	3	6
				1	1
	1		1	1	1
	1	1	1	1	
	1	1	1		
	1	1			
	1				

Comparison of the values $s_2 \cdot s_6 = 1$, $s_1 \cdot s_6 = 0$, $s_3 \cdot s_6 = 2$ with the numbers of common ones again supports the admissibility of the ordering.

Finally, column 5 is added, for example, from the triplet 645, satisfying condition (3). This gives:

	2	1	4	3	6	5
				1	1	
	1		1	1	1	
	1	1	1	1		
	1	1	1			1
	1	1				1
	1					

Having convinced ourselves of the admissibility of this configuration we construct the desired row ordering. From the examination of columns 1 and 2 the first position must be occupied by the row whose number is common to the two sets {4,5} and {3,4}, i.e. 4; then 5; and 3 last. From column 4 it is clear that row 1 is next-to-last. Information about columns 3 and 6 shows how to place the remaining rows. The final ordering (permutation) of the rows is 4,5,2,6,1,3 (of course the reverse: 3,1,6,2,5,4 is also admissible).

2. **The uniqueness theorem**. We first note that the process just described provides a proof for Theorem 3.2, since it uses only information about the scalar products of columns (including scalar squares, giving the number of ones in a column), on which formulas (3) and (4) are also based; and these uniquely determine (except for reversals) the ordering of the columns k for a component A_s ($s=1,\ldots,p$).

From this it follows that in the linear case identical matrices $\|s_i \cdot s_j\|$ of scalar products must lead to identical configurations on ones, so that the initial matrices can differ in the number of zero rows and in the ordering of the other rows. If, however, a matrix $\|s_i \cdot s_j\|$ does not allow an admissible configuration this signifies that, whatever the initial hypergraph that led to it, it is not linearizable.

The uniqueness properties of the Fulkerson-Gross method permit us to conclude the solution of the uniqueness problem for complementation mapping.

Theorem 2. If Γ is a minimal linear hypergraph for which the overlap graph contains p components, then there are exactly 2^p distinct admissible orderings (depending on selection of one of two mutually inverse orderings of the vertices corresponding to each component).

Proof. Suppose, as in Theorem 1, the graph $G(\Gamma)$ has p components $A_1, \ldots, A_p$ with a corresponding partition of the matrix $\Gamma = (\Gamma_1, \ldots, \Gamma_p)$, where the enumeration of the components corresponds to the partial ordering given by the graph $D(\Gamma)$.

Let $\Delta = (\Delta_1, \ldots, \Delta_p)$ be a (column)-linear form of Γ obtained by permuting the rows i. As in the proof of Theorem 6, we argue that the non-zero rows of the matrix Δ_p form an interval $[i_1, i_2]$ of the ordering which "covers" a set of identical rows of the matrix $\Delta^{p-1} = (\Delta_1, \ldots, \Delta_{p-1})$. This means that in Δ_p all the non-zero rows are distinct, since from the minimality of Γ no two rows coincide. As a consequence of the Fulkerson-Gross algorithm, this means the admissible ordering of the interval $[i_1, i_2]$ of Δ_p is uniquely defined up to reversals. Actually, in the case when all rows are different an arbitrary admissible configuration of the ones in the intervals uniquely determines the range of the vertex associated with each of the rows obtained. Therefore a unique ordering of the rows corresponds to each of the two possible mutual dispositions of the intervals.

We now examine the matrix $\overline{\Delta} = (\overline{\Delta}_1, \ldots, \overline{\Delta}_{p-1})$ obtained by eliminating from Δ all rows of the half-open interval $(i_1, i_2]$ (excepting i_1), as well as the trivial submatrix $\overline{\Delta}_p$ with the single remaining non-zero row i_1. This matrix is clearly linear. Moreover, in $\overline{\Delta}$ all rows are distinct. Identical rows of $\overline{\Delta}$ not equal to i_1, cannot occur, since such rows would also be identical in Δ, because of their zero extensions in Δ_p. But rows which coincide with i_1 in $\overline{\Delta}$ also cannot occur, since this contradicts the minimality of the initial hypergraph. In fact, every interval $S_k (k \notin A_p)$, because of Lemma 1 and the minimality of A_p in $D(\Gamma)$, either contains all $S_l (l \in A_p)$ (among them also the row i_1), or does not intersect at all with them and does not contain i_1. The intervals S_k $(k \notin A_p)$ must behave in just this way also with respect to a row i which coincides in $\overline{\Delta}$ with i_1. Moreover, in Δ_p this row i is zero, since it cannot coincide with i_1 in Δ. This means the row i can be removed from Δ without affecting the intersection relation of the sets $S_k (k \in A)$, but this contradicts Corollary 2 on the minimality of Γ.

The components of the graph $G(\overline{\Delta})$ coincide with components of the graph $G(\Delta)$ $A_1, \ldots, A_{p-1}$, and the graph $D(\overline{\Delta})$ coincides with $D(\Delta)$ with the exclusion of A_p. In fact, the intersection relation of columns is unchanged if (as was done) we exclude duplicate rows.

Thus, for the matrix $\bar{A}$ we can repeat the argument that only two mutually inverse permutations of the interval $[i_1, i_2]$ are possible, and that these are controlled by the minimal matrix, in the sense of $D(\Gamma^{p-1})$ (in this case Γ_{p-1}). Repeating this procedure p times proves the theorem.

The above proof provides not only the number of possible orderings of a minimal graph, but also a means of obtaining them from any specific ordering.

To do this it is necessary to select from s given minimal map those intervals $S=(k, n_k)$ (not necessarily map intervals) having the property that each map interval which intersects with S either is contained in it or contains it. Such an interval S corresponds to a component of the intersection graph, as is required by the proof of Theorem 2. Therefore, reversing the order of the complons of S leads to a new minimal map.

All the possible minimal maps can be obtained from a given minimal map by successive applications of this technique.

3. **Admissible orderings of linear matrices.** In this section we examine linear matrices. We referred to these matrices in connection with the recombination mapping problem (page 18). However, such matrices can appear naturally also in other situations -- anytime connections between objects reflect their linear order. Connections between objects that are near one another in the ordering must exceed connections between objects more distant from one another. In particular, linear matrices of connections arise repeatedly in the study of complementation maps. Thus, the complementation matrix for non-overlapping mutational distortions (the graph of non-covering intervals) is linear, as is the matrix of connections between separate sections of a map, i.e. complons (Theorem 3.1).

As it turns out, the theory of linear matrices [13] falls naturally within the theory of graphs and hypergraphs of intervals.

Thus, suppose that $\|a_{ij}\|$ is a matrix of connections between objects. We will say that it is *linear* (with respect to rows) if and only if

$$a_{ij} \geqslant a_{ik} \text{ for } i \leqslant j \leqslant k \text{ or } i \geqslant j \geqslant k. \tag{5}$$

In some situations the values a_{ij} characterize the distance between i and j: The nearer i and j are, the smaller a_{ij} is, since the interpretation of a_{ij} as a connection leads to a reverse dependence (smaller for a reduced connection between i and j). In such situations the notion of linearity (5) is naturally modified by substitution of the inequality $a_{ij} \leq a_{ik}$ for $a_{ij} \geq a_{ik}$. We will carry out all subsequent constructions using the relation (5), but their modifications in the case of distance is trivial. It is necessary only to use the relation "less-than" in place of the relation "greater-than."

An ordering $p = (i_1, \ldots, i_N)$ of objects, which brings a matrix to linear form will be said to be *admissible*. We will formulate admissibility critera for orderings.

Suppose the elements of the ith row of a matrix $\|a_{ij}\|$ take on m_i distinct values $a_1, \ldots, a_{m_i}$, with $a_1 > a_2 > \ldots > a_{m_i}$. We define

$$R^i_1 = \{j | a_{ij} = a_1\}, \quad R^i_2 = \{j | a_{ij} = a_2\}, \quad \ldots, \quad R^i_{m_i} = \{j | a_{ij} = a_{m_i}\}.$$

The set R^i_k ($i \in A$, $k=1,\ldots,m_i$) thus consists of those objects having identical connections with i (monotonically decreasing with increase of k). Consequently, the ith row of the matrix $\|a_{ij}\|$ induces an ordered partition (ranking) $R^i = (R^i_1, \ldots, R^i_{m_i})$ of the set A.

Theorem 3. An ordering $p=(i_1,\ldots,i_N)$ is admissible if and only if all the sets

$$S^i_k = \bigcup_{j=1}^{k} R^i_j \quad (i \in A, \; k=1, \ldots, m_i) \tag{6}$$

are its intervals.

Proof. From (6) $S^i_k = \{j | a_{ij} \geq a_k\}$, so that S^i_k consists of objects with which i has the largest connection. But in the linear form of the matrix columns corresponding to the largest elements of the row are situated consecutively. Therefore from the admissibility of the ordering it follows that the sets S^i_k are its intervals.

If, on the other hand, for some ordering p all the S^i_k are intervals then the simultaneous permutation of rows and columns of the matrix $\|a_{ij}\|$ in accord with p brings it to linear form. Thus in the ith row the maximal elements a_{ij} with $j \in R^i_1$ are situated next to a_{ii} (since by agreement $S^i_1 = R^i_1$ is the interval around them), followed (in order of size) by the elements a_{ij} with $j \in R^i_2$ (since $S^i_2 = R^i_1 \cup R^i_2$ is the interval ...), etc. The theorem is proved.

Corollary 1. The matrix $\|a_{ij}\|$ is linearizable if and only if the system $(A, \{S^i_k\}_{i \in A, \; k=1,\ldots,m_i})$ is an interval hypergraph.

Thus the problem of constructing an admissible ordering is transformed into the problem of ordering an interval hypergraph, and it may be solved by the method of Fulkerson and Gross.

It is useful to consider another construction, connected with a specific characteristic of swuare linear matrices. For example, in the first row of a linear matrix the elements are decreasing, while in the last they are in creasing. In order to formulate such properties we veral notions [13].

We say that an ordering p *includes* an ordered partition R if and only if $iRj \longrightarrow ipj$, i.e. p is obtained from R by the additional ordering of the objects internal to the classes of R. A set S is called an *interval of the ordered partition R*, if and only if it is an interval of some ordering contained in R.

The set S is said to *cover* an object k in the ordered partition R, if there exist $i, j \varepsilon S$, such that iRk and kRj. Then an interval of an ordered partition is a set which contains all objects covered by it.

In other words, a set S is an interval of an ordered partition $R = (R_1, \ldots, R_m)$, if and only if for some s,t $(1 \leqslant s \leqslant t \leqslant m)$

$$S = P_s \cup R_{s+1} \cup \ldots \cup R_t \cup P_{t+1},$$

where $P_s \subset R_s$, $P_{t+1} \subset R_{t+1}$. We will say that the ordered partition

$$R' = (R_1, \ldots, R_{s-1}, R_s - P_s, P_s, R_{s+1}, \ldots, R_t, P_{t+1}, R_{t+1} - P_{t+1}, R_{t+2}, \ldots, R_m)$$

is obtained by *superimposing the interval S on R*.

Property 1. If the ordering $p = (i_1, \ldots, i_N)$ is admissible and $i_1 \varepsilon R_1^l$ for some $l \varepsilon A$, then p includes the ordered partition R^l. In particular, p includes the ordered partition R^{i_1} induced by the i_1th row of the matrix $\|a_{ij}\|$.

Property 2. An admissible ordering includes an ordered partition R if and only if it includes the ordered partition R', obtained by superimposing on R its interval S which coincides with one of the sets (6).

Property 3. Suppose $R = (R_1, \ldots, R_m)$ is obtained from the ordered partition $R^l = (R_1^l, \ldots, R_m^l)$ (for some $l \varepsilon A$) by successive superimposings of some of its intervals (6) (it is possible that $R = R^l$). Suppose further that for some $i \varepsilon A$, $k = 1, \ldots, m_i$ the set S_k^i is not an interval of R, so that there exists an object $s \notin S_k^i$ which is covered by S_k^i. Then there do not exist admissible orderings beginning with any object which precedes s in R.

Proof 13. To prove property 1 it is sufficient to note that if $l = i_k$, then $a_{li_1} \leqslant a_{li_2} \ldots \leqslant a_{li_{k-1}}, a_{li_{k+1}} \geqslant \ldots \geqslant a_{li_N}$, but since $i_1 \varepsilon R_1^l$ then also $i_{k-1} \varepsilon R_1^l$, so that in fact

$$a_{li_1} = \ldots = a_{li_{k-1}} \geqslant a_{li_{k+1}}$$

Thus the elements $a_{li_1}, \ldots, a_{li_N}$ form a non-increasing sequence, and this means that R^l is contained in $p = (i_1, \ldots, i_N)$.

We now prove property 2. If the admissible ordering includes R' then clearly it includes R also. If, on the other hand, the admissible ordering p includes R but does not include R' then this means that in p some element of the set P_s precedes some element of the set $R_s - P_s$, or some element of the set $R_{t+1}-P_{t+1}$ precedes at least one element of P_{t+1}. But then the set S is not an interval of the ordering p and, by Theorem 1 p is not admissible.

We now prove property 3. First, note that the sets $S_k = \bigcup_{j=1}^{k} R_j$ ($k=1,\ldots,m$) are intervals of the admissible ordering. Suppose that R is obtained from R^l by superimposing some intervals of the form (6). Consider R', obtained from R by superimposing still another interval S of the form (6). Clearly only the sets

$$S'_s = \bigcup_{j=1}^{s-1} R_j \cup (R_s - P_s), \quad S'_{t+1} = \bigcup_{j=1}^{t} R_j \cup P_{t+1}$$

are not unions of the first classes of the ranking R, so that it is necessary to verify whether they are intervals of the admissible ordering. But

$$S'_s = \bigcup_{j=1}^{s} R_j - S = S_s - S, \quad S'_{t+1} = \bigcup_{j=1}^{t+1} R_j - S = S_{t+1} - S,$$

with S_s and S_{t+1} and S all being intervals of the admissible ordering. But then S'_s and S'_{t+1} are also intervals, as the differences of intervals that are not included in each other. Thus we have shown that the sets $S_k = \bigcup_{j=1}^{k} R_j$ ($k=1,\ldots,m$) are intervals of the admissible ordering (if it exists).

Now suppose that in R the object t precedes the object s, for which $s \not\in S_k^i$ for some $i \in A$, $k=1,\ldots,m_i$, with $s_1, s_2 \in S_k^i$ such that $s_1 R s$ and $s R s_2$. We suppose that an admissible ordering p, beginning with t exists.

There are two possible cases for the arrangement of the objects t, s and l in p: a) $p = (t\ldots l\ldots s\ldots)$ or b) $p = (t\ldots s\ldots l\ldots)$. Consider case a). All of the sets $S_k = \bigcup_{j=1}^{k} R_j$ are intervals of p, so that $l, t \in S_{k_1}$; $l, t, s_1 \in S_{k_2}$; $l, t, s_1, s \in S_{k_3}$; $l, t, s_1, s, s_2 \in S_{k_4}$, with $k_1 < k_2 < k_3 < k_4$ by assumption. From this it follows that the objects s_1 and s_2 are situated in p as follows: $p = (t\ldots l\ldots s_1\ldots s\ldots s_2\ldots)$. But then S_k^i is not an interval of p, so that p is not admissible, by Theorem 1.

Case b) goes similarly and leads to the same contradiction, so that property 3 is proved.

From properties 1-3 comes the following algorithm for discovering all admissible orderings. For each $l \in A$ the algorithm finds all admissible orderings which begin with l, or else shows that there are none.

Examined candidates for first place are stored in the set L. At first L is empty.

For each $l \notin L$ examine the ordered partition R^l and some arbitrary S_k^i of the form (6). If S_k^i is not an interval of R^l then by property 3 for any $s \notin S_k^i$ covered by R^l there does not exist an admissible ordering for any objects t which precede s. Place these t in L and return to examine a new R^l for $l \notin L$. If S_k^i is an interval of R^l, then proceed to examine $R^{l'}$, obtained by superimposing S_k^i on R^l, and a new set S of the form (6). By properties 1 and 2 all admissible orderings beginning with elements of R_1^l, if they exist, are contained in $R^{l'}$. Successively repeating this superpositioning procedure for all sets S_k^i ($i \varepsilon A$, $k=1,\ldots,m_i$) we either arrive at an ordered partition R, for which one of the sets of the form (6) is not an interval (in which case we place new objects t in L as above and proceed with a new $l \notin L$), or at an ordered partition R for which all the sets S_k^i of the form (6) are intervals, so that all the orderings contained in R are admissible[†]. In this last case place all elements of the set R_1 in L and return to the examination of R^l for $l \notin L$.

The process terminates when $L = A$. L may be filled more quickly if we keep in mind that the reverse ordering of any admissible ordering p is also admissible. Having obtained at least one admissible ordering, we can also make use of the uniqueness theorem (Theorem 2) to discover the remaining orderings.

Admissible orderings frequently do not exist. We consider the following generalization of this notion. Every ordered partition $R = (R_1, \ldots, R_m)$ of the set A induces a partition of the matrix $\|a_{ij}\|$ into submatrices (cells) $B_{st} = \|a_{ij}\|$ ($i \varepsilon R_s$, $j \varepsilon R_t$). We say that a cell B_{st} is *not greater than* a cell B_{sr}, if each element of each row of B_{st} does not exceed the minimal element of the corresponding row of the cell B_{sr}:

$$B_{st} \leq B_{sr} \leftrightarrow \forall i \varepsilon R_s \left[\max_{j \varepsilon R_t} a_{ij} \leq \min_{j \varepsilon R_r} a_{ij} \right].$$

The fact that $B_{st} \leq B_{sr}$, means that the closeness of the connections of objects of class R_s with objects of class R_t never exceeds the closeness of their connections with objects of R_r.

An ordered partition $R = (R_1, \ldots, R_m)$ of a set A is said to be *admissible*, if the matrix of cells $\|B_{st}\|$ has linear form, in the sense of the relation just introduced:

$$B_{st} \leq B_{sr}, \text{ iff } s \leq r \leq t \text{ or } s \geq r \geq t. \tag{7}$$

A set $S \subseteq A$ is said to be *closed* (with respect to the system of subsets S_k^i of form (6), if it contains, along with all $i \varepsilon S$, all sets S_k^i ($k=1,\ldots,m_i$) which do not contain S.

Theorem 4. An ordered partition is admissible if and only if all its classes are closed and all the sets S_k^i of the form (6) are its intervals.

[†]Of course, as in Section 1, only one component of the overlap graph of sets S_k^i is being considered. The other components correspond to subsets of objects, and are considered separately.

The proof of this theorem and of the algorithm (based on it) for finding admissible ordered partitions with a maximal number of classes are given in [13].

This result is connected with the problem of approximating real data by means of linear matrices. If an admissible ordering does not exist then an ordered partition with a maximal number of classes may be considered as an approximating structure, which characterizes the desired ordering up to permutations of objects within classes.

Probably a more natural theory for approximating data by graphs and hypergraphs of intervals can be developed within the framework of the graph approximation idea presented in Chapter 2. However we will not present any theoretical or experimental results in this direction. The particular problem of approximating a partial order by an interval graph contained in it, was examined recently by F. Aleskerov [1a].

§5. Examples of structural analysis of genetic systems

1. **The use of deletions and polar mutations**. The majority of mutations, both spontaneous and induced, occur as simple replacements of nucleotides in DNA, leading to corresponding replacements of amino acids in proteins.

The use of only such point mutations for constructing finely-detailed maps is not suitable in recombination mapping because of the necessity of producing a very large number of crosses, nor in complementation mapping because they do not provide information about the meshing of text fragments.

Very frequently mapping is made easier by the use of deletions, the loss of genome fragments of various lengths. Operationally (in genetic analysis), <u>deletions</u> are characterized by the following two properties: a) unlike point replacements, which permit the reverse mutations (reversions) to return the organism to the norm, deletions are stable and do not revert; b) deletions, depending on their lengths, are not able to recombine with some point mutations [30,43,48]. At the molecular level property a) is explained by the infinitesimally small probability of making up for the loss of an entire series of nucleotides, at times of great length, by inserting the lost text. Property b), at the molecular level, means that a deletion which overlaps a point mutation cannot recombine with it.

These properties make for wide (and highly successful) use of deletions both in recombination and complementation gene mapping.

The classic example of the use of deletions in constructing a recombination map is the work of S. Benzer [3,55] (see also page 18) in analyzing the fine structure of the rII locus in the bacteriophage T4, a virus of the intestinal bacterium *E. coli*. The bacteriophage destroys an *E. coli* cell within an hour, but mutations of this locus can significantly shorten this period (hence, this locus and its mutations are called *rapid-lysis*).

The collection of S. Benzer included more than 2400 mutations of this locus. The usual recombination approach would require an astronomical number of paired crossings. Therefore Benzer chose mutations able to revert to the norm (a priori, point mutations). He did recombination mapping by the usual paired crossings for several dozens of these mutations. Then he was able to select 32 deletions (according to their inability to revert) and to cross them with the point mutations which were already located on the map. Each deletion was associated with its list of point mutations with which it gave no recombinant progeny. In other words, the point mutations gave the vertices, and the deletions, the edges of the interval hypergraph. It turned out that this graph was divided into a system of imbedded hypergraphs Γ_s, i.e. the graph $G(\Gamma)$ had several components A_s (see Section 4.1) (Fig. 16a). Benzer first examined seven of the longest deletions, which showed a step-like arrangement. Then, crossing the remaining point mutations with these deletions, it was not difficult to establish on which "step" (i.e. on which of the distinguishing segments for neighboring deletions) their defects were located. A mutation is located in the segment distinguishing deletion i from deletion $i+1$ if it is not recombinant with the former, but is with the latter.

In this way all the mutations were divided among the seven segments formed by the deletions.

In succeeding stages Benzer carried out a similar procedure, working with deletions at succeeding "levels", localized within individual segments at higher levels (see Fig. 16a).

As a result he was able in three steps to divide the rII region into 47 deletion segments, "equivalent" to complons of the deletion intersection graph (see Section 3.2), and also was able to distribute the whole collection of mutations of the rII locus among those segments (see Fig. 16b). The arrangement of the point defects within these segments was determined by the usual recombination analysis.

The work of Benzer is a model of the use of qualitative analysis in mapping. It is no accident that it has attracted the interest of mathematicians to graphs and hypergraphs of intervals. Owing to the successful selection of deletions Benzer did not have to solve the algorithmic problems of ordering graphs and hypergraphs of intervals, but the stimu-

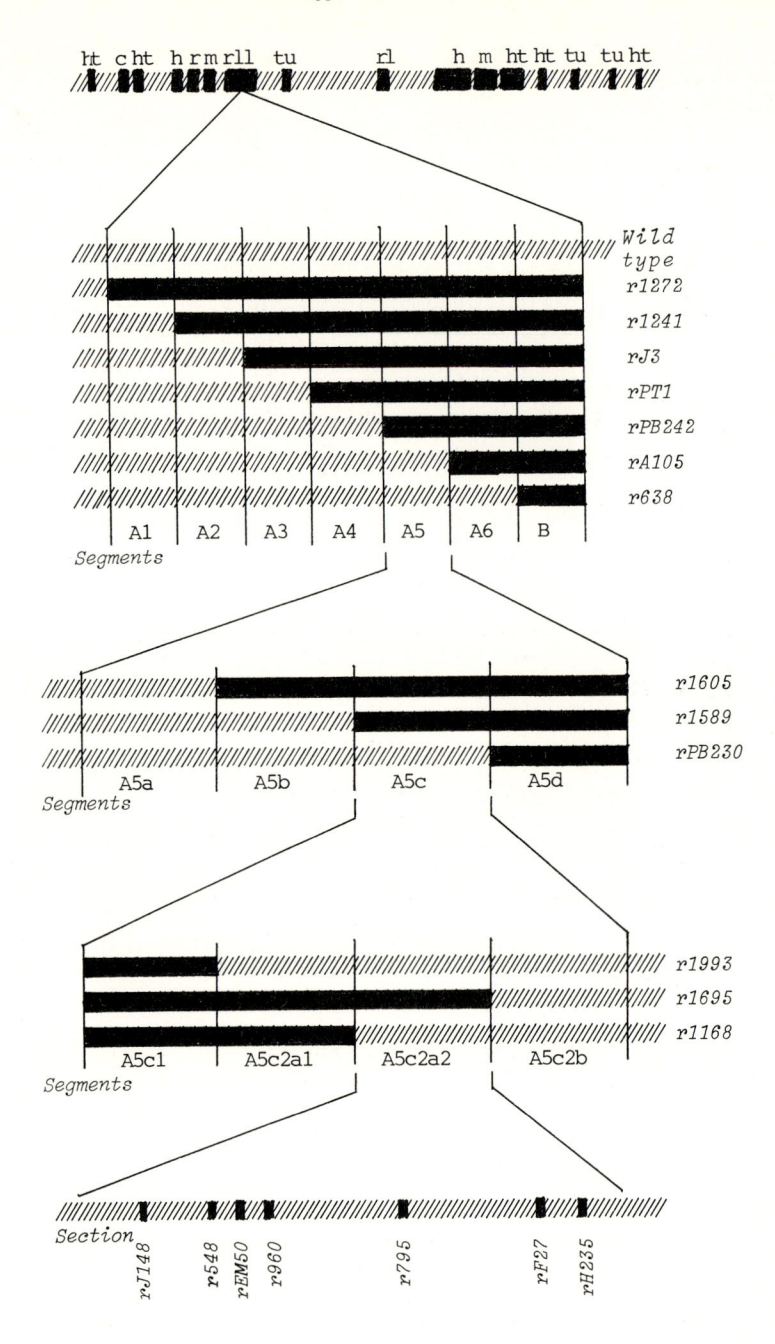

Figure 16a. The mapping of mutations in the rII locus of the bacteriophage T4 [3,55]. The three-stage division of the locus into segments by means of overlapping deletions.

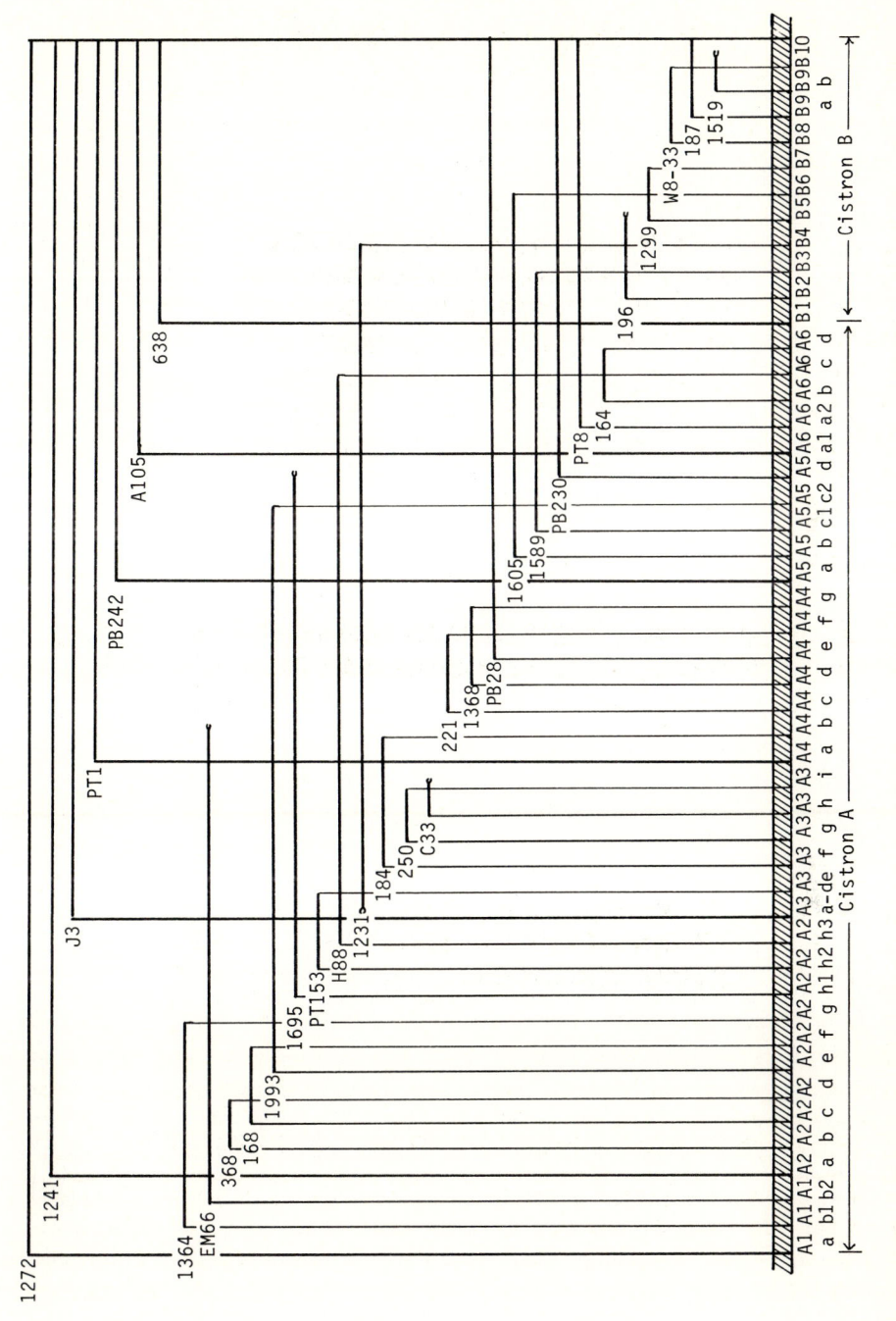

Figure 16b. The mapping of mutations in the rII locus of the bacteriophage T4 [3,55]. The deletion map of the locus, consisting of 47 segments (the right ends of five deletions which did not participate in determining the segments are shown with cavities).

lating influence of his investigations on the work reported in [67] and [80] is unquestionable.

Benzer also carried out a functional analysis of the rII locus, allowing him to fill the classical operational definition of a gene with molecular content.

It turned out that the complementation test divided the collection of rII mutations into two groups, within which mutations are noncomplementary, while any two mutations from different groups are complementary. The groups were compactly projected onto Benzer's map corresponding to the sections he designated as *cistrons*. This term comes from Benzer's modification of the complementation test to the so-called *cis-trans test* (details of his description can be found in [30,43,48]). In Fig. 16b cistron A occupies the left six segments of the "top layer", while cistron B occupies the remaining segment. Thus, according to Benzer, a cistron is a functional unit, having, at the same time, a complex mutational and recombinational structure. By means of some indirect data Benzer was able to closely estimate the physical dimensions of the basic components of his map. It turns out that the mutational and recombinational granularity of a gene ranges down to individual nucleotides, while the functional unit, the cistron, may consist of hundreds or even thousands of such nucleotides. The work of Benzer is one of the most fundamental in the growth of concepts about the genetic systems we described in Section 1.

We now investigate the possibility of using deletions in complementation mapping. We already remarked in Section 1.3 that recombination maps are collinearly recombinational, i.e. they reflect the structure of a locus in case there are mutations, among those being studied, that involve several cistrons at once. Deletions are among the more simple multi-cistron mutations, with respect to their functional manifestation: with them, a loss of function is directly connected with an insufficiency of the corresponding genetic material. The complementation matrix of a collection of mutations, among which are multi-cistron deletions, can be ordered by the methods described above.

As an example we consider the structure of the operon which controls the synthesis of the amino acid histidine in *Salmonella* bacteria. This is one of the most complex operons described up to the present. On its genetic map more than a thousand different mutations have been localized, which produce an insufficiency of histidine in the bacteria [71,81].

The histidine operon numbers nine cistrons, controlling 10 successive stages in the synthesis of the amino acid from biochemical precursors (Fig. 17,a). Benzer's methods described above were used to construct the map (Fig. 17) for preliminary distribution of mutational defects. Moreover, biochemical and complementation data were also used. Thus, the position of cistron H was determined at the beginning, by means of only the complementation matrix [48,81].

The first summaries of complementation data allowed the complete elucidation of the intercistronic topography of the operon by means of deletions. Thus, up to 1964, deletions at the supercistron level had formed the following complementation matrix:

	1	2	3	4	5	6	7	8	9	10	11	12	13	14	15	16
1	1	1	1	1	1											
2	1	1	1	1	1	1	1	1								
3	1	1	1	1	1	1	1	1	1	1	1					
4	1	1	1	1	1	1	1	1	1	1	1	1	1	1		
5	1	1	1	1	1	1	1	1								
6		1	1	1	1	1	1	1								
7		1	1	1	1	1	1	1	1							
8		1	1	1	1	1	1	1	1	1	1	1	1			
9			1	1			1	1	1	1						
10			1	1			1	1	1	1	1					
11			1	1				1	1	1	1	1				
12			1	1				1			1	1				
13				1				1					1	1		
14				1				1					1	1	1	1
15				1				1					1	1	1	1
16														1	1	1

The matrix is represented in quasidiagonal form. By means of the algorithm of Section 3.1 it is easy to conclude that there are only seven complons in this matrix, one of which (H) is ficticious. However, the linear order of the complons, determined by extended deletions, is collinear with the order of the associated cistrons on the genetic map (Fig. 17,b). The bounds of the deletions are such that the neighboring cistrons E, I and C, D are not different on the complementation map; they are represented by single complons. To separate them it is sufficient to include in the matrix complementation interactions of the deletions with point mutations. If each cistron is represented in the matrix by a point mutation, the corresponding map contains only **real complons**.

Also not represented in the matrix are many cases of intracistronic complementation discovered in the cistrons B, F and E. Their identification was accomplished biochemically. The inclusion of intracistronic

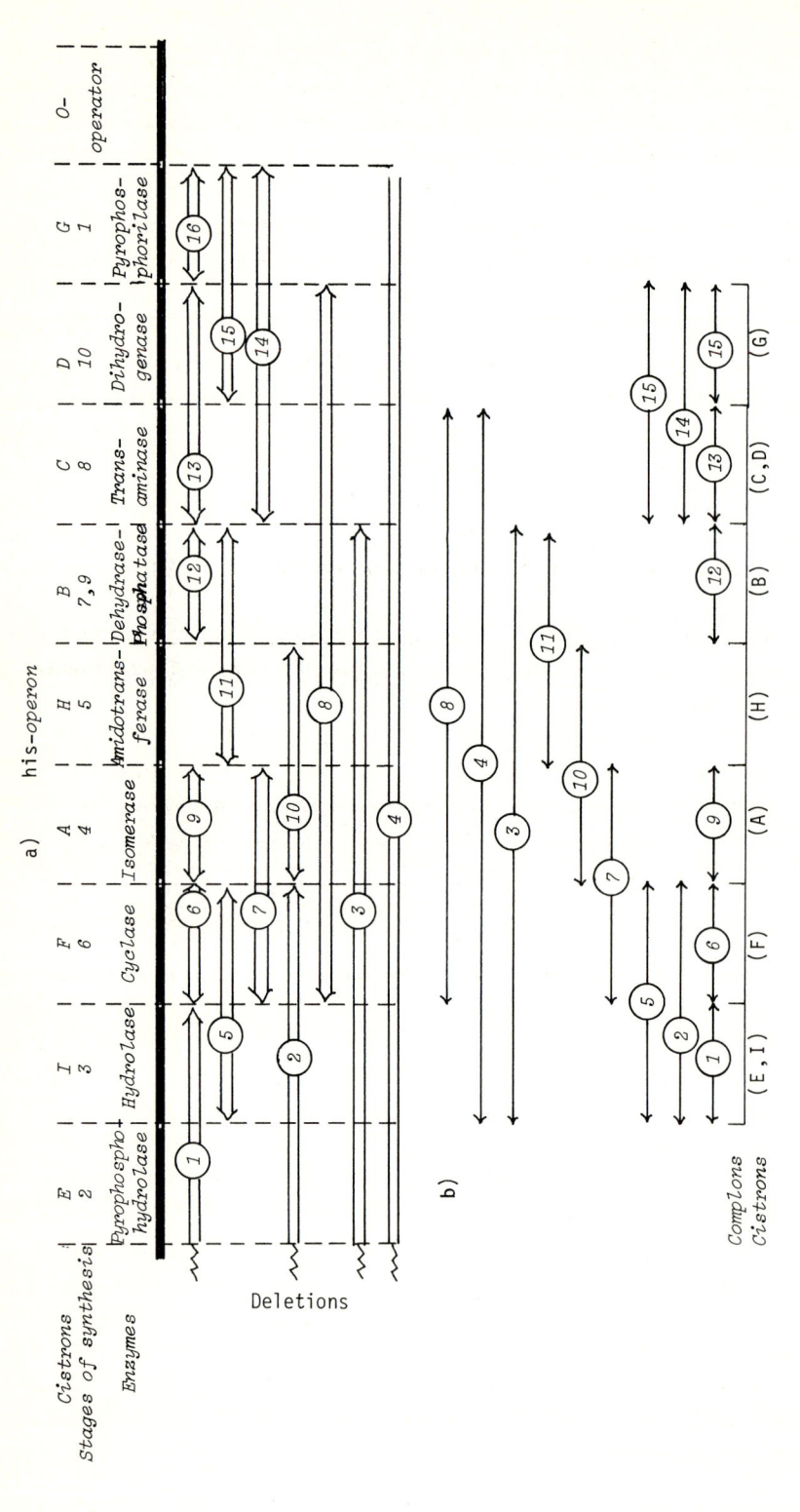

Figure 17. The his-operon map of the bacteria *Salmonella typhimurium*: a) At the top of the map are shown the separate structural cistrons of the operon and the names of the enzymes (0-operator) coded for by them. Below them are several deletions used by Loper et. al. [81] to order the cistrons in the operon (the bounds of the deletions in the cistrons are not shown); b) The complementation map constructed according to the matrix of complementation reactions of the deletions with one another.

complementation in the analysis leads to the impossibility of a linear ordering of the complons (see Section 2.1). However, in this analysis characteristic "concentrations" arise, deviations from linearity, which help determine the cistron groups.

Besides deletions, another class of extensive intercistronic mutations has been successfully applied to the complementation mapping of transcription units in bacteria and viruses. These are the *polar* mutations.

To describe their manifestations it is necessary to briefly dwell on the features of the transcription and translation processes in bacteria. Translation begins before transcription has even been completed. Ribosomes seat themselves on the initial codon of an already-transcribed mRNA fragment and perform the translation process by moving from one codon to the next. As a result, several ribosomes will be seated on an mRNA fragment, each with its partially formed polypeptide chain. On reaching the terminator, the ribosomes release the mRNA with the finished protein molecules.

Suppose that in some cistron a mutation of the type "codon→nonsense" occurs. On reaching the meaningless codon the ribosome, as a rule, cannot proceed because of the absence of the corresponding tRNA. Meanwhile, transcription continues, so that a section of the cistron is formed which is not occupied by ribosomes. Such "free" sections are subject to the action of nuclease enzymes which destroy them as well as the entire remaining chains [30,31]. Therefore, mutations of the "codon→nonsense" type lead to the possible loss of translation of following cistrons.

Thus, although the mutation arises in one cistron, its negative effect is felt in all succeeding ones. Such mutations are said to be *polar*.

We emphasize that polar mutations are possible only in scriptons, where the units of translation (cistrons) are joined at the mRNA level. Besides nonsense mutations, polar effects are frequently associated with the so-called *frameshift* mutations, which result from the loss or insertion of nucleotides. Such mutations change the content (meaning) of all succeeding triplets. Because of this there is a high probability that a meaningless codon will be encountered in the subsequent text.

The nearer the nonsense replacement occurs to the beginning of transcription, the greater will be the extent of the polar defect. A diagnostic feature of all polar mutations is the inactivation of the last structural cistron in the scripton. This is not necessarily so for deletions, although in Benzer's work (see Fig. 16) seven deletions at the top level did have this feature.

Polar mutations permit the easy determination of the linear order of the cistrons within a scripton.

Thus, for example, if among the initial mutations there are only point mutations and polar mutations (Table 2 and Fig. 18), then clearly the point mutations characterize the complons of the map (the cistrons), it being unnecessary to use a general algorithm to order them in this case (page 55). To find the order of the cistrons it is sufficient to find a satisfactory ordering of the polar mutations, according to the numbers of the cistrons (point mutations) they cover.

This method of mapping has been applied to a whole series of bacterial operons: lactose, tryptophan, galactose and others.

An analogous map can be observed for the locus his-3 in the fungus *Neurospora* [54]. In the opinion of some geneticists this is one example supporting the existence of operons in organisms of more complex structure than bacteria. Shown schematically in Fig. 19 are the complementation and recombination maps of this gene. The recom-

Table 2

A complementation matrix[†]

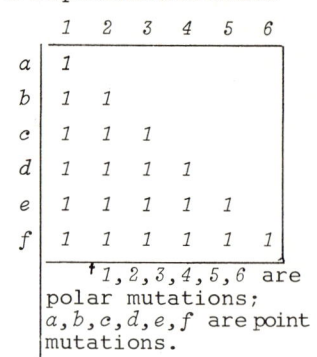

	1	2	3	4	5	6
a	1					
b	1	1				
c	1	1	1			
d	1	1	1	1		
e	1	1	1	1	1	
f	1	1	1	1	1	1

[†] $1,2,3,4,5,6$ are polar mutations; a,b,c,d,e,f are point mutations.

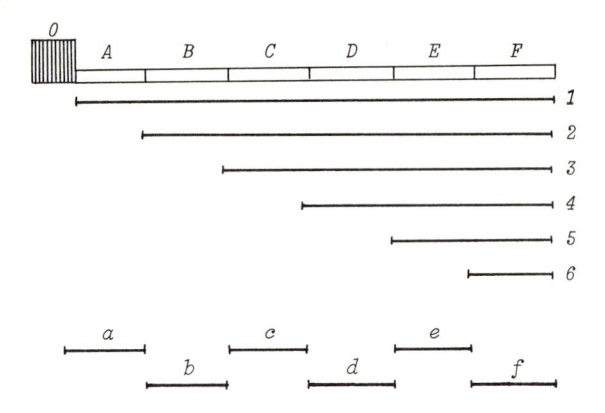

Figure 18. The complementation map of the operon corresponding to the complementation matrix of Table 2.

bination map is divided into five parts: section 0 (possibly an operator) and four structural cistrons (A, B, C and D), which code for corresponding enzymes, three of which are already known. The complementation map contains five polarly complementary mutations, probably due to nonsense replacements. The locations of the mutations on both maps is evidence of their collinearity (the correspondences in position are shown by the vertical arrows).

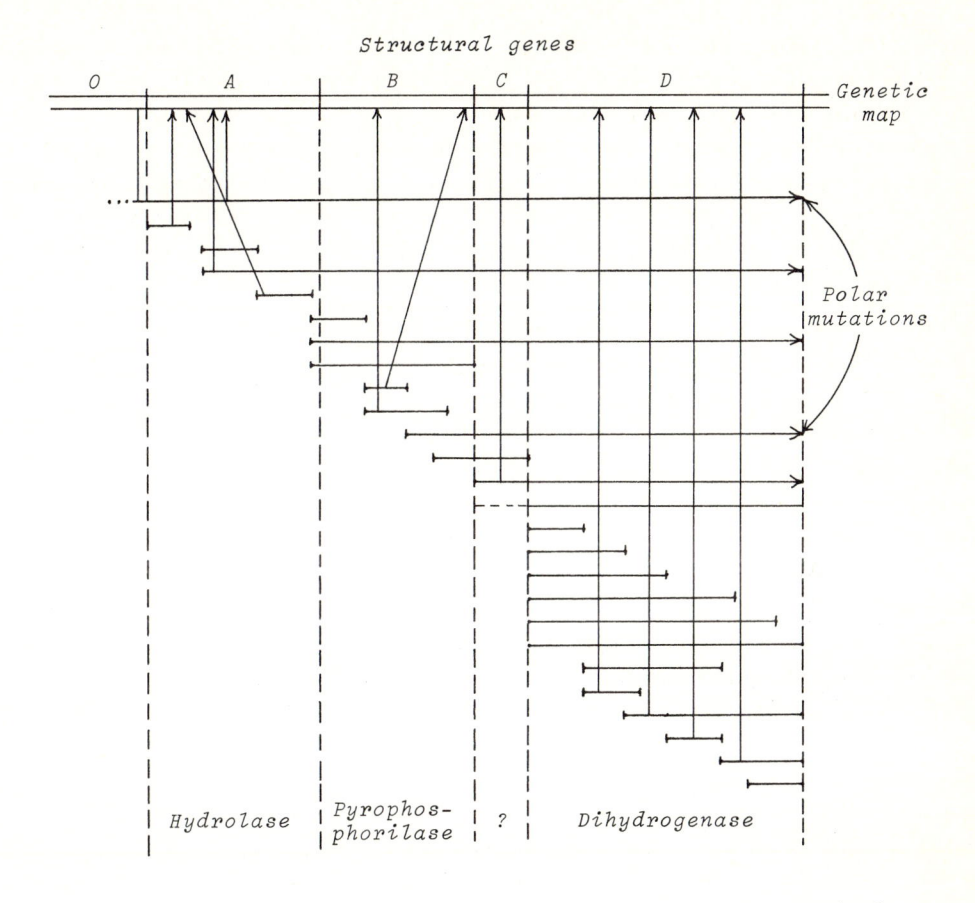

Figure 19. The map of the his-3 locus in *Neurospora crassa* [54]; the complementation map has a sharply dilineated polar character.

2. **The complementation maps of multiple mutational defects**. There are situations in which it is not possible to construct a linear map by complementation. First of all, the impossibility of constructing unbroken intervals on a linear map (even when all the complons are determined) may be connected with an actual discontinuity of the defects associated with a single mutation. A more exotic type of nonlinearity may involve special chromosome reconstructions called *inversions*, in which the order of the genes in a chromosome section is reversed. Under crosses of two homozygotes with normal and inverted gene positions, progeny heterozygous for the inversion are produced. Suppose, now, that an "inverted" chromosome in one of the parents has sustained a deletion involving two genes which are normally not neighbors (Fig. 20a).

Similar multicistron deletions in this region may arise also in normal (for gene order) parents. Then the progeny, which are heterozygous for inversion and deletion, may provide an intergene complementation matrix (Fig. 20,b) which cannot be represented by a linear map, although all the deletional defects shown in Fig. 20,a are unbroken. This matrix corresponds to a cyclic interval graph (Fig. 20,c).

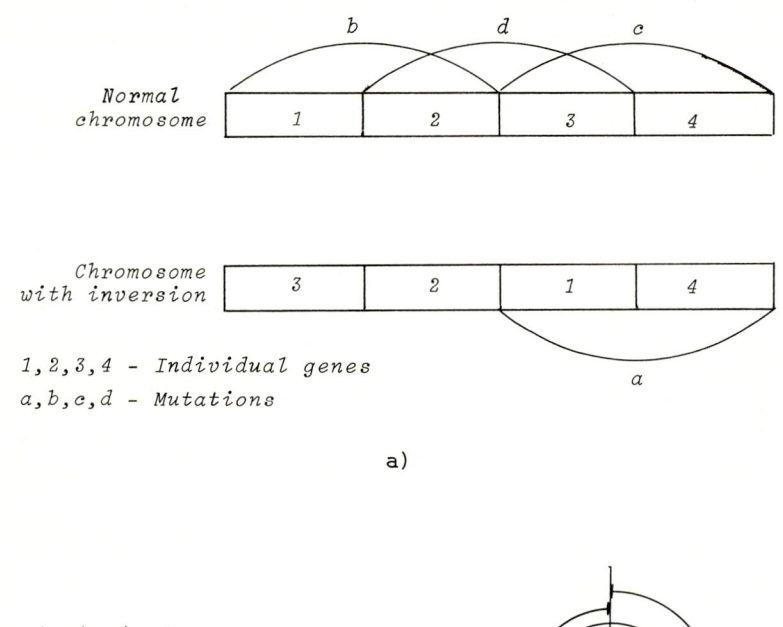

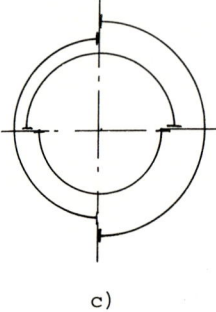

Figure 20

Naturally, the most simple sources of deviations from linearity are errors in the initial data. Such errors may be connected both with features of the object of study and with low resolution of the genetic technique (not to mention subjective aspects). Therefore, when nonlinearity appears one must first be convinced of the absence of systematic errors of this type.

TABLE 3

Basic complementational "parameters" of recessive lethals, induced in *Drosophila* by foreign DNA's and several viruses.

Series	Mutagen nature	No. of Lethals	Total	Real	Fictitious	Isolated
				Number of complons		
CIV	Cranefly iridescent virus	16	11	4	6	1
DA	DNA of blue-green algae	16	12	4	8	0
C-5	Coxsackie virus, type 5	17	12	2	9	1
C-3	Coxsackie virus, type 3	18	16	3	11	2
IV	Influenza virus (through food)	19	19	2	14	3
DH	DNA of herring	21	20	6	14	0
IVI	Influenza virus (injection)	22	15	2	8	5
DCVD	DNA of CIV + DNAase	22	21	4	16	1
C-1	Coxsackie virus, type 1	24	21	3	16	2
RV	Rous sarcoma virus	25	21	4	8	9
PV	Poliomyelitus virus	27	27	5	21	1
DC	DNA of chick erythrocytes	32	38	4	33	1
DCV	DNA of CIV	41	61	7	53	1
DCT	DNA of calf thymus	66	99	5	90	4

However, the basic, regular source of deviations from linearity in the mapping of a complementation matrix are the cases of interallelic, intracistronic complementation, the molecular events of which do not satisfy the initial premises of linear mapping (see Section 2.1).

We have also concluded the impossibility of linear mapping from the analysis of complementation interactions of mutations induced in *Drosophila* by foreign molecules of DNA, RNA and several viruses [4,37].

These mutations have a recessive lethal action, i.e. in crosses with normal individuals they are not manifest, while the corresponding mutant homozygote flies do not survive. Unlike the usual mutagens (such as X-ray irradiation or chemically active substances), foreign nucleic acids and viruses act on the genetic apparatus of *Drosophila* to some extent directionally: all the mutations elicited by them are localized in a small section of the right arm of one of the chromosomes (the second) [4]. Fourteen different mutagens from this class were examined, to which were matched complementation matrices for the mutations they induced, having dimensions from 16 × 16 to 66 × 66 (Table 3).

All the complons were constructed by means of the algorithm given in Section 3.3, but a continuous linear map was not obtained for any of the matrices. Moreover, the continuous variants which we constructed from the maps had necessarily complex topologies: The mutations appeared as G-intervals of the graph G of extremely complex cyclic structures.

What causes such complex structure in the maps? Chromosome rearrangements of the inversion type cannot be the reason. An investigation of the corresponding chromosomes by S. M. Gershenzon showed that inversions and other structural aberrations were encountered very rarely, and even when encountered they were weakly connected with those regions in which complex interval structure was found.

Perhaps the source of nonlinearity is the fact that the mutational defects examined fall within one cistron. The following facts refute this. The chromosome region in which the mutations lie encompasses several dozen (or more) cistrons, and several mutations of particularly complex meshing are located at opposite ends of the region, i.e. they probably belong to different cistrons. Suppose, nevertheless, that some of these mutations belong to the same cistron. Then the cyclic character of the associated map sections implies that we have a situ-

---

These data are the results of many years of experiments on induced mutations, carried out under the direction of Academician S. M. Gershenzon of the Ukrainian Academy of Sciences (the Institute of Molecular Biology and Genetics) [4]. The authors express deep appreciation to him for the opportunity to analyze the data.

ation of the type "defect corrects defect" (the corresponding molecular mechanism is discussed in Chapter 2). But such mutual correction is rare and characterizes specific pairs of defects, while in this case the mutations are complementary to many others.

Thus, there remains the quite simple and natural explanation, that such mutations affect the chromosome in several places at once.

This conclusion is supported by the following regularity (see Table 3). For the smaller matrices the number of complons is slightly less than the number of mutants; for the intermediate-sized matrices these numbers are comparable; and for the larger matrices (the last three mutagens) the number of complons exceeds the number of mutants. An analogous result is obtained when one examines randomly selected mutants from the larger matrices.

In fact, those sections of the chromosome (separated by testing) which are able to mutate individually as well as jointly are the real pre-images of complons. We recall that the complementation map at the level above cistrons, characterizes the mutual disposition of defects associated with mutants. Therefore, if the number of complons equals or exceeds the number of mutants it means that some mutants carry several non-adjacent defects in the chromosome.

How can one nevertheless find a purely linear order for the defects? It is sufficient to merely isolate the groups of mutants representing cyclic map intervals, and carry out direct recombination analysis for them. As a result, those mutants which carry several defects will be determined, and subsequent complementation mapping of these defects (not the mutants) will provide the desired map.

As a basis for this prognosis S. M. Gershenzon undertook, with his colleagues, experiments which completely supported these expectations. The recombination map clearly fixed the desired order of the mutational defects and their multiplicity for specific mutants. It turned out, moreover, that the defect multiplicities obtained are connected with more general and fundamental aspects of the organization of genetic systems. We will discuss this in more detail. What genetic mechanisms have a similar appearance? The distinguishing feature of foreign nucleic acids and viruses, as mutagens, is the specificity of location of their effects in the genetic apparatus of the host. Therefore, understanding of the molecular mechanisms of mutagenic actions is extremely important, since it contributes to the solution of one of the leading problems of genetics — control of the mutational process in multicellular organisms. A number of hypotheses have been expressed that the basis of mutagen action of observed agents lies in their capacity for being taken into the chromosome host-recipient by means of segments homologous to segments of the chromosome [4]. In this process genes near the introduced sgements are destabilized and begin to mutate more frequently.

The multiplicities of induced defects we have revealed provide indirect support for such a mechanism, for the observed mutagens are of the most diverse origins (from RNA of viruses to DNA of the thymus of a calf), and consequently, the segments of homologous identification probably also vary greatly. And in the final analysis, this also leads to multiple defects in a *Drosophila* chromosome: Mutagens of a given type may be present either in a free state, in which case they replicate autonomously, or they may be bound to definite loci of a chromosome, causing mutations in those and in nearby regions. Thus, although local specificity of these mutagens is apparent, it is hardly restricted to one cistron.

Intensive genetic and biochemical investigations in the last ten years have shown that the traditional belief in the invincible stability of a specific genome is nothing more than an antiquated dogma. In the genomes of both higher and lower organisms a large number of varied, mobile genetic elements have been determined, described and partially studied. These are the so-called *insertosomes*, *transposons* (saltatory genes) and other disperse, repeating DNA sequences [69a]. All these elements have a common distinguishing feature: They may change their location in the genome and cause mutational instability in the regions of their intrusion. The frequency of such jumps is several orders of magnitude higher than the average frequency of the usual spontaneous mutation. As a result, multiple defects arise in the genome. Apparently it is such elements that are responsible for the multiple mutagenic effects of foreign DNA and viruses. Moreover, multiplicity of mutational defects itself may be used as an indirect indicator of the presence of mobile elements in the mutational screening of natural populations.

The particular significance of the results described above, in conformity with human populations, adds to the fact that viruses figure here as a mutagenic source (more accurately, the nucleic acids contained in them) [4]. Therefore they should be considered not only as a source of disease, but also as a mutation-causing factor, i.e. as an evolutionary factor. They play this role independently of whether they are infectious. If one considers how widely dispersed viruses are in man's environment and how frequently the two come in contact, the problem of viral mutational action becomes particularly serious in a practical as well as theoretical sense.

3. **Complex traits and the loci which control them**. It is usual to use recombination analysis (which directly reflects the arrangement of mutational markers) as the means of eliciting the structure of genetic loci. In Section 1 we saw how it is used to localize defects not only on the supercistron level, but also within cistrons. However, recombination analysis of small sections of a chromosome, commensurate with the size of a cistron, requires the examination of tremendous numbers of progeny, since recombinations of nearby markers are so rare. Consequently, investigations similar to those which have been successful for microorganisms are extremely difficult for higher organisms.

The question arises whether the more easily obtainable results of complementation testing can be used in analyzing the structures of such loci. In Section 1 we showed how this could be done when the mutations have a supercistronic character. Although complementation data contain information not about the interactions of actual nucleotide texts, but rather their protein products, i.e. they only indirectly reflect the structure of the genetic material, the success of such endeavors is ensured by the collinearity of DNA, RNA and the primary structure of proteins, functioning as wholes.

However, many phenotypes of higher organisms are variable expression, and binary characteristics of their manifestations in terms of "yes - no" (as was done in the preceding examples) looks very coarse: one can only say that "this thing is and that one is not." In this case the results of complementation testing are not representable as complementation graphs, and the notion of a map is not directly applicable.

Nevertheless, one can try to estimate the similarity of variable expression phenotypes for various mutations quantitatively, and afterwards consider the matrix of indicators of the functional nearness of examined mutations as a "portrait" of the locus. To do this one must step back from the position that structural nearness of defects within the locus being examined determines, to a certain degree, the similarity of their functional manifestations. In other words, the matrix of functional similarity must be reducible to a linear form (Section 4.3) which determines the internal topography of the mutations in the locus (when, of course, the initial data do not include cases of interallelic complementation).

This idea was used in analyzing the functional manifestations of mutations of the *scute* locus in *Drosophila*.

This locus controls the appearance of bristles on the head and thorax of the flies (Fig. 21). Mutations of this locus lead to the loss of growth of some bristles, in which a strict dichotomy is discernible with regard to each specific bristle on a specific fly: the bristle is either normal or non-existent (reduced). Thus, in this case each separate mutant individual is characterized not by one, but by a whole set of binary manifestations — for each of 20 pairs of investigated bristles, i.e. for a 40-element boolean vector. Is each bristle controlled by its own gene? No. Recombination analysis has shown that all these mutations affect the same very small, recombinationally indivisible section of the X chromosome in *Drosophila* — the *scute* locus.

We note that individuals of one genotype carrying the same scute mutation may, generally speaking, have different sets of bristles, so that for a specific mutation it only makes sense to speak of the probabilities of reduction for each bristle. In other words, each mutation of scute is characterized by a 40-element vector of probabilities of reduction in the macrochaeta. The probabilities are estimated according to frequencies of reduction which are quite reliably produced in genetic experiments.

The *scute* locus was first investigated in the 1920's and 1930's by the Soviet geneticists A. S. Serebrovsky N. P. Dubinin et. al. [5,38]. Their work dealt the first serious blow to the classical concept of an indivisible gene. The gene was divisible, as witness the sectional mutability of the *scute* locus. Under the union of different mutations in a heterozygote, local dominance of norms over reductions was observed with respect to separate bristles. Namely, in each type of heterozygote only those bristles were formed which were in at least one of the original mutations in a homozygote.

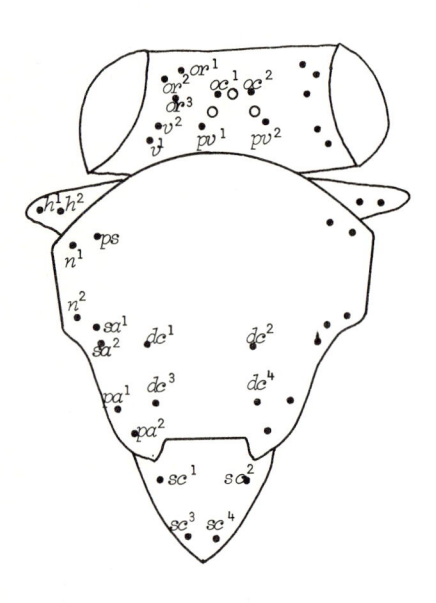

Figure 21. The arrangement of bristles on the body of *Drosophila melanogaster*: *oc—ocellar, pv—post-vertical, or—orbital v—vertical, h—humeral, n—notopleural, ps— presutural, sa—supra-alar, pa— post-alar, dc— dorsocentral, sc—scutellar.*

Of course the correctness of this assertion depends on the frequency of reduction taken as the threshold, above which a manifestation is considered normal, and otherwise, mutant. However, with real data mutant and normal reduction frequencies are quite accurately separable (Fig. 22).

By that time it was already known that different mutations affect compact, unbroken sections of a chromosome (the most simple example is a deletion). Therefore, the next logical step was the premise that for each bristle there is a special section (center) in the *scute* gene that

a)

	pa^1		n^1		$pv^{1\ 2}$		or^1		$oc^{1\ 2}$		or^2		$sc^{3\ 4}$		$sc^{1\ 2}$	
	l	r	l	r	l	r	l	r	l	r	l	r	l	r	l	r
sc9	52	52	100	100	95	93	100	100	100	100	96	95	100	100	96	96
scD1	44	47	100	100	92	92	100	100	88	88	83	81	99	98	99	99
sc1	5	6	100	100	97	96	100	100	100	100	98	98	84	88	68	60
sc7	1	2	100	100	94	95	100	100	98	99	78	80	99	98	94	96
scD2			100	99	89	88	100	100	95	96	84	85	64	68	55	56
sc6	1	1	93	89	73	79	100	100	99	99	99	99	2	3	1	
sc3B	1	1			34	38	100	100	76	76	74	71	93	92	62	60
sc2B	2	3	1	1	1	2							99	99	97	94
sc260-22					1	2							100	100	100	100
sc5													69	68	18	17
sc8	3	3	2	1	1	1	2	2			5	4	19	19	37	40
scV₂			1	4	4	2	3	6	3			8	9	24	24	23	22

b)

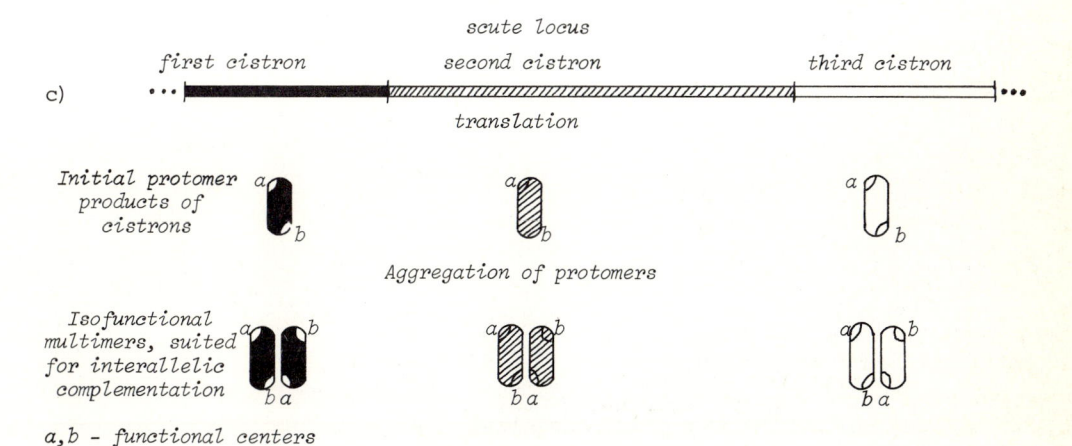

c)
scute locus
first cistron   second cistron   third cistron
translation

Initial protomer products of cistrons

Aggregation of protomers

Isofunctional multimers, suited for interallelic complementation

a, b - functional centers

<u>Figure 22</u>. A map of the *scute* locus in males of *Drosophila melanogaster* at 14°C: a) linear map of bristle indices, b) map of mutational defects (the numbers give the percentage reductions in the corresponding bristles), c) a model of the structure and functioning of the locus.

is responsible for its formation, with each mutation affecting neighboring centers. To test this hypothesis A. S. Serebrovsky, N. P. DuBinin et. al. performed a formal procedure which anticipated the method of ordering interval hypergraphs: They found a linear ordering of the bristles for which the indices of those bristles reduced by a given mutation were situated together.

It seemed that the described constructions not only filled a "gap" in the classical theory, but also permitted quite realistic description of the internal topographic structure of the *scute* locus. However, tests of the scute phenomena in various laboratories showed that the phenotypic manifestations of its mutations are very susceptible to changes in temperature and a whole series of other factors [56,68,95]. This contradicted a basic property of genetic material — the stability necessary for faultless transmission of hereditary information from generation to generation.

Since the character of reduction changes with change in temperature it raises doubt whether this phenomenon has a real connection with the structure of a stable locus.

For example, the noted geneticist R. Goldschmidt [68] felt that such a map reflects a sequence of events in the formation of bristles over time, the course of which permits dependence on such factors as temperature. As a result of this criticism work on the genetic analysis of the scute system ceased.

Recently, complementation phenomena in the *scute* locus were investigated anew with dependence on temperature [33,45].

Bristle reduction frequencies for 468 genetic combinations of 12 scute mutations were analyzed. Calculations of the frequencies for each combination were produced for a set of 400-500 flies. Crossing data for 12 × 11 = 132 heterozygotes and 24 homozygotes (by twelves separately for male and female, since the trait being investigated is coupled with sex, and its appearance differs accordingly) were examined. In all, 156 combinations of the genotype were examined, each one for three temperatures (14°, 22° and 30° C), constituting a total of 468 experiments.

To estimate the functional nearness of mutations i and j ($i,j = 1, 2, \ldots, 12$) we calculate the distance between the 40-element vector of reduction frequencies corresponding to the homozygote for mutation i, and the analogous vector corresponding to the homozygote for mutation j, by the formula

$$f_{ij} = \sum_{s=1}^{40} |r_{is} - r_{js}|,$$

where i, j are the indices of mutations (alleles), and r_{is} is the reduction frequency of the sth bristle for the ith allele.

Thus twelve distance matrices $\|f_{ij}\|_1^{12}$ were obtained (separately for males and females for each of the temperatures).

One of these matrices is presented in Table 4. It is not difficult to convince oneself that this matrix cannot be reduced to linear form. However, with the ordering given, the number of deviations from linearity is not great, as one can distinguish its block structure with the "unaided eye": The first and last sets of five mutations form groups which are not connected with one another, since the intragroup distances are noticeably less than those between groups. A similar picture holds for the remaining five matrices: The same three groups of mutations are discernible, with the "almost linear" forms of the matrices obtained by ordering the alleles differing from this one in only one case, in which three permutations of neighboring mutations within groups occur. This supports the existence of substructure in the *scute* locus corresponding to the three groups obtained. These groups are invariant with respect to changes in temperature, while the proteins (unfortunately, as yet unidentified), which evidently also define the observed phenotype, must significantly change their activity with temperature fluctuations. Consequently, the invariance of these groups is explained by the structural stability of the gene, not by the corresponding proteins. Moreover one can conjecture that the substructures of the locus corresponding to these mutational groups are cistrons, because it is they that code for separate proteins. With a change of temperature the order of the mutations changes only within the groups, which easily explains the changes in protein activity.

We will now trace the extent to which our speculative construction is connected with the real process of bristle formation. Do the invariant orderings of mutations we have obtained correspond to some kind of regularities in bristle reduction? To answer this question consider the table of reduction frequencies for homozygotes (see Fig. 22). Columns (bristle indices) were ordered so that for each mutation (row) bristles with the largest reduction values fell together (i.e. the matrix was row linear), while the ordering of rows was obtained from the preceding distance matrix analysis.

As is evident from Fig. 22, the form of the table obtained is nearly linear (for rows). Consequently the order of bristles also reflects the locus structure.

We see that mutations of the various groups correspond to different types of rows: those of the first group correspond to long rows, those of the third to short rows, and the second group to intermediate-length

TABLE 4

An ordered matrix of the functional differences of scute mutations, constructed according to percentage reductions of bristles in mutant homozygous females at 14°C (the separated blocks are of mutations functionally most similar to one another).

	scD1	sc9	sc7	sc1	scD2	sc6	sc3B	sc28	sc260	sc5	sc8	scV2
scD1	0	2694	3418	4078	5178	6563	7434	11966	11162	13648	15850	16971
sc9	2694	0	2506	3136	3388	5877	5484	10374	10276	11714	13528	14925
sc7	3418	2506	0	4076	3324	6033	5432	10208	9584	11490	13712	14601
sc1	4078	3136	4076	0	2528	2863	4744	11124	11904	10330	12508	13825
scD2	5178	3388	3324	2528	0	2841	3176	9320	10344	8742	10716	12025
sc6	6563	5877	6033	2863	2841	0	5095	11243	12271	9361	10481	11922
sc3B	7434	5484	5432	4744	3176	5095	0	6420	7232	6358	8584	10237
sc28	11966	10374	10208	11124	9320	11243	6420	0	1056	1890	4064	6083
sc260	11162	10276	9584	11904	10344	12271	7232	1056	0	2918	5092	7107
sc5	13648	11714	11490	10330	8749	9361	6358	1890	2918	0	2334	4248
sc8	15850	13528	13712	12508	10716	10481	8584	4064	5092	2334	0	2643
scV2	16971	14925	14601	13825	12025	11922	10237	6083	7107	4248	2643	0

rows.

The columns of the table of Fig. 22 also lend themselves to natural division into three groups. The right-most of these groups is engendered by the short rows of the third mutation group. The exactness of division of the remaining columns into two groups flows from the constitution of analogous tables for different temperatures. In the columns of the left group reduction frequencies change more or less monotonically with change in temperature in all the mutants, without exception. The same regularity holds for columns of the third group, while in the columns of the middle group dependence of reduction on temperature has a complex, irregular character (for more details see [45]).

We not only confirmed the preceding conclusions, but also obtained a number of additional ones. The principle one is that individual mutations can be interpreted in terms of the linear ordering of bristles, so that the bristle groups which have been determined correspond to separate cistrons, resulting from analysis of the matrix of intermutation distances. In this the structure of the table points out the possible polar character of the mutations peculiar to scriptons and operons.

The above considerations allow one to propose the following model of structural organization in the *scute* locus (see Fig. 22).

It is conjectured that the *scute* locus is a gene of operon type, consisting of a small number (3-5) of structural cistrons coding for proteins which fulfill the same function, but in different parts of the fly body, forming different bristles. In accord with the model, the first five mutations affect all three cistrons. The defects corresponding to them may either be the usual deletions of all the cistrons, or begin in the first cistron, affecting all information subsequently read out, as in the case of polar mutations of the nonsense or frameshift type in operons. The last five mutations are point mutations and affect only the final cistron. Several fall in the general picture of mutations sc6 and sc3B. Their nature is unknown, so that their position on the map cannot be accurately determined. In Fig. 22 the bristles controlled by each cistron are evident. In this connection it is interesting that each cistron controls its own compact portion of the fly body (compare Figs. 21 and 22), although there is no correlation between the order of the cistrons in the map and corresponding locations of bristle formation on the body.

How can this model be verified? Direct methods for investigating the structure and functioning of loci like *scute* are still not possible. We therefore revert to the usual method of verification: We draw from the model consequences concerning phenomena directly observable in experiments. If the observations contradict the consequences the model is declared to be groundless, otherwise the model is supported (although indirectly). We will attempt to take advantage of this method, bearing in mind that up to this time we have no material in any way reflective of the behavior of scute mutations in the heterozygous state.

Thanks to the elegant experiments of C. Stern [94] we know that the plan of arrangement of the bristles on the fly body (called prestructure) is not under control of the *scute* locus; other genes are responsible for it. The *scute* locus plays the role of a distinctive "commutator", triggering the bristle formation process in response to a specific factor ("signal") from the prestructure. Our model implies that such a triggering occurs at the post-translation level: From among all the isofunctional proteins synthesized by the *scute* locus some one is activated by a factor specific to it, and afterward the formation process begins. The specific "identification" of prestructure factors with protein products of the *scute* locus means that they possess allosteric properties (we will discuss these in Chapter 2) and therefore they must have quaternary structure.

The analysis of scute mutation behavior in heterozygotes can serve as indirect support for this conclusion. If homo- and heterozygous mutant manifestations are compared as such, before their reduction to the binary values (0,1), then it is found that besides the local dominance of reductions by the norm, in heterozygotes reduction on some bristles may be either equal, lower (complementation), or higher (anticomplementation), than in each of the mutant homozygotes. But as we will see in Chapter 2, such complementation effects are inherent only for proteins with quaternary structure. In passing, we remark that in the model of an operon with a complex structure in its operative section, suggested by V. A. Ratner for the *scute* locus [29,31,33], such effects are theoretically unexplainable.

We now consider what kind of picture one can expect for functional nearness of mutations, if nearness is estimated for heterozygotes under the assumption that our operon-like model holds.

We fix some mutation i and consider all 11 heterozygotes of type (i,j), where $i \neq j$. We will call mutation i the reference for this collection of heterozygotes. What does our model say about the qualitative features of functional nearness in heterozygotes?

First consider the case when the reference mutation is from the first group. These mutations affect (inactivate) all three cistrons, so that bristle formation is determined by the second allele in the heterozygote. Thus, the nearness of heterozygotes (i,j) and (i,k) is determined by the nearness of the second components j and k. This means

TABLE 5

The ordered matrix of functional differences of scute mutations, constructed according to percentage reductions of bristles in mutant heterozygotes: sc1 × sci at 14°C (blocks of mutants which are functionally most similar are the same as for the homozygous case.

	scD1	sc9	sc7	sc1	scD2	sc6	sc3B	sc28	sc260	sc5	scV2	sc8
scD1	0	1719	1839	2058	2867	3582	4717	7836	7447	9147	5750	8649
sc9	1719	0	1930	1437	2282	3253	4178	7557	7482	8688	5619	8290
sc7	1839	1930	0	1283	1676	3221	3436	6611	6206	7892	4755	7410
sc1	2058	1437	1283	0	1559	2518	3375	7288	7211	7861	4788	7357
scD2	2867	2282	1676	1559	0	1753	2176	6511	6416	6460	4931	6148
sc6	3582	3253	3221	2518	1753	0	2537	7480	7385	6271	5700	5921
sc3B	4717	4178	3436	3375	2176	2537	0	5301	5704	4558	5435	4672
sc28	7836	7557	6611	7288	6511	7480	5301	0	563	1379	2986	1657
sc260	7447	7482	6206	7211	6416	7385	5704	563	0	1770	2891	1886
sc5	9147	8688	7892	7861	6460	6271	4558	1379	1770	0	3807	664
scV2	5750	5619	4755	4788	4931	5700	5435	2986	2891	3807	0	3323
sc8	8649	8290	7410	7357	6148	5921	4672	1657	1886	664	3323	0

TABLE 6

The ordered matrix of functional differences of scute mutations, constructed according to percentage reductions of bristles in mutant heterozygotes: sc8 × sci at 14°C (the matrix is not reducible to linear form, mutants cannot be separated into blocks).

	scD1	sc9	sc7	sc1	scD2	sc6	sc3B	sc28	sc260	sc5	scV2	sc8
scD1	0	380	385	304	368	439	339	280	388	332	509	1364
sc9	380	0	369	362	432	537	549	370	562	446	535	1464
sc7	385	369	0	483	377	600	648	521	665	543	672	1613
sc1	304	362	483	0	448	337	271	288	316	214	335	1288
scD2	368	432	377	448	0	633	669	562	688	608	695	1648
sc6	439	537	600	337	633	0	174	323	261	221	368	1229
sc3B	339	549	648	271	669	174	0	229	135	129	348	1137
sc28	280	370	521	288	562	323	229	0	248	168	403	1216
sc260	388	562	665	316	688	261	135	248	0	164	319	1152
sc5	332	446	543	214	608	221	129	168	164	0	297	1168
scV2	509	535	672	335	695	368	348	403	319	297	0	1209
sc8	1364	1464	1613	1288	1648	1229	1137	1216	1152	1168	1209	0

that the matrix of functional distances must have a near-linear form for the same ordering of mutations as in the homozygous case.

If the reference mutation is from the third group then elements of the matrix of functional distances of second components must be small, and the matrix must be far from linear and from a block structured form. The fact is that in this case the reference mutation does not affect the first cistrons, so that in all combinations of heterozygotes the qualitative composition of functioning proteins is the same (although their concentrations may vary, depending on how many cistrons affect the second component).

The data at hand completely confirm these conclusions. In Tables 5 and 6 both cases are shown. The form of the matrix for the reference mutant sc1 from the first group is the same as the one for homozygotes (nearly linear, with a similar ordering of mutations, the same blocks of mutations near to one another).

For the matrix of functional distances of heterozygotes for the reference mutation sc8, from the third group, our prediction also holds: matrix elements are small, the matrix itself is far from linear, the mutations are not divided into blocks. The same is true for other mutations (considered as reference mutations) from the first and third groups, respectively.

This provides the desired experimental confirmation. From this model a number of more subtle conclusions also follow, concerning the specific functioning of protein-multimers (the products of structured cistrons) in corresponding sections of the fly body, which also are amenable to experimental genetic testing (and sharpening of the model).

Thus, the results obtained indicate that the initial linearity of structure of the genetic material is nevertheless sufficiently stable to be manifest even after a long chain of intermediary influences from various ontogenetic factors.

Of course final confirmation of this model must await direct biophysical investigation of this system.*)

---

*) Very recently the scute-region of the X chromosome has been cloned and mapped by the "direct" restriction method 55b . The comparison of these new biochemical results with ours is premature because the number of scute mutations investigated at DNA level is too small. It should be noted, however, that DNA localization of mutations sc1, scD1, scD2, sc6 and sc3B *correlates with our cluster-polar map.*

Chapter 2. Graphs in the Analysis of Gene Semantics

§1. Interallelic complementation and the functioning of protein multimers.

1. <u>Interallelic complementation</u>. The complementation test, unlike the recombination test, has an initial functional nature: in it the differences in mutant functions are considered. However, at the intercistronic level of complementation events (by the analysis of scriptons and operons (1.5.1)) this test permits the characterization not only of the functioning of separate cistrons, but also the structural topography of a system as a whole. The molecular-genetic basis of this is, on the one hand, the independence of cistron translation and the subsequent functioning of the proteins which correspond to it, and on the other hand, the presence of polar mutations which, having appeared in a single cistron, block the action of all subsequent ones. Because of this, polar mutations lead to the violation of the classical rule that mutations of different cistrons are complementary to one another.

In this chapter we discuss a different kind of violation of the rule, intracistronic complementation, in which the complementary mutations are within the bounds of a single cistron. This phenomenon, called *interallelic complementation* was discovered at mid-century. Later, it was established, that interallelic complementation is characteristic for those cistrons which code for proteins having quaternary structure, i.e. which are complex aggregates of separate polypeptide chains, produced by the cistron [43,57]. Thus the test for interallelic complementation is a genetic method of finding out whether a protein has quaternary structure. If there is no quaternary structure then complementation within the corresponding cistron is ruled out.

Here, the objects to be crossed are mutant homozygous individuals. The offspring of such crosses then are heterozygous for the given cistron: one defect of the cistron has paternal origin and the other maternal origin. Each of these cistrons repeatedly synthesizes the corresponding form of the modified polypeptide chain. It is assumed that the aggregation of individual instances of these chains (monomers), during the formation of quaternary (multimer) structure, takes place randomly.

If a multimer is formed by the union of identically altered monomers (of one type or another), then the proteins obtained are identical to the parental ones. However, the aggregation of differently altered monomers gives rise to a new hybrid form of the protein multimer which is absent in the parents. It is known that such hybrid proteins are responsible for new functional properties of the heterozygous offstring, which is the manifestation of interallelic complementation.

Usually the phenomenon of interallelic complementation is studied in microorganisms (viruses, bacteria, fungi etc.). Initial mutant parents unable to grow in a well-defined medium due to insufficient activity of the corresponding mutant protein are used. If, as a result of crossing of these mutants, growth is observed in the descendant colonies (complementation), then this is clearly due to a hybrid fraction of the protein multimer, since it alone distinguishes the offspring from the parents. This means that the hybrid protein multimer, consisting of homologous but differently altered monomers, displays greater activity than that found in the parents (which contain identically defective subunits).

These phenomenological peculiarities point out the fact that complementary mutations of a single cistron must be sufficiently profound to inactivate the protein in a homozygous situation, but at the same time, not too serious, so that renewed function (even if only partial) is ensured in the heterozygote [35,40,88].

Thus, to understand the molecular mechanisms of such renewals-of-function it is necessary to discuss the principles of the construction and functioning of protein multimers.

2. <u>The fundamental principles of organization of protein multimers</u>. Proteins are the most complex macromolecules of cells. Direct physicochemical investigations of their structures are very laborious, and for proteins occuring in the cell in insignificant concentrations (and there are many such) it is simply impossible, with current methods.

To the present time, detailed atomic structure has been determined for only a few proteins: globins, lysosomes, dehydrogenase and others [49,53,86,87].

From a functional point of view proteins are constructed as follows. On the tertiary and quaternary levels proteins form special spatial structures called *functional centers*, each of which fulfills a definite operation, connected with overall protein function. Among the kinds of functional centers in proteins are: *catalytic centers*, which directly process substrates; *contact centers*, those parts of a monomer at which it comes in contact with other monomers of the multimer; *allo-*

steric centers, through which the activity of the protein is regulated by external, low-molecular cell substances; and a number of others.

As an example we consider the functional organization of hemoglobin, a protein which has been studied rather thoroughly in recent years. In particular, in the last five years the structure of the gene which codes for hemoglobin in various species has been completely deciphered by means of DNA sequencing techniques (see below, page 159 and Fig. 43'). *Hemoglobin*, a transport protein, participates in a most important function in animals, oxygen respiration. The basic results concerning the molecular structure of this protein were obtained by M. Perutz and co-workers as the result of many years of investigation [87]. Hemoglobin is of interest in this book not only as a thoroughly studied molecular object for demonstrating the principles of semantic analysis of genetic texts, in this chapter, but also as a specific object for genetic discussion in Chapter 3.

Normally a mammalian hemoglobin molecule appears as a *tetramer*, consisting of four monomer subunits of two types, designated α and β. The conventional arrangement of the tetramer is shown in Fig. 23, including the contact points between separate subunits.

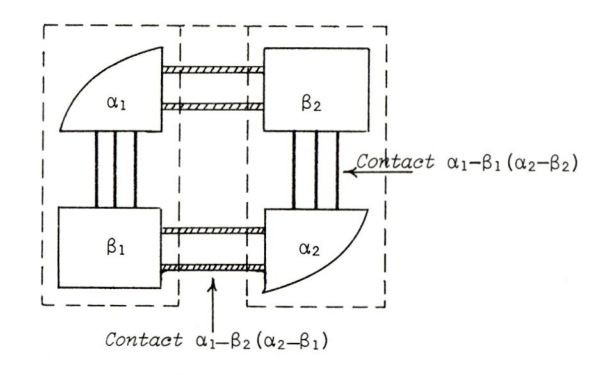

Figure 23

Each chain of the tetramer has a complex spatial packing, and connects to two chains of the opposite type, as shown in Fig. 23. Under chemical dissociation the molecule at first separates into the two dimers $\alpha_1\beta_1$ and $\alpha_2\beta_2$ outlined in Fig. 23 by dashed lines. The centers of mutual identification (contact) of the subunits are designated $\alpha_1-\beta_1$ and $\alpha_2-\beta_2$ within the dimers, and $\alpha_1-\beta_2$ and $\alpha_2-\beta_1$ between the dimers. Although the α and β subunits are coded by different cistrons, their structures are very similar, which clearly points to the common character of their origin in the evolutionary process.

Protein multimers consisting of identical, or homologous (as in hemoglobin) chains are symmetrically structured. The hemoglobin tetramer has three axes of symmetry, with the contacts $\alpha_1-\beta_1$ and $\alpha_1-\beta_2$ being realized near these axes [1,49,86,87].

The most important semantic part of any protein molecule is its active center. In the transport molecule, hemoglobin, the organization of the active heme-specific center includes, aside from the set of amino acids, a pigment molecule called *heme* (containing an iron ion), which binds a molecule of oxygen. Thus the heme-specific center recognizes as a substrate an oxygen molecule.

It is curious that the amino-acid segment of the heme-specific center is also involved in one of the contact centers (in the chain α_1 it is the contact center $\alpha_1-\beta_2$). This plays a decisive role in the functions of binding and releasing oxygen molecules, performed by the hemoglobin tetramer. An oxygen molecule bound to one chain changes the conformation of its heme-specific cavity, and consequently, of the contact center between the dimers, thus transmitting the change to the other subunit. As a result the probability of the other subunit taking up the next oxygen molecule is increased by two orders of magnitude. Thus, because of the quaternary nature of its structure, the binding of even one oxygen molecule automatically leads to a sharp acceleration in the absorbtion of subsequent molecules (by a restructuring of the recognizing apparatus).

Interactions of this kind, between different centers of proteins, are said to be *allosteric*. Clearly, the appearance of quaternary structure in proteins is due primarily to the advantages connected with allosteric interactions, which provide cooperative functional effects [16,49].

The role of the contact centers $\alpha_1-\beta_2$ and $\alpha_2-\beta_1$ in hemoglobin is thus of particular importance, since they also facilitate allosteric interactions. In a number of vertebrates, notably mammals, an important allosteric function is carried out by a small center situated in the $\beta_1-\beta_2$ contact zone (consisting of only four amino acids), which binds to 2,3-diphosphoglycerate. The bound organic phosphate is an intermediate product of carbohydrate decomposition (glycolysis); and when bound, it lowers the affinity of hemoglobin for oxygen. In other words, by this means the process of blood saturation by oxygen is regulated, depending on the intensity of the metabolic processes of the body. However, in general, contact centers and allosteric centers can be spatially separate, as they are in a number of enzymes which have been studied.

Thus the functioning of a single hemoglobin chain, for example α_1, may be represented in the following way. A newly-translated α-chain includes heme in its tertiary structure (after identifying its heme-specific cavity), and is then aggregated with a β-chain to form a dimer (by means of the contact $α_1-β_1$), which then combines with another dimer (contact $α_1-β_2$). The binding of an oxygen molecule by a chain is possible in one of two states, determined by the contact $α_1-β_2$, by which the chain "knows" whether there is an oxygen molecule on the other dimer. In tissues with a lowered concentration of oxygen the subunit gives up its oxygen molecule.

Analyzing the details of the structure and interaction of functional centers, V. A. Ratner reached the conclusion that the cistron instructions which code for the α-chain of hemoglobin can be represented in the following form [31]:

If heme is recognized then
If the center $α_1(β_1)$ is recognized then
If the center $α_1(β_2)$ is recognized then
If the center $α_1(β_2)$ is not restructured then
Bind a molecule of O_2 with reduced affinity
Otherwise
Bind a molecule of O_2 with increased affinity;
If the concentration of oxygen is below the threshhold level then
Release a molecule of O_2.

This phrase illustrates very well a basic property of the genetic language: the imperative form of the protein sentences, which consist of "cases" or conditions for the execution of various commands. In this there is a strong similarity between genetic information and the "machine" texts of algorithmic languages.

As we see, the semantic analysis of cistrons requires knowledge of the functional centers of the corresponding proteins, including knowledge of the molecular character of their functional interactions. Hemoglobin is practically the only example of a protein for which researchers have obtained a more or less detailed knowledge of this kind. Exact physico-chemical analysis of proteins by modern methods is so laborious and cumbersome that it requires years and years to decipher the semantics.

This is why attempts to determine the semantic properties of genes by some other method are so attractive. The analysis of interallelic (intracistronic) complementation interactions of mutations seems particularly promising for this approach, since it is complementation that is connected with the meaning of genetic information as manifest in protein functions.

3. **The molecular mechanisms of interallelic complementation.** It is clear that, at the intracistronic level, the possibility of complementation of mutations depends on the constitution of the functional centers in defective monomers, aggregated together into multimers.

Until recently the model generally used was that of F. Crick and L. Orgel [57]. In their model the initial molecular events of interallelic complementation take place in the contact centers of the hybrid protein multimer, since it is there that the defects of the various subunits are spatially near together and can have influences on one another (see also [40,43]).

Mutational replacements of separate amino acids (and it is only these mutations we are concerned with) lead, at the level of tertiary structures, to the alteration of the conformation of the polypeptide chain. Generally speaking the sizes of these regions differ, depending on the character and the position of the replacement, since the tertiary configuration of a protein is completely determined by its primary structure.

The fact that certain proteins are only active in heterozygotes, and not in homozygotes, leads one to notions of the "norm correcting for the defect" and to principles of covering or non-covering of defects similar to those for intercistronic complementation (but on the protein level). According to Crick and Orgel, if the hybrid multimer is not active, then it is as if the altered zones of conformation in the mutants are partially overlapping in the region of contact of the subunits. On the other hand, even a partial renewal of activity corresponds to non-overlap of the altered zones. This aspect justifies the importance of mapping data about interallelic complementation by the methods of interval graphs (Chap. 1): If the mutations alter homologous contacting portions of subunits (as is always the case with mutant homozygotes), then they are non-complementary; otherwise the normal part "corrects" the defect homologous to it.

It is important to emphasize that such a mechanism for homologous correction of a defect by the norm works near the axis of symmetry of the multimer. Therefore, according to Crick and Orgel, the mutual disposition of the sections of the homologous contacting subunits in the multimer must be reflected in the complementation map.

However in recent years there has accumulated a considerable amount of data, both of genetic and purely biochemical kinds, which does not agree with this model.

We have already mentioned the fact that the functional centers of a protein are not represented, in the protein's primary structure, by compact fragments. Similarly a mutational deformation of the centers is also not compact, but "spread" over the whole amino acid sequence.

But then the first postulate of complementation mapping has already been violated, since such a mutation cannot be represented by an interval on a line. Therefore, in particular, it is too much to expect any appreciable coordination of the positions of mutational defects in complementation (functional) and recombination (structural) gene maps. Moreover, considering the latest data concerning the multi-domain organization of a number of proteins (see for example the immunoglobulins on page 7), in which domains are sequentially projected onto the primary protein structure (and, correspondingly, onto the recombination map), one can expect a correlation between the genetic and complementation maps to within the occurrence of domains. In accord with the hypothesis that domains are coded for by separate exons, the situation here is completely analogous with intercistronic complementation, ex-

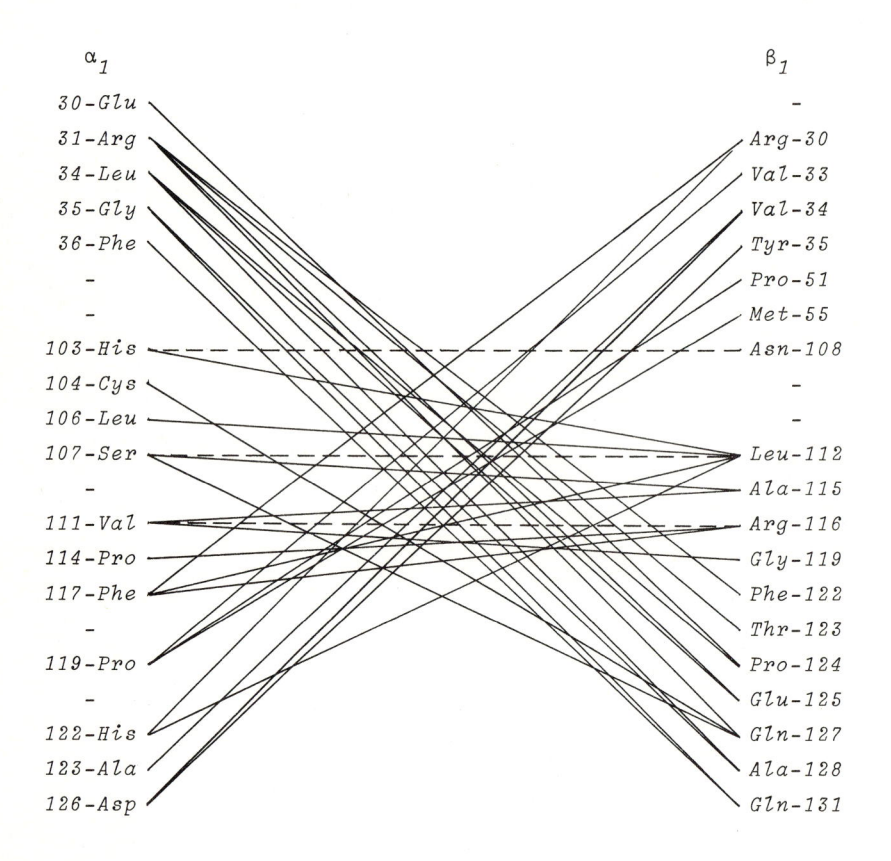

Figure 24. The structure of the contact centers α_1-β_1 in hemoglobin: Opposite each position in the α-chain is its homologous position in the β-chain; connections between amino acids in homologous positions are designated by dashed lines; those between non-homologous positions, by solid lines.

cept that exons replace cistrons.

The notion of mutually homologous correction of protein defects has also been shown to be faulty, since it turns out that the subunits of a multimer are situated in "antiparallel" positions [53,87], so that as a rule the homologous parts of different subunits are distant from one another, and not in contact (Figs. 24 and 25). Consequently, there is simply no place that the mechanism of Crick and Orgel works.

It is known that the phenomenon of intracistronic complementation is "tied to" the basic semantic substructures of proteins: the func-

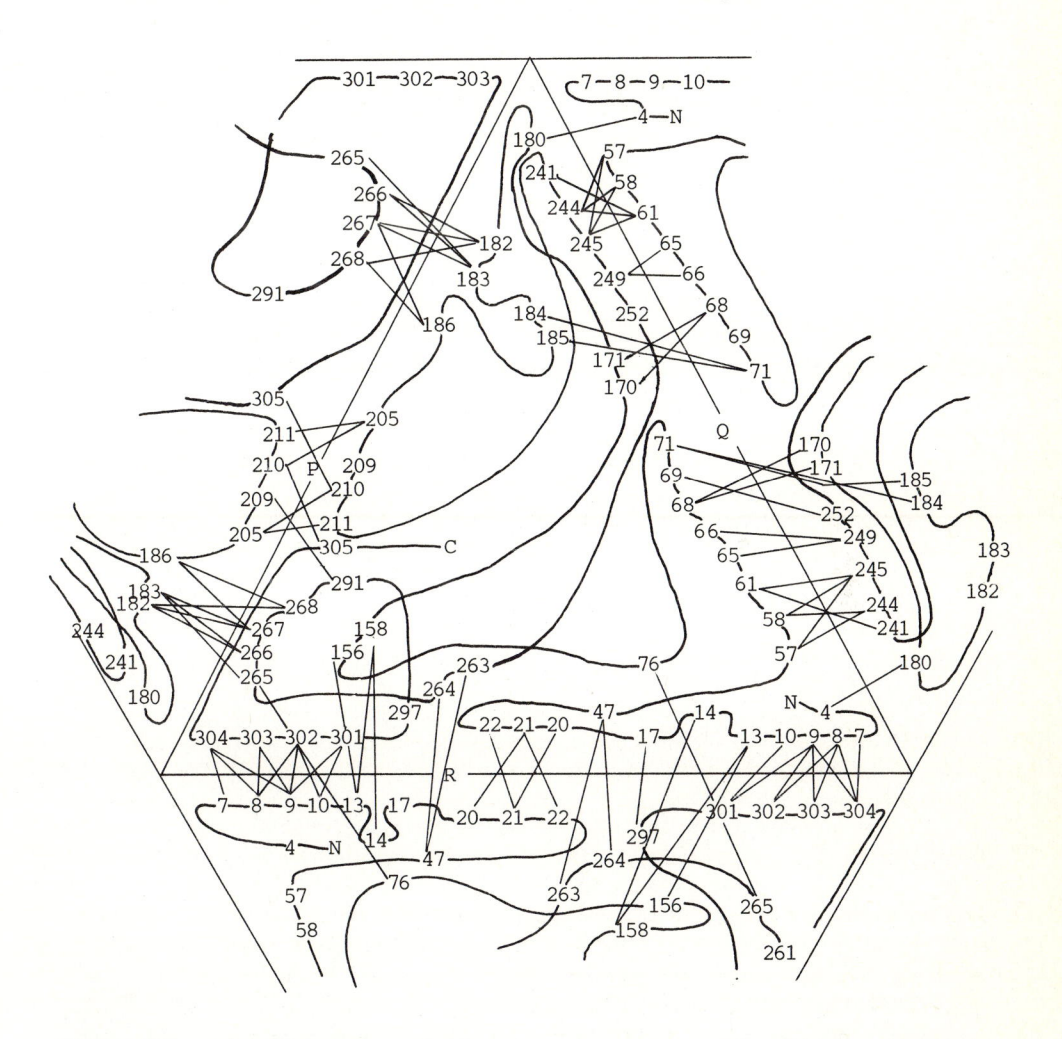

Figure 25. The structure of the contact centers of subunits in an isoenzyme lactatedehydrogenase of a shark [53]: P, Q and R are the axes of symmetry of the multimer; the numbers designate the order of the amino acids in the primary structure; lines join the amino acids which are in contact.

tional centers. Therefore any model of the mechanism of interallelic complementation must also be based on the organizational features of these centers.

In a paper by Ratner, Rodin and Shenderov [35] a model is examined which satisfactorily explains the fundamental known facts (see also [43,59,88]). The basis of this model is the fact that each subunit of a multimer contains a complete set of its functional centers, so that in a multimer each type of center is represented by several copies, depending on the number of subunits. Each mutation is able to block one or more centers of separate chains of the multimer. We will consider that a multimer is able to fulfill its function if it contains even one correctly-working copy of a functional center of each type, i.e. in the hybrid multimer there are no common affected centers (complementarity). If in a hybrid multimer identical centers are affected then these centers are blocked in all the subunits of the multimer, and as a result it is inactivated (non-complementarity).

Contact centers occupy a critical position. Effects on contact centers frequently lead to the "break down" of the multimer, independently of whether the mutations are homozygous or heterozygous. However if both mutations affect the same contact center, then it is possible that they are complementary, i.e. mutually-correcting. This is connected with the fact that in the contact zone the amino acids of different subunits interact directly (physico-chemically), and this ensures fulfillment of the contact function. The normal character of interactions in a contact zone may be preserved by the "corrective" influences of the mutually-contacting defects, which leads to complementation. Such mutual corrections of the amino acid replacements participating directly in contact centers must be very specific, and consequently, infrequent. This is because of the physico-chemical individuality of connections between amino acids in the contact zones of subunits. It is another matter, if the mutations affect the contact centers indirectly, by means of their nearby environments, so that the amino acids participating directly in the centers are preserved and only the conformation of the centers is changed. It is easy to understand in this case that for mutual correction of defects it is not necessary to have stringent, mutual, physico-chemical specificity of replacements but only that the stereometry of that part of the multimer be maintained [35,88].

It may appear that the mechanism we have described reiterates the model of Crick and Orgel. This is not so, however: With Crick and Orgel "the good corrects the bad", while here "the bad corrects the good." this is possible because the subunit contacts are nonhomologous by their very nature.

This model can be called a *mosaic* model, since complementation effects are defined by a distinctive mosaic of normal copies of the functional centers of a protein multimer.

In realizing the function of a protein the centers interact with one another, as we discussed in the example of hemoglobin. This mutual interaction is reflected in the behavior of mutations which affect more than one center, in particular through their complementation reactions. If the mosaic model is correct, then mutations which are similar in their complementation manifestations must correspond to the same functional center. In other words, the structure of the basic complementation relations of a mutant system must correspond with the structure of the interactions of the functional centers, and not with the "mythical" structure of the arrangement of mutational defects. Actually, for a number of well-studied proteins (of the hemoglobin type) one notices that mutations which effect the same center are similar in their functional manifestations; running all the way from local molecular changes of protein function to the pathology at the level of the whole organism [30,46,70,86].

It is clear from the foregoing that the analysis of the structure of intracistronic complementation matrices presupposes the identification of the functional centers and the structure of their interactions.

The formulation and investigation of mathematical problems of structural analysis in relation to the mechanism described above, is given in the following section. It must be mentioned that, as will become clear, these formulations are applicable in the much wider context of the analysis of arbitrary matrices of connections (see, for example, Section 2.2).

§2. The approximation of graphs

1. **Approximation problems in a space of relations.** In the preceding section we established that to analyze interallelic complementation it is necessary to understand the structure of the initial complementation matrix. At first glance there is nothing at all new in this assertion; the entire preceding chapter was dedicated to the analysis of such matrices. However, then we had an exact idea of the type of structures to be expected in the real system: a linear or near-linear ordering of its elements. Since the mosaic model does not provide any information about the structure, a more general mathematical language is required.

Before giving an exact statement of the problem we will briefly discuss the general notion of structure in a system of interconnected objects.

The analysis of a system's structure is usually understood to mean the representation of its objects and their basic connections in an aggregate form. For example, structures of an ordered type are possible (such as were studied in the preceding chapter), when connections are defined according to the nearness of objects in some ordering. Also possible are structures associated with partitions, in which the system is defined in the form of a collection of relatively unconnected groups of objects, with the basic connections of the system being concentrated within the groups. Structures of more complex forms are possible, involving cycles, trees etc.

This brings us to the following more precise notion. We say that a *partition* $R = \{R_1, \ldots, R_m\}$ on a set A is *structured* with structure $\varkappa$, if $\varkappa$ is an oriented graph (relation) on the set $\{1, \ldots, m\}$ of indices of the classes of the partition R. The fact that $(s,t) \in \varkappa$ is interpreted as the existence of an essential (though not necessarily bidirectional!) connection between the classes R_s and R_t.

The pair $(R, \varkappa)$ is sometimes called the *macrostructure* of the set A, since it is an aggregated description of the system.

If $\varkappa = \phi$, then $(R, \varkappa)$ is nothing more than the usual unordered partition of the set A. If $\varkappa$ is a linear order then $(R, \varkappa)$ is an ordered partition. If $\varkappa$ is a cycle then $(R, \varkappa)$ is a cyclic partition, etc.

To each partition with structure $(R, \varkappa)$ there corresponds a relation $R \subseteq A \times A$, defined by the condition

$$(i,j) \in R_\varkappa \longleftrightarrow \exists s, t = 1, \ldots, m$$

$$\left[i \in R_s \text{ and } j \in R_t \text{ and } (s,t) \in \varkappa \right].$$

It is clear that the relation R on A coincides with $R_\varkappa$ if and only if there exists a homomorphism of the relation $R \subseteq A \times A$ onto the relation $\varkappa \subseteq \{1, \ldots, m\}^2$. This homomorphism is specified by associating with each object $i \in A$ the number of the class of R which contains it.

We consider the inverse problem of reconstructing the macrostructure corresponding to a given relation $R \subseteq A \times A$. Clearly, its solution is the partition into groups of objects R_s with identical images and pre-images, specified by the canonical homomorphism. The corresponding structure is defined by the condition: $(s,t) \in \varkappa \longleftrightarrow i \in R_s, j \in R_t \left[(i,j) \in R \right]$, which is correct, since if there exist in R_s and R_t objects i, j such that $(i,j) \in R$, then $R_s \times R_t \subseteq R$, i.e. every pair of objects $(i,j) \in R_s \times R_t$ is contained in R.

However this solution is not unique, since any partition more detailed than R, and with similarly defined structure is also generated by the homomorphism, and consequently forms a macrostructure corresponding to R.

The matrix corresponding to the relation $R_\varkappa$ (i.e. the partition with structure $(R, \varkappa)$), is defined as the $N \times N$ matrix consisting of $m \times m$

cells $r_{st} = \|r_{ij}\|_{i \in R_s, j \in R_t}$, all elements of which have the same value: 0, if $(s,t) \notin \varkappa$, and 1, if $(s,t) \in \varkappa$.

In particular, the matrix associated with an unordered partition has unit cells on the main diagonal (and only there), and the matrix associated with an ordered partition has as its only zero blocks those situated below the main diagonal. The matrix associated with a strict ordering of the objects can be reduced, by a simultaneous permutation of rows and columns, to a triangular form, in which all elements above the main diagonal are ones and all elements below it are zeros.

Thus any type of structure which can be described in terms of partitions with structure $(R, \varkappa)$ is characterized by binary relations (graphs) on A, or equivalently, by $N \times N$ Boolean matrices.

Let E be the class of relations (matrices) on A, which characterize the hypothetical types of structures for a given system. For instance it could be the class of rankings (linear or interval orderings), as was the case in the preceding chapter, when we spoke of a known, linear type of structure, with only the mutual disposition of elements being unknown. It could also be the class E_m of matrices corresponding to all possible macrostructures whose number of classes does not exceed m, if, as was the case with interallelic complementation of mutations for a specific gene, there is no basis for preferring one type of structure over another, but only the confidence that the number of functional centers does not exceed m.

Our task is to find in E that relation (matrix) which approximates the given complementation matrix (or the relation corresponding to it) closely enough. In the most simple situation E contains a macrostructure $(R, \varkappa)$ with a relation $\varkappa$ which is homomorphic to the original complementation relation I: it characterizes the macrostructure of the matrix of the complementation data, since the corresponding matrix $R_\varkappa$ coincides with the original one. It would seem to be exactly this situation for interallelic complementation, if the postulated mosaic mechanism is correct. For if the centers interact, then any mutation affecting one of them automatically alters the others. Therefore if $(i,j) \in I$ for i and j from different centers, this will also be true for any other i, j from these centers.

However this is not so. Even though a mutation affects a center as a whole, whether the effect is transmitted to another center which interacts with the first is not clear. It all depends on the extent to which the defect influences the interaction of the centers. Therefore a real complementation matrix is not homomorphic to the structure of the interactions among its functional centers. One can at least say that it must be nearly so.

Therefore to select a "representative" of the initial matrix in the class E it is necessary to learn to measure the degree of nearness between matrices. A natural measure of closeness of two matrices $r = \|r_{ij}\|$ and $p = \|p_{ij}\|$ is the *Hamming distance*, the number of corresponding positions in which they differ:

$$d(r,p) = \sum_{i,j=1}^{N} |r_{ij} - p_{ij}| = \sum_{i,j=1}^{N} (r_{ij} - p_{ij})^2. \quad (1)$$

The last equality in formula (1) follows from the important property of Boolean quantities, that they coincide with their squares: $0^2 = 0$ and $1^2 = 1$.

In terms of relations the distance (1) is calculated as the number of pairs (i,j) which are contained in one and only one of them. In terms of diagrams the distance is the minimal number of operations of adding and deleting arcs necessary to change one graph into the other. Several properties of this distance measure are studied in the books [21,22], where it is used in the analysis of qualitative phenomena. Its importance for us lies in its possible use for selecting from E that relation which is closest to the initial matrix.

In the next section we will also consider the case when a relation is sought in E which is nearest to several given relations.† We therefore examine the general formal problem.

Let $R^1, \ldots, R^n$ be n given relations on the set A. It is required to approximate this system of relations by a relation from the class E, i.e. to find a relation $R \varepsilon E$ for which the sum

$$f(R) = \sum_{k=1}^{n} d(R, R^k) \quad (2)$$

is minimal, for all $R \varepsilon E$.

This approximating relation is naturally called the *median* of the system $\{R^1, \ldots, R^n\}$ in the class E.

The following assertion is fundamental to the material we will be discussing [20].

Theorem 1. A relation R is the median of the system $\{R^1, \ldots, R^n\}$ in the class E if and only if it maximizes

$$g(R) = \sum_{(i,j) \varepsilon R} b_{ij} = \sum_{s,j=1}^{N} b_{ij} r_{ij} \quad (R \varepsilon E),$$

where

$$b_{ij} = a_{ij} - \frac{n}{2}, \quad a_{ij} = \sum_{k=1}^{n} r_{ij}^{k}.$$

---

†These situations also arise frequently in such data analysis tasks as the mutual reduction of expert orderings or the construction of classifications based on qualitative traits.

Proof. We consider the function $f(R)$ which, from (1), is equal to

$$f(R) = \sum_{i,j=1}^{N} \sum_{k=1}^{n} (r_{ij} + r_{ij}^k - 2 r_{ij} r_{ij}^k).$$

Noticing that $\sum_{k=1}^{n} r_{ij} = n r_{ij}$, $\sum_{k=1}^{n} r_{ij}^k = a_{ij}$, we have

$$f(R) = \sum_{i,j=1}^{N} a_{ij} - 2 \sum_{i,j=1}^{N} \left(a_{ij} - \frac{n}{2}\right) r_{ij}.$$

Since $\sum_{i,j=1}^{N} a_{ij}$ is fixed for given $R^1, \ldots, R^n$ the minimization of $f(R)$ is equivalent to the maximization of the second member of the sum. Q.E.D.

2. **The optimal partition problem** [15]. We consider the approximation problem in the specific case when E is the set of all (unordered) partitions $R = \{R_1, \ldots, R_m\}$ of the set A (for arbitrary $m \geq 1$). For a matrix r corresponding to the partition $R = \{R_1, \ldots, R_m\}$, an element r_{ij} equals 1 if and only if for some s ($s=1,\ldots,m$) the objects i and j both belong to R_s. Consequently, in this case

$$g(R) = \sum_{s=1}^{m} \sum_{i,j \in R_s} \left(a_{ij} - \frac{n}{2}\right). \qquad (3)$$

Expanding the bracketed part of (3) we have

$$g(R) = \sum_{s=1}^{m} \sum_{i,j \in R_s} a_{ij} - \frac{n}{2} \sum_{s=1}^{m} N_s^2, \qquad (4)$$

where N_s is the number of objects in the class R_s.

Equation (3) has a simple interpretation. The quantities $a_{ij} = \sum_{k=1}^{n} r_{ij}^k$ are naturally interpreted as indicators of connection of the objects i and j in the system $R^1, \ldots, R^n$. Therefore $g(R)$ in (3) is simply the sum of the "internal" connections (under subtraction from each of them of the significance "threshhold" $\frac{n}{2}$) in the partition R. Solving the approximation problem then reduces to constructing the partition for which the sum (3) of internal connections with subtracted significance threshhold is maximized. The meaning of the threshhold $\frac{n}{2}$ becomes clear in formula (4): To maximize $g(R)$ in agreement with (4) it is necessary to increase the sum of internal connections $\sum \sum a_{ij}$, and to decrease the sum of squares of class sizes $\sum N$. Maximizing the first member of (4) increases the "compactness" of the partition R, while minimizing the second member increases the degree of uniformity of its distribution.

Actually, the minimum $\sum N_s^2$ is reached for $N_1 = \ldots = N_m$, since $\sum_s N_s = N$ is fixed. Thus the size of the threshhold $\frac{n}{2}$ characterizes the degree of compromise between the two non-coincident goals of "compactness" and "uniformity" of the desired classification.

For $n=1$, when a single initial relation R^1 with matrix $\|r_{ij}^1\|$ is being approximated, the elements of the matrix $\|a_{ij} - \frac{n}{2}\|$ are the oppositely placed numbers $1/2$ and $-1/2$. Then the elements of the matrix $\|2a_{ij} - \frac{n}{2}\|$ are 1 and -1. The approximation problem in a class of equivalence relations is solved by maximizing $2g(R)$, the sum of the internal "connections" of the form $2a_{ij}-1$ in the partition R. A symmetric matrix with elements 1 and -1 is frequently associated with the so-called *signed graph* [10], a complete non-oriented graph, each edge of which is marked with the sign "+" or "-". Thus in order to approximate a given graph it necessary to pass over to the corresponding signed graph, in which an edge is "positive" if it appears in the initial graph, and "negative" otherwise, and then introduce an optimal cut of the signed graph, so that the edges are "cut" in such a way as to minimize the algebraic sum of the losses.

The exact solution of this problem is extremely laborious, even for a small number of objects. We introduce without proof some results of G. S. Friedman[44] which help to limit the number of trials, in some cases.

If a regular graph does not contain a triangle, then any maximal pair-combination of the graph [7] solves the approximation problem.

If the maximal length of a non-repeating chain (cycle) of the initial graph does not exceed $k < N$, then the cardinalities of the classes of an optimal partition do not exceed k (respectively $k+1$).

The number of classes of any optimal partition is not less than two-thirds of the radius of the initial graph. If the initial graph has a k-element clique then any optimal partition has a class with no fewer than $\left[\frac{k}{2}\right] + 1$ elements.

In connection with the complementation matrix $\|a_{ij}\|$, keeping in mind that for the mosaic mechanism a single class of a partition contains those mutations which are "more or less" similar in their complementation reactions, one should consider as the indicators of a given collection of mutatheir complementarities with all of the mutations under consideration. More formally, the indicators of a given collection (of mutations) are the columns of the complementation matrix: For any mutation i the jth column tells whether mutation i is complementary to mutation j, according to whether a_{ij} is zero or not. Thus functional similarity of mutations is characterized by N binary indicators corresponding to the columns of the complementation matrix. For a particular mutation each such indicator partitions the set A into two classes of mutations (in the one class are those mutations complementary to it, and in the other, those which are not).

By Theorem 1 the approximation of these N indicators can be replaced by the classification task for the matrix of connections $a_{ij} = \sum_{k=1}^{N} r_{ij}^k$. What does the quantity a_{ij} mean in this case? The value of r_{ij}^k is 1 if the mutations i and j are identically complementary with the kth mutation, i.e. if the numbers in positions i and j of the kth column of the complementation matrix are identical. Consequently a_{ij} is none other than the number of identical positions in rows i and j of the complementation matrix, viewed as N-element vectors. We will call these quantities a_{ij} *indicators of functional similarity* of mutations i and j.

Curiously, it is usual in experimental investigations involving objects represented by binary indicators, to measure similarity by the number of identical positions (or non-identical ones). We see from the above that this measure also shows up in the framework of an approximation problem.

Experiments have shown that in analyzing specific data it is useful to vary the value of the significance threshhold of individual connections (it is the indicator of the degree of compromise between the requirements of "compactness" and uniformity).

In the book [22] it is shown how to interpret the value of the threshhold when it is not necessarily $\frac{n}{2}$, in geometric terms (see also [25]). We consider the modified problem of constructing an optimal partition $R = \{R_1, \ldots, R_m\}$ according to the criterion

$$g(a,R) = \sum_{s=1}^{m} \sum_{i,j \in R_s} (a_{ij} - a) = \sum_{s=1}^{m} \sum_{i,j \in R_s} a_{ij} - a \sum_{s=1}^{m} N_s^2, \quad (5)$$

where a is an arbitrary real number and R is one of a fixed class E of partitions on the set A [15,22].

We note that in this formulation the problem has meaning not only in the framework of approximations, but can also be applied to connection matrices $\|a_{ij}\|$ of arbitrary form.

We first convince ourselves that a partition, optimal in the sense of (5), is also "good" in a meaningful sense of that word unconnected with any indicator of optimality.

In a "good" partition each class is "good", i.e. "compact". To sharpen this aphorism we introduce the following definitions.

A partition $R = \{R_1, \ldots, R_m\}$ is *compact* if the average internal connection between objects in each class exceeds the average external connection between two classes:

$$\frac{1}{N_s(N_s-1)} \sum_{i,j \in R_s} a_{ij} \geq \frac{1}{N_q N_t} \sum_{i \in R_q} \sum_{j \in R_t} a_{ij} \qquad (6)$$

$$(q,s,t=1,\ldots,m),$$

where N_s is the number of objects in class R_s ($s=1, \ldots, m$).

A set $S \subseteq A$ is a *concentrate* if the average connection between objects of S exceeds the average connection of S with the remaining objects:

$$\frac{1}{|S|(|S|-1)} \sum_{i,j \in S} a_{ij} \geq \frac{1}{|S|(N-|S|)} \sum_{i \in S} \sum_{j \notin S} a_{ij}. \qquad (7)$$

Although the notions introduced also have no connection with any indicator of the quality of the partition, it will be shown in what follows that a partition, optimal in the sense of (5), must be compact and that its classes must be concentrates.

We note that for the case when R_s and S contain only single elements the left parts of the inequalities (6) and (7) are meaningless. From the preceding footnote they should be taken to be 1.

As a preliminary, note that in analyzing an optimal partition in the sense of (5) for an arbitrary connection matrix, we can limit our attention to the case when the connections are symmetric ($a_{ij}=a_{ji}$). For if $\|a_{ij}\|$ is not symmetric the passage to a new symmetric matrix $\|a'_{ij}\|$ with $a'_{ij} = \frac{a_{ij}+a_{ji}}{2}$ is straight-forward, and does not change the values in (5) or the optimal partition.

Now for a given partition $R = \{R_1, \ldots, R_m\}$ and number $a \geq 0$ let

$$A_{st} = \sum_{i \in R_s} \sum_{j \in R_t} (a_{ij}-a).$$

The quantity A_{st} is the sum of elements $a_{ij}-a$ of the connection matrix "cell" $a_{ij}-a(i \in R_s, j \in R_t)$ corresponding to classes R_s and R_t of the

---

Here and in the remainder of this chapter a connection a_{ii} of an object with itself is not considered. This means that our discussion is about approximating data in the class of Boolean matrices with all zeros on the main diagonal (graphs without loops). Such restrictions will not be stated formally. In particular, removing the parentheses in (3) we obtain in (4) the expression $\sum N_s(N_s-1)$ in place of $\sum N_s^2$, since these quantities differ by a constant (equal to N), we may, as before, consider (4), and more generally (5), as optimization indicators. Similarly, since for $r_{ii}=0$ $a_{ii}r_{ii}=0$, in the sums $\sum a_{ij}$ one need not write the restriction $i \neq j$, but consider that $a_{ii}=0$. This corresponds to the fact that in the initial matrices r^k one considers elements on the main diagonal to be zeros.

partition R. Since we are not considering the connection of an object with itself it is natural to take $A_{ss}=0$ for a single-element class R_s.

The following necessary condition for the optimality of a partition holds [15]:

Theorem 2. If a partition $R = \{R_1, \ldots, R_m\}$ is optimal according to (5) in the class of all possible partitions of the set A, then for $s \neq t$ $(s,t=1,\ldots,m)$ $A_{st} \leq 0$, while $A_{ss} \geq 0$ for all s.

Proof. Suppose that R is optimal, and for some s $A_{ss}<0$. In this case we move from R to a better partition R' by "separating" the objects of class R_s into single-element classes $\{i\}$ for $i \in R_s$. But then $\sum_{i \in R_s} A_{ii}=0>A_{ss}$, so that $g(a,R')$ is obtained from $g(a,R) = \sum_{t=1}^{m} A_{tt}$ by the exclusion of $A_{ss}<0$ from the number of terms, and thus $g(a,R')>g(a,R)$. But this is contrary to the optimality of R.

On the other hand, if $A_{st}>0$ for $s \neq t$, then an increase in the value of (5) can be obtained by combining classes R_s and R_t. In this case the quantity $A_{st}+A_{ts}=2A_{st}>0$ is added to the sum, which again contradicts the optimality of R. This proves the theorem.

The required result now follows easily [15]:

Theorem 3. A partition, optimal in the sense of (5), is compact, and each of its classes is a concentrate.

Proof. From Theorem 2 the following inequalities hold for an optimal R:

$$A_{ss} = \sum_{i,j \in R_s} a_{ij} - aN_s(N_s-1) \geq 0,$$

$$A_{qt} = \sum_{i \in R_q} \sum_{j \in R_t} a_{ij} - aN_q N_t \leq 0.$$

From this it follows that

$$A_q = \sum_{t \neq q} A_{qt} = \sum_{i \in R_q} \sum_{j \notin R_q} (a_{ij}-a) = \sum_{i \in R_q} \sum_{j \notin R_q} a_{ij} - aN_q(N-N_q) \leq 0.$$

But then

$$\frac{1}{N_s(N_s-1)} \sum_{i,j \in R_s} a_{ij} \geq a,$$

$$\frac{1}{N_q N_t} \sum_{i \in R_q} \sum_{j \in R_t} a_{ij} \leq a,$$

$$\frac{1}{N_q(N-N_q)} \sum_{i \in R_q} \sum_{j \notin R_q} a_{ij} \leq a.$$

Finally, inequalities (6) and (7) follow from this, completing the proof.

We now consider the question of the relationship between an optimal partition and the value of the threshhold a [12].

We first remark that if $a_{ij} \geq a$ for all i,j i.e. $a_{ij} - a \geq 0$ for all i,j, then the largest value of (5) is reached for the partition consisting of a single class containing all the objects. On the other hand, if for all i,j $a_{ij} \leq a$, i.e. $a_{ij} - a \leq 0$, then the optimal (zero) value of the quantity $g(a,R)$ is obtained for the trivial partition which divides the set A into N single-element classes.

Changing a from the first situation $(a_{ij} \geq a)$ to the second $(a_{ij} \leq a)$ corresponds to changing the number of classes of the optimal partition from 1 to N. The value of the threshhold a, as well as the number of classes m, reflects the degree of "coarseness" of aggregation of the initial information.

Can we assert that the number of classes of an optimal partition is monotonically dependent on a? The answer is "no". It is not difficult to find examples in which increasing the threshhold decreases the number of classes of the optimal partition. Moreover, there exist connection matrices for which m does not take on all the values from 0 to N, for all possible values of a [12].

However, there is a second, more refined characteristic of the optimal partition, which *is* dependent on a. It is the quantity $\Delta(R) = \sum_{s=1}^{m} N_s^2$ which, as was remarked earlier, characterizes the degree of non-uniformity of the distribution of objects to the classes of R. This quantity is closely connected with the well-known measure of the non-uniformity of a distribution, entropy. It is obtained by taking the quadratic part of the Taylor expansion of the entropy (see also [22]).

Theorem 4 [12]. For a partition R, optimal according to $g(a,R)$, the value $\Delta(R)$ is monotonically non-increasing for increasing a.

Proof. Let R' be optimal for $a = a'$ and R'' be optimal for $a = a''$, with $a' > a''$. We must show that $\Delta(R') \leq \Delta(R'')$.

From (5) we have

$$g(a,R) = g(0,R) - a\Delta(R).$$

By the definition of R' and R'',

$$g(a',R') \geq g(a',R''),$$

$$g(a'',R'') \geq g(a'',R').$$

Substituting the expression $g(a,R)$ in these inequalities we obtain

$$g(0,R') - a'\Delta(R') \geq g(0,R'') - a'\Delta(R''),$$

$$g(0,R'') - a''\Delta(R'') \geq g(0,R') - a''\Delta(R').$$

Hence

$$a''(\Delta(R') - \Delta(R'')) \geq g(0,R') - g(0,R'') \geq a'(\Delta(R') - \Delta(R'')).$$

Consequently,

$$a''(\Delta(R') - \Delta(R'')) \geq a'(\Delta(R') - \Delta(R'')),$$

i.e.

$$(a' - a'')(\Delta(R') - \Delta(R'')) \leq 0.$$

Since $a' - a'' > 0$, this is possible only for

$$\Delta(R') - \Delta(R'') \leq 0,$$

Q.E.D.

Notice that for the proof only the optimality of R' and R'' was used, independently of the class E of admissible relations. This means that Theorem 4 is in fact true for approximative relations from any class of relations E by utilization of the quantity $\sum_{(i,j) \in P}(a_{ij} - a)$ $(P \in E)$: with increase of a the value $\Delta(P) = |P|$ decreases. This is true in particular in the case when E is a collection of partitions with a previously fixed number of classes.

Searching for an optimal partition (in the class of all possible partitions), which is minimal in the sense of (5), is a combinatorily complex task, and in general, it cannot be solved by any simple methods. In these situations a frequent approach is to build a simple algorithm which gives an exact solution in the large majority of cases, but not necessarily in all cases. Unfortunately, in seeking an optimal partition this approach is not suitable, as the following result shows.

We will say that a property of systems of relations $\{R^1, \ldots, R^n\}$ holds *almost always* if the fraction of collections $\{R^1, \ldots, R^n\}$ which satisfy the property approaches 1 as N approaches ∞.

Theorem 5. The summed distances $f(R)$ from a system of relations $\{R^1, \ldots, R^n\}$ to an optimal equivalence R is asymptotically equal to $\frac{nN^2}{2}$ almost always, i.e.

$$\frac{f(R)}{\frac{N^2 n}{2}} \longrightarrow 1 \quad (N \to \infty).$$

The proof of the theorem follows from a result of G. S. Friedman, who showed that the distance from an arbitrary relation to the equivalence nearest it is almost always asymptotically equal to $\frac{N^2}{2}$. This means that for given N the collection of relations for which the distance from the nearest equivalence is not less than $(1-\delta_N)\frac{N^2}{2}$ contains a fraction not less than $(1-\varepsilon_N)$ of the whole quantity of relations, with $\delta_N, \varepsilon_N \to 0$ for $N \to \infty$ [44]. Designate the size of this class by $\overline{N}$, and consider all possible n-tuples of relations from this class. The number of these is $C^n_{\overline{N}+n-1}$ from the well-known formula for the number of combinations of things with repetitions. From the above,

$$C^n_{\overline{N}+n-1} > C^n_{(1-\varepsilon_N)2^{N^2}+n-1},$$

since the entire number of relations is 2^{N^2}. But then the fraction of the n-tuples being considered exceeds

$$\frac{C^n_{(1-\varepsilon_N)2^{N^2}+n-1}}{C^n_{2^{N^2}+n-1}}$$

and, clearly approaches 1 as $N \to \infty$ for fixed (or even slowly increasing) n.

For each relation from these n-tuples the distance to the nearest equivalence is asymptotically equal to $\frac{N^2}{2}$, so that the sum of the distances is asymptotically equal to $\frac{nN^2}{2}$. Q.E.D.

We note in addition that there are $\frac{N^2}{2}$ arcs in the median relation, and for $N \to \infty$ almost all the relations have asymptotically the same number of arcs. Consequently, from Theorem 5 it follows that for almost all collections $\{R^1, \ldots, R^n\}$ the median in the class of partitions may be obtained by rejection of all arcs in $\{R^1, \ldots, R^n\}$, i.e. by taking the trivial partition, or vice versa, adding arcs necessary to make the universal relation. This means that any relation is almost always asymptotically optimal, including the trivial relation consisting of N classes, and the universal relation consisting of only one class. Thus, in the problem being considered, it is necessary to build locally optimal algorithms having purely experimental bases.

These algorithms work as follows. At step k some partition R of objects is considered, along with some collection of partitions $O(R)$, which constitute a "neighborhood" of R. An optimal partition is selected from this neighborhood according to the given criterion, and this becomes the new R for the next step. The algorithm terminates when the new partition coincides with the preceding one.

We will now describe two of these algorithms [22,15].

The *"unification" algorithm*. Let $R(l) = \{R_1, \ldots, R_l\}$ be an arbitrary partition of the set A into l classes. Take as the neighborhood

$O(R(l))$ all possible partitions R^{st} with $l-1$ classes, obtained from $R(l)$ by unifying the classes R_s and R_t ($s,t=1,\ldots,l$). The optimal partition from among these becomes $R(l-1)$.

The algorithm, beginning with the trivial partition of A into N single-element classes, produces a sequence of partitions $R(N-1)$, $R(N-2)$, ...

To make use of the indicator (5), calculations are organized on the basis of the fact that $g(a,R^{st})$ is large when the sum of connections $A_{st} = \sum_{j \in R_s} \sum_{j \in R_t} (a_{ij}-a)$ is large, as follows.

For the partition $R(l) = \{R_1, \ldots, R_l\}$ consider the matrix of summed connections $\|A_{st}\|$ between its classes (at the first step, for $R(N)$ the initial matrix $\|A_{ij}=a_{ij}-a\|$ is taken). The maximal element A_{st} of this matrix is selected. If it is positive then the classes R_s and R_t are combined, and the transition to the $(l-1)\times(l-1)$ matrix is made, by the component-by-component composition of the rows (and columns) s and t. From Theorem 3, the calculations are stopped when all of the A_{st} ($s \neq t$) are non-positive.

The partition thus obtained satisfies the necessary optimality condition, and consequently it is compact and its classes are concentrates.

The unification algorithm, in connection with the criterion (5) is one of a widely-known group of so-called *agglomerative algorithms* of a taxonomy, consisting of successive unifications of classes near to one another. In the given case nearness of classes is estimated quantitatively. In other cases nearness of classes is estimated by: a) the value of the maximal connection between their elements (the nearest-neighbor method), b) the value of the minimal connection between their elements (the farthest-neighbor method), c) the value of the average, etc. In addition to the final partition, the whole sequence of partitions is often used in the analysis, and is viewed as a hierarchical tree (see Section 3.1.3).

The *"transfer" algorithm*. For any partition R, the neighborhood considered is the collection of all partitions R^{is} ($i=1,\ldots,N; s=1,\ldots,m$), obtained by transferring the object i into the class R_s. In particular, if $i \in R_s$ then $R^{is} = R$.

At each step i is fixed, so that the optimal variant of the transfer of i into some class of R can be selected. To make use of the indicator (5) for this, it is sufficient to calculate the summed connections of i with each of the existing classes. Objects are selected in the order of their indices, with the Nth one being followed by the first. The number of these cycles can be fixed in advance.

As a rule we use a combination of the "unification" algorithm and the "transfer" algorithm in practical computations, as follows [15].

By means of the unification algorithm obtain a partition $R(l)$ with a previously fixed (and sufficiently large) number of classes. This partition is then improved by means of the transfer algorithm. The necessary condition $A_{ss} \geq 0$ for optimality may be violated in this process. Any such classes R_s are divided into single-element subclasses, for which the internal connections A_{ss} are zero. After this, unification produces a new partition into $l-1$ classes, which is then improved by the transfer algorithm with subsequent testing of the condition $A_{ss} \geq 0$ and the "dividing" of "bad" classes. Thus a sequence of alternate unifications and transfers leads to a final partition with $A_{st} \leq 0$ ($s \neq t$), $A_{ss} \geq 0$.

3. **Detecting macrostructure.** From Theorem 1 the problem of detecting the macrostructure of a given $N \times N$ connection matrix $\|a_{ij}\|$ (in the simplest case $\|a_{ij}\|$ is an $N \times N$ Boolean matrix) may be formulated as follows.

Find that macrostructure $(R, \varkappa)$ in a given class of macrostructures, which maximizes the following quality indicator (for given a):

$$F(R, \varkappa) = \sum_{(s,t) \in \varkappa} \sum_{i \in R_s} \sum_{j \in R_t} (a_{ij} - a). \qquad (8)$$

If $\|a_{ij}\|$ is a Boolean matrix, then by Theorem 1 $a = \frac{1}{2}$ (in the approximation problem).

For a fixed partition $R = \{R_1, \ldots, R_m\}$ a structure $\varkappa$, optimal in the sense of (8), is determined as follows. As in the preceding section, we define the quantity A_{st} to be the total connection of the classes R_s and R_t ($s,t=1,\ldots,m$), and the structure $\varkappa(R)$ is defined by the condition:

$$(s,t) \in \varkappa(R) \leftrightarrow A_{st} > 0.$$

Clearly $F(R, \varkappa(R))$ is maximal (for all $\varkappa$). Consequently, if a pair $(R^*, \varkappa^*)$ maximizes (8), then $\varkappa^* = \varkappa(R^*)$ (to within pairs (s,t) for which $A_{st}=0$). This means that criterion (8), if all structures $\varkappa$ are admissible, forces the maximization of the sum of the positive connections of A_{st} ($s,t=1,\ldots,m$). Since the sum of all the numbers $a_{ij}-a$ ($i,j=1,\ldots,m$) is fixed, then during this process the maximum of the sum of the absolute values of the negative connections of A_{st} is reached.

We have thus proved that the maximization of the indicator F in the case when for every R all structures are admissible, is equivalent to the construction of a partition R for which the maximum sum of the modulus of the connections A_{st} is reached [14]:

$$F'(R) = \sum_{s,t=1}^{m} \left| \sum_{i \in R_s} \sum_{j \in R_t} (a_{ij}-a) \right|. \qquad (9)$$

In other words, in optimizing the indicator (8) a partition R is obtained for which the matrix of "cells" $\|a_{ij}\|$ ($i \varepsilon R_s, j \varepsilon R_t$) has the most contrast: The positive connections $a_{ij}-a$ are located basically in certain cells, and the negative ones in other cells. Thus objects which are more or less identical in their interactions with other objects fall into one class. Therefore the detection of macrostructure may be considered as the aggregation of information in such a way that objects which possess identical interactions are combined [14].

We designate by F_m^* the optimal value of the indicator (8) in the set of all macrostructures $(R, \varkappa)$ with a fixed number m of classes in the partition R.

Clearly,

$$F_m^* \leq F_{m+1}^* \quad (m=1, \ldots, N-1). \quad (10)$$

For if $F(R, \varkappa) = F_m$ and $R = \{R_1, \ldots, R_m\}$, then it is not difficult to construct a partition with $m+1$ classes, by dividing any class R into two subclasses. For this new partition the structure $\varkappa$ is redefined only for the new classes, in such a way that these classes are bi-connected with each other, and also with those classes of R with which the initial class was connected. The macrostructure thus obtained clearly corresponds to a value of F equal to F_m^*, which proves inequality (10).

From (10) it follows that in the class of all possible macrostructures the optimal macrostructure is defined simply by the trivial partition R into single-element classes, with

$$(\{i\}, \{j\}) \varepsilon \varkappa \leftrightarrow a_{ij} \geq a.$$

In this case $\varkappa$ is known as the *a-similarity graph* for the connection matrix $\|a_{ij}\|$.

The inequality (10) shows that, unlike the optimal partition problem, besides the threshhold a it is necessary to fix in advance the number of classes, for otherwise the a-similarity graph will be obtained. As a rule, this graph is not of interest, since it starts out with N orders [15, 20].

To maximize (8) one can use the *"Structure" algorithm* [14], which is a combination of the unification and transfer algorithms described in Section 4.

In the unification algorithm, as in the maximization of (5), calculations are organized according to the matrix $\|A_{st}\|$ of the total connections between classes. However, in this case the unification of the classes R_s and R_t is realized not according to the maximum A_{st}, but by the minimum of the function

$$\Phi(s,t) = \sum_{\substack{k=1 \\ k \neq s,t}}^{l} \left[\left| \frac{\text{sign}A_{sk} - \text{sign}A_{tk}}{2} \right| \min(|A_{sk}|, |A_{tk}|) + \right.$$
$$\left. + \left| \frac{\text{sign}A_{ks} - \text{sign}A_{kt}}{2} \right| \min(|A_{ks}|, |A_{kt}|) \right] + \min(A_+, |A_-|), \quad (11)$$

where A_+ (A_-) is the sum of all positive (negative) numbers from the quadruple A_{ss}, A_{st}, A_{ts}, A_{tt}.

The reasonableness of using this particular $\Phi(s,t)$ is ensured by the following equality [14]:

$$F'(R) - F'(R^{st}) = 2\Phi(s,t). \quad (12)$$

From (12) it follows that a minimal decrease in the value of the indicator (9) as a result of unification, corresponds to the minimum of $\Phi(s,t)$, so that the optimal R^{st} is actually defined by this minimum.

We now prove (12). First note that for two quantities x and y having opposite signs ($xy<0$), the following equality holds:

$$|x+y| = |x| + |y| - 2\min(|x|, |y|). \quad (13)$$

In fact, if, for example, $|x|<|y|$, then

$$|x+y| = |y| - |x| = |x + y| - 2\min(|x|, |y|).$$

We continue the proof of (12) by considering the absolute values of the total connections A_{pq}^{st} between classes of the partition R^{st}. For ease of working, the class $R_s \cup R_t$ of the partition R^{st} will be designated by $R_{s \cup t}$. This permits the use, for the remaining classes of R^{st}, of the indices p, q, which they had in the partition R. We consider all possible cases:

1. $p \neq s \cup t$, $q \neq s \cup t$;
2. $p = s \cup t$, $q \neq s \cup t$;
3. $p \neq s \cup t$, $q = s \cup t$;
4. $p = s \cup t$, $q = s \cup t$.

In case 1 the classes with numbers p, q in the partitions R and R^{st} coincide, so that A_{pq}^{st} is equal to A_{pq} (the total connection of the pth and qth classes in the partition R), and of course $|A_{pq}^{st}| = |A_{pq}|$. The quantity A_{pq} does not appear in $\Phi(s,t)$, so that the "contribution" of the cell (p,q) in $F'(R)$ and $F'(R^{st})$ satisfies (12).

In case 2 $p = s \cup t$, so that $A_{pq}^{st} = A_{sq} + A_{tq}$. If the signs of A_{sq} and A_{tq} agree, then $|A_{pq}^{st}| = |A_{sq} + A_{tq}| = |A_{sq}| + |A_{tq}|$, and consequently the "contribution" of the cell (p,q) in the left part of (12) is zero. This cell also does not introduce anything into $\Phi(s,t)$, since $\text{sign}A_{sq} - \text{sign}A_{at} = 0$, i.e. the coefficient for the corresponding term $\min_i(A_{sq}, A_{tq})$ is zero. On the other hand, if $\text{sign}A_{sq} \neq \text{sign}A_{tq}$, then from (13) $|A_{pq}^{st}| = |A_{sq}| + |A_{tq}| - 2\min(|A_{sq}|, |A_{tq}|)$. Consequently $|A_{sq}| + |A_{tq}| - |A_{pq}^{st}| = 2\min(|A_{sq}|, |A_{tq}|) = |\text{sign}A_{sq} - \text{sign}A_{tq}| \min(|A_{sq}|, |A_{tq}|)$. So that in this case also the "contribution" of the cell (p,q) in the partition R^{st} is identical in the left and right parts of (12).

Case 3 is handled similarly.

In case 4 $p = s \cup t$, $q = s \cup t$, so that $A_{pq}^{st} = A_{ss} + A_{st} + A_{ts} + A_{tt}$. In accord with (11) we designate by A_+ the sum of the positive numbers among the quadruple $A_{ss}, A_{st}, A_{ts}, A_{tt}$, and by A_- the sum of those that are negative. We know that $|A_{ss}| + |A_{st}| + |A_{ts}| + |A_{tt}| = |A_+| + |A_-|$. On the other hand, $|A_{pq}^{st}| = |A_+ + A_-| = |A_+| + |A_-| - 2\min(A_+, |A_-|)$, i.e. $|A_{ss}| + |A_{st}| + |A_{ts}| + |A_{tt}| - |A_{pq}^{st}| = 2\min(A_+, |A_-|)$, so that again the contributions of the cell (p,q) to the left and right sides of (12) are equal.

This concludes the proof of the equality (12).

Thus, beginning with the trivial partition into N single-element classes, it is possible to successively combine pairs of classes by the unification algorithm in accordance with the minimum $\Phi(s,t)$.

The transfer algorithm is applied just as described on page 113, except that the valuation of the partition R^{is} obtained from R by transferring object i from R^t ($t=1,\ldots,m$) into class s, is not produced according to the sum of the connections of i with the objects of R^s, but according to the maximum value of

$$f(i,t,s) = \sum_{k \in \mathcal{X}(s) \cup \mathcal{X}-1(s)} \sum_{j \in R_k} (a_{ij}-a) - \sum_{k \in \mathcal{X}(t) \cup \mathcal{X}-1(t)} \sum_{j \in R_k} (a_{ij}-a). \quad (14)$$

Here $\mathcal{X} = \mathcal{X}(R)$ and the first summand characterizes the connection of the object i within the macrostructure after its transfer from R_t to R_s, and the second summand, the connection of i before the transfer.

As in the search for a partition, the best results are obtained by a combination of the unification algorithm and the transfer algorithm.

The value of the threshhold a can be determined from considerations of interest. In [25] the method of least squares is also used to determine the value of a for which the desired macrostructure of the "best form" approximates a given connection matrix $\|a_{ij}\|$, whether or not it is Boolean.

§3. **Analyzing the spatio-functional organization
of specific genetic systems**

1. **Complex protein organization.** The goal of this section is, on the one hand, to examine the validity of the mosaic mechanism of interallelic complementation, and, on the other, to obtain individual descriptions of the functional organization of specific systems of "gene-enzymes."

The "structure" algorithm was used in analyzing several specific cistrons. We will examine here the results of only one of these. This cistron is studied both genetically and biochemically more throughly than the others in order to form a possible molecular interpretation of the constructed graphs.

The discussion will be about the B cistron of the his operon, which controls the synthesis of the amino acid histidine in *Salmonella* bacteria (see Fig. 17). A peculiarity of this cistron is that its protein product, the enzyme, dehydratase-phosphatase, is bifunctional: it catalyzes two non-adjacent reactions (the 7th and 9th) in the biosynthesis of histidine. The reactions are: the splitting off of water molecules (*dehydratase function*) and the adjoining of a phosphoric-acid residue (*phosphatase function*). Correspondingly, some mutations may be defective for the dehydratase function, some for the phosphatase function, and some for both functions. However, in spite of the large number of complementary mutations which have been described in this cistron, up to this time only the defects for the phosphatase activity have not been located [71,81,48]. The results below allow, in particular, an explanation of this peculiarity (see page 125).

The initial complementation data were taken from the work of Loper et. al. [81]. Out of 64 his B mutants which were studied, some coincided in their complementation behavior, i.e. in the image and preimage of the corresponding relation, so that the complementation matrix (Fig. 26), with which all further operations were carried out, consisted of 36 different groups of mutants. (Clearly, these groups are none other than the classes of the canonical partition; see the Appendix.)

In the given collection of mutations there are various types of changes: The greater part are the usual point replacements "codon⟶codon", replacements of the type "codon⟶nonsense", frameshift (i.e. the loss and insertion of individual bases), and deletions of varying extents. Some of these are defective only for the dehydratase activity (the monofunctional ones underlined in Fig. 27,b), some for both activities (the bifunctional ones written in parentheses in Fig. 27,b), and for the remaining mutations the functions affected by them are still not known.

The results of aggregating mutations by means of the "structure" algorithm are shown in Figure 27. It was natural to select as the value

	12,...	40,...	53,...	56	59	65	116	136,...	138	167,488	206,...	217	234	241	243	262	286	289	353	355	380	391,836	470	480	542	562,812	569	578	612	641	662	669,672	824	865	902	923
12,...	1														1							1					1		1						1	
40,...		1				1						1	1	1		1		1							1	1			1							
53,...			1						1		1		1		1	1							1							1						1
56				1	1			1	1	1			1		1	1	1				1		1									1				
59				1	1			1	1	1			1		1	1	1			1	1		1			1						1		1		
65,...		1				1			1	1		1	1	1	1	1	1	1	1	1	1			1	1			1	1			1		1		
116							1	1	1	1			1		1	1	1			1	1		1			1						1		1		
136,...				1	1		1	1	1	1			1		1	1	1			1	1		1			1						1		1		
138			1	1	1	1	1	1	1	1		1	1	1	1	1	1	1	1	1	1		1			1						1		1		1
167,488				1	1	1	1	1	1	1			1	1	1	1	1		1	1	1		1			1						1		1		
206,...			1								1		1		1	1													1	1						1
217		1				1			1			1	1	1	1	1		1	1		1			1	1			1	1			1				
234		1	1	1	1	1	1	1	1	1	1	1	1	1	1	1	1	1	1	1	1		1			1		1	1	1		1		1		1
241		1				1			1	1		1	1	1	1	1	1	1	1	1	1			1	1			1	1			1	1	1	1	
243	1		1	1	1	1	1	1	1	1	1	1	1	1	1	1	1		1	1	1		1			1			1	1		1		1	1	1
262		1	1	1	1	1	1	1	1	1	1	1	1	1	1	1	1	1	1	1	1		1	1		1		1	1	1		1		1		1
286				1	1	1	1	1	1	1			1	1	1	1	1			1	1		1			1						1		1		
289		1				1			1			1	1	1		1		1	1		1			1	1			1	1				1		1	
353						1			1	1		1	1	1	1	1		1	1		1							1				1		1		
355					1	1	1	1	1	1			1	1	1	1	1			1	1		1			1						1		1		
380				1	1	1	1	1	1	1		1	1	1	1	1	1	1	1	1	1		1			1						1		1		
391,836	1																					1					1		1						1	
470			1	1	1		1	1	1	1			1		1	1	1			1	1		1			1						1		1		1
480		1				1						1		1		1		1						1	1			1	1							
542		1				1						1		1				1						1	1			1	1							
562,812					1		1	1	1	1			1		1	1	1			1	1		1			1						1		1		
569	1																					1					1		1		1				1	
578		1				1						1	1	1		1		1	1					1	1			1								
612	1					1					1	1	1	1	1	1		1				1		1	1		1		1		1		1		1	1
641			1								1		1		1	1														1						1
662																											1		1		1				1	
669,672				1	1	1	1	1	1	1		1	1	1	1	1	1		1	1	1		1			1						1		1		
824														1				1											1				1		1	
865					1	1	1	1	1	1			1	1	1	1	1		1	1	1		1			1						1		1		
902	1													1	1			1				1					1		1		1		1		1	
923			1						1		1		1		1	1							1						1	1						1

Figure 26. The matrix of interallelic complementation for cistron B of the his-operon in *Salmonella typhimurium*. Mutants are numbered as in the original paper of Loper et. al. [81]. Each row (column) corresponds to a separate complementation group which is the union of mutants with identical complementation reactions. For groups containing more than one mutant, only one of them is designated (non-complementarity is designated by 1, complementarity by a blank).

of the threshhold $a=\frac{1}{2}$, so that, formally, the problem solved was that of approximating a relation given by the matrix of Fig. 26, in the class E_m of partitions with structure. Since in this case a value for a was not known in advance, the problem was solved for various a. Beginning with seven classes, it was found that each successive partition for larger a was contained in the preceding one. In other words, the transition to a larger number of classes simply produces a subdivision of classes, not their rearrangement, with the closest "neighborhood" of any mutation in the structure being unchanged. This permitted stopping at $m=7$.

		391,836	12,...	569	662	902	612	824	289	578	40,...	480	542	217	353	65,...	241	380	669,672	167,488	865	138	286	355	136,...	59	562,812	116	56	470	262	234	243	53,...	923	206,...	641
1.	391,836	1	1	1		1	1																														
2.	12,...	1	1	1		1	1																											1			
3.	569	1	1	1	1	1	1																														
4.	662				1	1	1	1																													
5.	902	1	1	1	1	1	1	1	1									1													1						
6.	612	1	1	1	1	1	1	1	1		1	1	1			1	1														1	1	1			1	1
7.	824				1	1	1	1	1									1																			
8.	289					1	1	1	1	1	1	1	1	1	1	1	1				1										1	1					
9.	578								1	1	1	1	1	1	1	1	1														1	1					
10.	40,...								1	1	1	1	1	1		1	1														1	1					
11.	480				1				1	1	1	1	1	1		1	1														1						
12.	542				1				1	1	1	1	1	1		1	1																				
13.	217				1				1	1	1	1	1	1	1	1	1	1	1												1	1	1				
14.	353									1	1			1	1	1	1	1	1	1											1	1	1				
15.	65,...				1				1	1	1	1	1	1	1	1	1	1	1	1	1	1									1	1	1				
16.	241			1	1	1	1	1	1	1	1	1	1	1	1	1	1	1	1	1	1	1									1	1	1				
17.	380								1					1	1	1	1	1	1	1	1	1	1	1	1	1	1	1	1	1	1	1	1				
18.	669,672													1	1	1	1	1	1	1	1	1	1	1	1	1	1	1	1	1	1	1	1				
19.	167,488													1	1	1		1	1	1	1	1	1	1	1	1	1	1	1	1	1	1	1				
20.	865													1	1	1	1	1	1	1	1	1	1	1	1	1	1				1	1	1				
21.	138								1					1	1	1	1	1	1	1	1	1	1	1	1	1	1	1	1	1	1	1	1	1	1		
22.	286														1	1	1	1	1	1	1	1	1	1	1	1	1	1			1	1	1				
23.	355														1	1	1	1	1	1	1	1	1	1	1		1			1	1	1	1				
24.	126,...																	1	1	1	1	1	1	1	1	1	1	1		1	1	1	1				
25.	59																	1	1	1	1	1	1	1	1				1	1	1	1	1				
26.	562,812																	1	1	1	1	1	1	1	1	1	1				1	1	1				
27.	116																	1	1	1	1	1	1	1			1		1		1	1	1				
28.	56																	1	1	1		1			1	1			1		1	1	1				
29.	470																	1	1	1	1	1	1	1	1	1	1	1	1	1	1	1	1	1			
30.	262						1		1	1	1	1		1	1	1	1	1	1	1	1	1	1	1	1	1	1	1	1	1	1	1	1	1	1	1	1
31.	234						1		1	1	1			1	1	1	1	1	1	1	1	1	1	1	1	1	1	1	1	1	1	1	1	1	1	1	1
32.	243	1			1	1								1	1	1	1	1	1	1	1	1	1	1	1	1	1	1	1	1	1	1	1	1	1	1	1
33.	53,...																					1									1	1	1	1	1	1	1
34.	923																										1				1	1	1	1	1	1	1
35.	206,...																					1									1	1	1	1	1	1	1
36.	641																														1	1	1	1	1	1	1

Figure 27a. Classes of complementationally close his B mutants. Reconstruction of the complementation matrix with the classes singled out.

Almost the identical partition was obtained from the construction of the optimal partition for the matrix of functional similarities of mutants with a threshhold $a=24$. This value of the threshhold corresponds to the assumption that mutations which affect the same functional centers must be identically complementary in at least 2/3 of the cases. The combination of the unification and transfer algorithms described in Section 1.2 led to a partition into seven classes, distinguished from the one shown in Fig. 27 only by the interchange of two objects and the division of class 5 into two subclasses. We will only consider further the structure of Figure 27, since the matrix of functional similarities does not precisely reflect the specific character of local paired mutational interactions, and it also forces the selection of a threshhold a for which, unfortunately, there is still no meaningful basis [14].

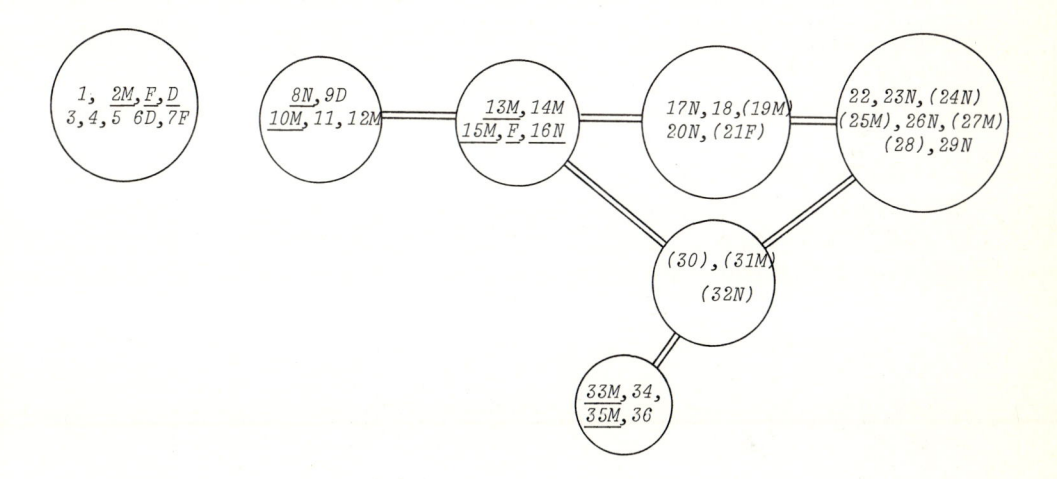

```
( 1,  2M,F,D )   ( 8N,9D       )   ( 13M,14M  )   ( 17N,18,(19M) )   ( 22,23N,(24N)   )
( 3,4,5 6D,7F)   ( 10M,11,12M  )   ( 15M,F,16N)   ( 20N,(21F)    )   ( (25M),26N,(27M) )
                                                                      ( (28),29N       )

                                   ( (30),(31M) )
                                   (   (32N)    )

                                   ( 33M,34,   )
                                   ( 35M,36    )
```

<u>Figure 27,b</u>. The classes of complementationally similar his B mutants, with inter-class connections. Most mutants are given with a two-part designation: The number is the row from Fig. 27,a and the letter indicates the type of mutation involved in each complementation group: "M" means the group contains a missense mutation, i.e. a replacement of the "codon⟶codon" type, "N"— a nonsense mutation, "F"—a frameshift mutation and "D"—a deletion. The numbers of those groups containing monofunctional mutations are underlined, while those for groups of bifunctional mutations are enclosed in parentheses.

We will compare the structure which has been obtained with available independent data about the system.

As is evident in Fig. 27, there are no classes containing both mono-functional and bifunctional mutations in this structure. This fact is an indirect argument in support of the mosaic mechanism of interallelic complementation. Within the scope of this mechanism bifunctional mutations affect several functional centers, and consequently are identical in their complementation behavior with respect to monofunctional mutations associated with other centers. From this it is clear that classes of monofunctional mutations must be connected with one another through the class of bifunctional centers. Exactly this situation is observed in the structure of Fig. 27.

We will compare this structure with the genetic map for the his B cistron. All the classes are projected onto the cistron map (and this means onto the primary structure of the protein) non-compactly (for example, Fig. 29). This kind of disagreement between complementation and recombination has been observed in all the cistrons we have studied carefully. This is a serious argument against the model of Crick and Orgel, since in that model such discrepancies should be the exception (see also [40,43,88]). In addition, the functional centers of the protein are the results of complex spatial packing of chains and are therefore represented in the primary structure by disconnected sections.

		391,...	20M,...	569	902	662	542M	480	478D	40M,...	217M	65M,...	669, 286	562	53M,...	923	206M,...	641
I	391,836	1	1	1	1													
	20M,...	1	1	1	1													
	569	1	1	1	1	1												
	902	1	1	1	1	1												
	662			1	1	1												
II	542M						1	1	1	1	1	1						
	480						1	1	1	1	1	1						
	578D						1	1	1	1	1	1						
	40M,...						1	1	1	1	1	1						
	217M						1	1	1	1	1	1	1					
	65M,...						1	1	1	1	1	1	1	1				
III	669,672									1	1	1	1	1				
	286										1	1	1	1				
	562											1	1	1				
IV	53M,...														1	1	1	1
	923														1	1	1	1
	206M,...														1	1	1	1
	641														1	1	1	1

Figure 28

We will now attempt to examine in a more detailed way the direct description of the functional centers of the his B protein. For this it is necessary to exclude from the initial matrix those mutations known to affect several functional centers. These are primarily the bifunctional mutations, but also a group of deletions, nonsense mutations and frameshifts. Unlike the usual point replacements, mutations of this group lead to the modification or loss of relatively long segments of the polypeptide chain. Therefore, with respect to them, it is not suitable to speak of affecting an individual functional center.

Moreover, the result of complementation testing of such mutations in pairs with missense mutations depends primarily on the ability of the corresponding polypeptide fragments to aggregate with the polypeptides carrying the missense replacements, and also consequently on the level of activity of the "mosaic" hybrid multimer. This ability frequently turns out to be allele-specific [39,88]. In particular this is pointed out by the fact that the deletions and nonsense mutations do not form a single class (see Fig. 27).

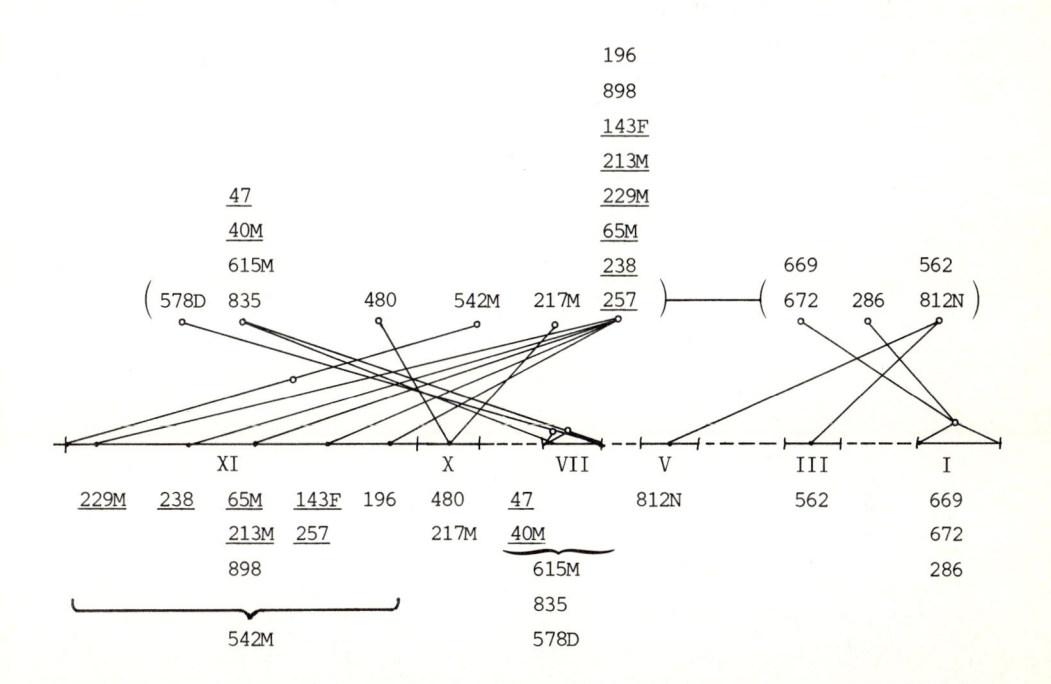

Figure 29. Projections of classes II and III (Fig. 28) on the genetic map of the cistron B: a) classes of complementationally-similar mutations, b) the genetic map of the cistron built by Hartman et. al. [71]. As is evident in this scheme, class II, which contains monofunctional mutations, is projected onto the distal (with respect to the beginning of translation in the gene) half of the map, as is also true of the other classes with monofunctional mutations. On the other hand, class III is projected onto the proximal half of the gene, where the overwhelming proportion of the bifunctional mutations are located.

As a result (Fig. 28) we obtain four almost isolated classes, three of which contain monofunctional mutations. One class (the third) may be "suspected" of bifunctionality, both from the presence in the matrix of Fig. 28 of a characteristic linkage, and from its behavior in the general structure (see Fig. 27).

The following peculiarity is associated with these classes. For monofunctional (defective for dehydratase activity) mutations, their projections onto the genetic map lie in the distant (*distal*) part of the gene, with respect to the beginning of translation. At the same time the bifunctional mutations are projected on the other, nearby (*proximal*) part of the map (see Fig. 29). Mutations of unknown nature, which are contained in a class with the known monofunctional ones, are also projected on the distal half of the map. On the other hand, the class suspected of being bifunctional is projected onto the proximal part of the map, which supports the assumption of bifunctionality of the mutations contained in it. These facts lead to the notion that the protein sections responsible for the dehydratase activity are a collection of amino-acid residues from the distal half of the cistron, while the phosphatase segments are associated with the proximal half.

Analyzing the projections of complementation classes on the genetic map, one can construct a boundary separating the sections which control the two functions even at the level of the primary structure of the protein (see Fig. 29).

These results, obtained from the complementation data [81] by means of the structure algorithm, are corroborated by data from biochemical investigations on a series of mutant forms of the his B protein, carried out by Houston [72,73]. Studying the distribution of a series of nonsense mutations in the map and comparing it with the activity of the corresponding proteins, he concluded that the his B cistron is divided into two functionally differing parts: for dehydratase and phosphatase. The boundary he found is practically the same as ours.

Such coincidence of results of two completely independent approaches inspires confidence that the entire preceding construction, connected with the detection of macrostructure in a complementation matrix, is correct, and does solve the postulated problem.

With this in mind, we can proceed to the description of the functional centers of this protein. First of all three sections are distinguishable, which are responsible for the dehydratase function. These sections (centers) correspond to classes of Fig. 28 which are projected onto the distal part of the cistron. The isolated nature of these classes shows that the corresponding centers are not directly connected with one another. However, since these centers must interact (they work on exactly the same function), this leads us to hypothesize the existence of still another functional center, an allosteric center, which is connected with all of these centers and is a "conductor" of all their interactions. Naturally, it is the presence of the catalytic phosphatase center which ensures the other protein function. Since there is no mutation defective only for the phosphatase center, this center must overlap considerably (perhaps completely) with the allosteric center. By the same token, any effect of this center must be transmitted to the dehydratase portion of the molecule. The possibility that this mechanism is realized in the synthesis process for the initial polypeptide chain cannot be ruled out. Actually, the phosphatase center is a collection of amino acid residues from the proximal part of the cistron. It is logical to suppose that in the process of translation, after obtaining a necessary length, the growing polypeptide chain is folded in such a way that it forms the phosphatase center. Any distortion of this portion of the chain may lead to such curtailment or disturbance of the subsequent growth and conformation of the protein that both its functions are lost. This quite easily explains the absence of monofunctional phosphatase mutations.

We now turn our attention to the class of Fig. 27 containing the bifunctional mutations 24, 25, 27 and 28. It almost fails to overlap any of the classes of monofunctional mutations. Moreover, despite the simultaneous loss of both functions in the homozygote, mutations within this class can complement one another in the heterozygote (see Fig. 27,a).

These two peculiarities cannot be simultaneously explained in terms only of dehydratase, phosphatase and allosteric centers. However, if we recall the necessity of the presence in each subunit of the multimer of a contact center, by means of which it aggregates with the other subunits, then both these peculiarities are easily interpreted. It is possible that mutations of this class affect the contact zone, which leads to the de-aggregation of the subunits in the homozygote, and consequently, to the loss of all the basic properties of the multimer. In the heterozygote specific mutual correction of such defects is possible (see Section 1.1), which also explains the possibility of complementary mutations within this class. On the other hand, the absence of direct connections of these mutations with those which are defective for the the dehydratase function shows that the contact center only partly interacts with the allosteric center.

All this may be summed up in the scheme of functional organization of this enzyme (Fig. 30).

The structure thus obtained characterizes the semantic structure of the cistron: the number of functional centers in the gene product, their nature and interconnections. In particular, both the structural and functional peculiarities of lightly studied mutations can be predicted on the basis of "which neighbors they have" in the constructed graphs. In the situation we are considering an example of this is the prediction of whether mutations are monofunctional or bifunctional, depending on which class they belong to.

In our view, the use of the mathematical techniques presented here for the analysis of structure may become a highly effective means in purely genetic studies of the semantic organization of systems of "gene-enzymes."

Figure 30. Scheme of functional organization of the his B cistron protein product.

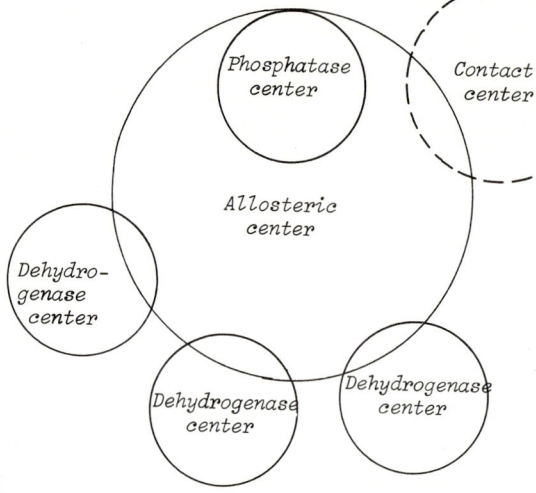

2. **The investigation of genome spatial structure**. In the preceding investigations of specific genetic systems we restricted ourselves primarily to the analysis of genes in organisms of relatively simple structure: bacteria, viruses, bacteriophages, fungi, etc. The single exception was the locus *scute* in *Drosophila*, which, however, was naturally modeled by the most simple linear gene system of the operon type, characteristic for microorganisms.

It should be emphasized that the existence of such genetic systems in higher organisms has not been rigorously shown, up to this time. Along with such systems, one can expect essentially different methods of organization of the functional connections of gene groups.

The determination and analysis of gene systems in higher organisms by application of the usual genetic and biochemical methods which have been so fruitful for microorganisms, meets with significant difficulties. On the one hand, the complexity of the systems themselves complicates the identification of separate genetic components and their connections. On the other hand, the existing methods of experimental analysis are of little value (by reason of the much longer life cycle, as well as a number of other factors).

One might think that success in investigating complex genetic systems in higher organisms comes "from the other end", as one of the results of studying the total integrated picture of the organization of the genetic apparatus.

We examine in this regard some results concerning key elements of the spatial organization of the genetic apparatus in *Drosophila*, obtained by V. A. Kulichkov and E. F. Zhimulev, using the structure algorithm [11].

We first remark that the basic genetic processes (replication of DNA, translation and synthesis of specific proteins, mutation, recombination etc.) run their course in the cells of higher organisms not continuously, but only in a definite period of their life cycle, called the *interphase* period.

The genetic apparatus of a cell is localized primarily in the nucleus. In the interphase period the chromosomes of the nucleus are found to be in working (uncoiled) form, so that their spatial structure in the interphase nucleus differs considerably from that found during other phases. These other phases consist of the actual process of cell division, as well as steps preparatory to it. Here the genetic material of the cell is inactive: structural genes do not function.

This is the reason for the importance of studying the spatial organization in this interphase period: It is here that the fundamental, vitally important sysnthesis functions of the cell are determined.

The "inactively arranged" nucleus lends itself immediately to direct observation, since the chromosomes are contracted (in spiral form) and are able to absorb specific stains which bring out their outlines clearly.

With present methods it is impossible to observe the intact, three-dimensional structure of the interphase nucleus. Cytologists study the internal structure of the cell in a "two-dimensional" form, using what are called *stressed preparations*. In even the best of these preparations we can only view the "fragments" of intra- and interchromosomal connections. By comparing such two-dimensional pictures it is hoped to be able to reconstruct the spatial arrangement of the chromosomes in the interphase nucleus. The detection of this structure helps us understand the meaning of the genetic processes described previously, as well as their interrelations.

Two extreme variants of spatial organization in the chromosome set can be given a priori:

1) When the nature of the contacts (both within and between chromosomes) is random, so that in different cells of the same type there are different structures;

2) When the nature of the contacts specifically determines the spatial organization uniquely for all cells of a given type (given tissue).

Information about the spatial organization of a chromosome can be obtained with the help of the matrix of paired contacts of its various regions (Fig. 31). This particular matrix contains the reduced results

	102F	101F	100A	98C	94D	94A	89E	86D	84D	84A	83D	81F	80ABC	75C	73F/74A	71C	70C	70AB	67D	67A	65B	64C	61A	60F	59D	57AB	56F	42B	39E	39A	36D	35F	35A	33A	25E	25A	22AB	21A	19E	18F/19A	16D	12EF	11A	9AB	8B	7ABC	4C	3C	1A
102F	1	1									1	1	1																										1								1		
101F		1										1	1	1														1	1										1									1	
100A			1	1				1													1	1					1													1			1						
98C				1	1	1		1									1				1	1		1			1																						
94D					1	1										1	1											1																1					
94A						1										1	1					1									1																		
89E							1	1											1		1		1	1								1		1				1											
86D								1	1	1	1						1		1		1		1					1									1					1							
84D									1	1	1																				1					1													
84A										1	1														1																							1	
83D											1	1				1							1					1																					
81F												1										1																				1							
80ABC													1														1	1																			1		
75C														1	1	1	1	1	1	1	1		1								1	1									1							1	
73F/74A															1		1	1															1	1															
71C																1	1	1				1				1		1																		1	1		
70C																	1		1	1		1																	1			1							
70AB																		1	1			1																	1				1						
67D																			1	1	1	1												1									1	1		1			
67A																				1	1	1																											
65B																					1	1																											
64C																						1	1	1	1		1					1						1				1			1				
61A																							1	1														1											1
60F																								1		1	1								1		1	1									1	1	1
59D																									1	1											1												
57AB																										1	1																				1		
56F																											1										1												
42B																												1				1																	
39E																													1		1	1																	
39A																														1	1															1			
36D																															1	1	1	1	1					1					1				
35F																																1	1	1		1				1	1								
35A																																	1	1		1					1	1				1			
33A																																		1					1							1			
25E																																			1	1	1	1							1				
25A																																				1	1				1				1				
22AB																																					1				1	1						1	
21A																																						1		1									1
19E																																							1		1	1	1				1		
18F/19A																																								1	1	1				1			
16D																																									1	1	1		1				
12EF																																										1	1		1	1		1	
11A																																											1	1		1	1	1	
9AB																																												1		1	1	1	
8B																																													1	1		1	
7ABC																																														1	1	1	1
4C																																															1	1	1
3C																																																1	1
1A																																																	1

Figure 31

obtained by a phase-contrast microscope from approximately 170 investigations of interphase cell nuclei from the salivary gland of *Drosophila*† [11].

†This type of cell is of particular interest to cytologists, since its chromosomes are *polynemic*, containing up to 1000 strands of DNA in cross-section. All these strands arise from an initial one through repetitive replication, without the subsequent separation of the daughter molecules. Thus these chromosomes are much thicker than usual, and are more easily stained and viewed in the interphase.

Of roughly a hundred of the chromosome segments (which were separated cytologically), only those 49 which participated in at least one contact are given in the table.

It must be said that the numbers contained in the initial experimental table specified the number of times the corresponding regions were observed to be in contact. Neighboring sections within individual chromosomes were found to be more frequently in contact. This is probably an artifact connected with the method of preparing the cells: In the destruction of the nucleus evidently the most easily destroyed connections are those between more distant regions of the chromosomes. The sample was sufficiently large that we may assume the basic contacts are determined. Therefore the initial matrix may be reformulated by replacing all the non-null elements with ones (see Fig. 31). A one means that the corresponding sections were observed to be in contact at least once.

To reconstruct the general picture of interchromosomal contacts we must partition the chosen regions into groups according to the similarities of their contacts and search for the structure of the connections among the groups.

If the spatial structure of the genes is random then the connections among the selected sections of the chromosome will surely be arranged in a complex form and will have a very complex macrostructure. On the other hand, if the spatial structure is fixed, the observed collections must consist of relatively isolated groups, corresponding to key knots in the structure.

The results of applying the structure algorithm support the latter supposition. The bounds for the degree of partition were set quite broad. The resulting graph has a complex form if the number m is greater than 11. However, beginning with 11 classes, practically all the regions of the chromosome were united into groups strongly connected within themselves and having no connections with each other (Fig. 32 and Table 7). Subsequent reductions in the number of classes had practically no effect on the form of the graph. Thus the given system actually represents a collection of isolated classes. However, one cannot suppose that the sections of the chromosome falling into one class necessarily form a single knot. We suppose that within the bounds of a particular group there may be several possible conjugations, i.e. in one nucleus A and B are in contact; in another, B and C; and in a third, A and C; but the three do not form a knot, and the real structure has the form A-B, C, or A-C, B, or B-C, A. In such a situation A, B and C may fall into one class as a result of classification.

An important feature of this graph is the presence of a connection between classes 3 and 6 in the absence of intraclass connections, and the presence of a connection between classes 8 and 9 in the absence of internal connections in class 8 (see Fig. 31). Apparently, this fact points out the method of formation of knots in the structure. We examine the simplest example of such a situation. Suppose that sections A, B and C form a knot, with the property that B is always connected

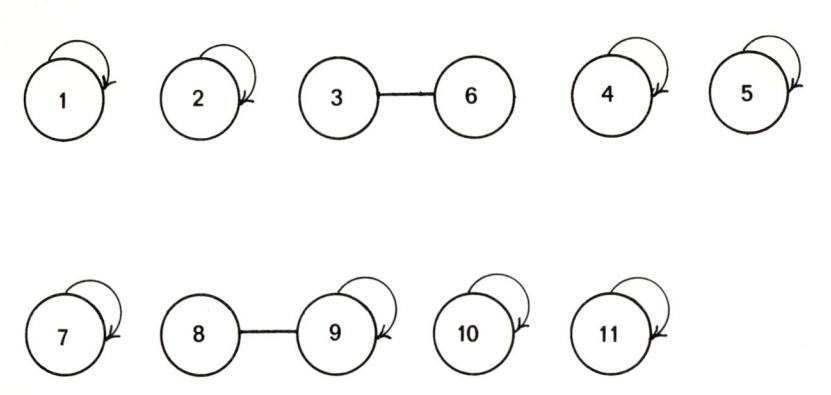

Figure 32. Classes of *Drosophila* chromosome regions according to the similarities of their non-homologous contacts (the result of a partitioning into 11 classes).

with A and C, but A and C are never in contact (A-B-C). Structures of this type have been encountered frequently in analysis. Then A and C are similar to one another, since they are both connected to B, though they do not contact each other. If a partition were given into two classes, then the graph would have this form:

Consequently, graphs of the form described reflect the formation of a knot around some center, in which the sections associated with the center do not directly interact with one another. Thus the structure algorithm permits, in this case, not only the determination of the number of knots and the arrangement of their connections, but also the regularity of interaction of components within a knot.

So the results of analyzing the structure of a contact matrix confirm the presence of a fixed spatial structure in a set of interphase chromosomes.

Recombinations probably occur more frequently within the knots found, since recombinations imply spatial nearness of exchanged genetic sections. The ends of sections involved in chromosomal rearrangements (for example, inversions) must also lie in regions which form knots, since rearrangements arise from recombinations. A further, most important characteristic of these sections is the hypothesis of their later replication, if only the spatial structure of the nucleus is maintained up to the very beginning of nuclear division, since for replication of DNA a linear unwinding of the corresponding chromosome segments is necessary.

TABLE 7
The Contents of the classes corresponding to Fig. 32

1	102F, 101F, 80ABC, 42B
2	100A, 98C, 86D, 65B, 64C
3	94D, 94A, 75C, 73F/74A, 71C, 67D, 67A, 19E,
4	89E, 61A, 60F, 21A, 1A
5	84D, 84A, 83D, 81F
6	70C, 70AB
7	59D, 57AB, 56F, 22AB
8	39E, 39A, 35A, 33A, 18F/19A
9	36D, 35F
10	25EF, 25A, 16D, 12EF, 8B
11	11A, 9AB, 7ABC, 4C, 3C

It turns out that all of these conjectured events actually occur [11]. In the early papers of N. P. Dubinin and V. V. Khvostova [6] it is conjectured that complex chromosomal rearrangements (for example, inversions), which have arisen at the same moment, are the result of recombination exchanges in knots formed by segments which may belong to the same or to different chromosomes. The analysis of matrices constituted on the basis of the distribution of the ends of restructurings would, of course, permit more exact building of the knot structure of the interphase nucleus. However, the data necessary for this are not yet available [11].

Moreover, other, previously unexplained facts are in agreement with this structure, such as the change in the intensity of the recombination process in certain sections of chromosomes under the influence of inversions, occurring in other chromosomes: The majority of the corresponding sections fall in some one of the constructed classes (knots) of the structure.

This structure also partially clears up the following phenomenon, usual for higher organisms. With some exceptions (with which, in all probability, the locus *scute* is associated), genes which constitute a single functional system "working" on a single autonomous trait (for example, biochemical), are frequently located in different chromosomes or are far from one another within the bounds of one chromosome. Can it be that these genes are neighbors in the spatial structure of the interphase nucleus?

These systems have not been studied sufficiently to permit a final answer. However, in *Drosophila*, for a number of known gene systems (a group of genes controlling the development of pigments, a group controlling the structure of various fractions of ribosomal RNA, and others) this is exactly the situation [11].

Thus this example was included in the chapter not just as an application of a formal device: The spatial structure of the interphase nucleus may determine essential features of the functioning of complex systems of "gene-enzymes" in higher organisms.

The whole situation is reminiscent of the interallelic complementation problem discussed in the previous section. In both of these situations the initial material is represented by linear structures (polypeptide chains and chromosomes), but the realization of their inherent functional possibilities occurs after spatial packing, ensuring the proximity of sections responsible for a single function. The thought automatically presents itself that here we have an example in which nature uses the same principles in ensuring the functioning of the genetic system at different levels. At the same time, in the case of interallelic complementation the constructed graphs characterize exactly those semantic connections of the most important functional sections of the protein macromolecule. But it is still early to talk of constructing a proper model of the protein's structure, based on this information. On the other hand, in the second case we are able to show the key knots in the spatial organization of the chromosome as a whole, and on the basis of this are able to attack the description of the semantic features of the system. From the perspective of knowledge of the spatial arrangement of genetic systems in the interphase nucleus one may be led to understand the principles of functional coordination of separate blocks of the genetic program of cells in the process of their individual growth (i.e. ontogenesis).

Chapter 3. Graphs in the Analysis of Gene Evolution

§1. Trees and phylogenetic trees

1. **The notion of a phylogenetic tree**. In the preceding chapters we examined the uses of graphs for analyzing the structure and semantics of "contemporary" genetic texts. However, graph-theoretic methods can also be used for investigating the evolution of these texts (both structural and semantic).

The recent growth of interest in the evolution of biological macromolecules (nucleic acids and proteins) is completely natural. Since molecular events at the level of separate genes (mutations, recombinations etc.) form the basis for evolution in living organisms, the direct study of the former contributes significantly to the understanding of the evolutionary process as a whole. In comparison with classical investigations of evolution in living things, which are based on the comparative analysis of "external" morphological features, the study of evolution at the molecular level relies, as we shall see below, on incomparably more exact initial data. It is well to bear in mind also that changes in morphological features depend, in the final analysis, on elementary changes in genetic texts. It is not without reason that the coding biopolymers (DNA, RNA and proteins) are known as "the molecular documents of evolution" [99].

For natural reasons any attempts to describe the rise and evolution of genetic systems can be based only on the analysis of contemporary genetic texts. It is in the structure of these texts that we will seek traces of their descent. But the significance of such analysis is not exhausted by this. The modeling of the evolutionary processes of genetic systems helps clarify the causes, methods and principles of structural and functional organization in presently existing systems. In what follows we will illustrate a concrete definition of the hackneyed truth that "studying the past, one finds better understanding of the present." This will be accomplished by analyzing the evolutionary path of quaternary protein organization in the family of globins.

The following statement will serve as the basic supporting premise for the study of evolution in genetic systems: Similarity of genetic texts implies their common ancestry. Actually, the potential number of variants of a genetic text of average length (say a hundred to a thousand characters: amino acids and nucleotides) is astronomical. However, a simple calculation will show [30,31,52], that such diversity could not be selected out in a reasonable evolutionary period. Therefore, the independent appearance of similar texts by evolutionary means is extremely improbable. In other words, in nature the chances of randomly meeting two or more identical variants of genetic information is insignificant.

Thus, if two genetic texts contain noticeably similar features this is not by accident, it clearly must come about by divergent evolution of the texts (See page 162). As far as the process of convergence is concerned, it is realized only in very short fragments of genetic text[31].

The beginning point for the investigation of evolution in texts is quantitative estimation of the similarity or difference between biopolymers taken from different species, but which are synonymous in the sense that they fulfill the same molecular function. The distance matrix obtained for the species being considered is used to construct the *phylogenetic tree*, the distinctive genealogy of the given family of synonymous biopolymers, which reflects the path of its, primarily divergent, evolution.

Nucleotide texts have been seldom used for this purpose, since the corresponding molecular data are fragmentary. More frequently, in order to analyze the differences in genetic information the investigator has turned to the secondary data of amino-acid sequences (the primary structure of the polypeptide language).

The comparison of synonymous polypeptide or polynucleotide sequences can be performed by counting the number of symbol-by-smybol differences in homologous positions. The fixing of the fact of coincidence or noncoincidence of symbols in corresponding positions in two genetic texts assumes that replacements of symbols one for another are equally likely. This is reasonable for comparison of nucleotide texts, but for protein sequences it may be applicable only over quite long evolutionary time intervals. Over relatively short time intervals a more adequate hypothesis is the equi-probability of replacements of corresponding nucleotide triplets, which, due to the non-uniform degeneracy of the genetic code, leads to unequal probabilities of replacements of amino acids. As a first approximation this situation is handled by the so-called *minimal mutational distance*: the minimal number of nucleotide replacements required to pass from the codon of one of the amino acids being compared to the codon of the other. The matrix of minimal mutational distances for all 20 amino acids follows:

	1	2	3	4	5	6	7	8	9	10	11	12	13	14	15	16	17	18	19	20
1. Asp	0	2	2	2	1	1	2	1	3	1	1	2	2	2	2	3	2	1	2	1
2. Cys	2	0	2	1	3	2	3	2	3	2	1	2	2	3	3	3	2	1	2	1
3. Thr	2	2	0	2	2	2	1	1	1	1	2	1	2	1	1	2	2	2	1	2
4. Phe	2	1	2	0	3	2	3	2	2	2	1	2	2	2	1	2	1	1	1	2
5. Glu	1	3	2	3	0	2	1	1	2	2	2	2	1	2	2	2	2	1	3	1
6. His	1	2	2	2	2	0	2	2	3	1	1	1	1	1	2	3	1	2	2	2
7. Lys	2	3	1	3	1	2	0	2	1	1	2	1	2	1	1	2	2	2	2	2
8. Ala	1	2	1	2	1	2	2	0	2	2	2	1	2	2	1	2	2	1	2	1
9. Met	3	3	1	2	2	3	1	2	0	2	3	2	2	1	2	2	1	1	1	2
10. Asn	1	2	1	2	2	1	1	2	2	0	1	2	2	2	1	3	2	2	1	2
11. Tyr	1	1	2	1	2	1	2	2	3	1	0	2	1	2	1	2	2	2	2	2
12. Pro	2	2	1	2	2	1	2	1	2	2	2	0	1	1	1	1	1	2	2	2
13. Gln	2	2	2	2	1	1	1	2	2	2	1	1	0	1	1	1	1	2	3	2
14. Arg	2	1	1	2	2	1	1	2	1	2	2	1	1	0	1	1	1	2	2	1
15. Ser	2	1	1	1	2	2	1	2	1	1	1	1	1	1	0	1	1	2	1	1
16. Trp	3	1	2	2	2	3	2	2	2	3	2	3	1	1	1	0	1	2	3	1
17. Leu	2	2	2	1	2	1	2	2	1	2	2	1	1	1	1	1	0	1	1	1
18. Val	1	1	2	1	1	2	2	1	1	2	2	2	2	2	2	2	1	0	1	1
19. Ile	2	2	1	1	3	2	2	2	1	1	2	2	3	2	1	3	1	1	0	2
20. Gly	1	1	2	2	1	2	2	1	2	2	2	2	2	1	1	1	1	1	2	0

The common distance between two polypeptides is obtained by summing the minimal mutational distances for all positions of their primary structure.

It should be said that this distance measure does not begin to take full account of the diversity of mutational differences between polypeptides. In the real process of evolution the principle of the minimality of mutational transitions is scarcely observed in all cases. Besides individual replacements, other kinds of changes dissimilar to them occur, such as the deletion and insertion of nucleotides, as witness the fact that the lengths of a number of synonymous polypeptides are not identical[†]. Therefore the connection of mutational distances with real (paleontologic) times requires some corrections (the corresponding methods are described in [99,31]).

Before proceeding to the precise mathematical formulation of problems associated with constructing phylogenetic trees of proteins, we remark on the appropriateness of considering evolution not only in its syntactic (structural) aspects, as is usually done, but also in its semantic (functional) aspect, since the uniformities of action of natural selection are connected in the first place with the estimation of the

---

[†]To estimate the distance between two polypeptides in this situation, a mutual positioning is sought, for which the number of positions containing identical amino acids is maximized. Discontinuities in the texts are allowed in the process [62].

semantics of genes.

To make this more clear we note that the distance measure defined above reflects only the structural aspect: in calculating the distance between polypeptides all positions are considered to have identical weights. This corresponds to the mutational mechanism, which has no connection with the meaning of the information; so that a mutational replacement may affect any nucleotide of the cistron with equal probability. However it is clear that the fixation of a mutation in a population depends on the position in the primary structure where it arises. Mutations affecting functional centers (the semantic parts of the molecule), are, of course, controlled by more strict selection. Therefore, in calculating the "semantic" distance between proteins, positions within functional centers must be considered to have significantly greater weights than the others.

2. **The metric generated by a tree**. We will consider *weighted trees*, i.e. trees in which each edge ij has an associated numeric "length" $f_{ij} > 0$. In such a tree it is natural to introduce a distance measure $d(i,j)$ between any two vertices as the sum of the lengths of the edges of the unique chain $[i,j]$, which joins them. This distance measure is clearly symmetric, and $d(i,i)$ is 0 by definition. Moreover $d(i,j)$ satisfies the triangle inequality $d(i,j) \leq d(i,k) + d(k,j)$.

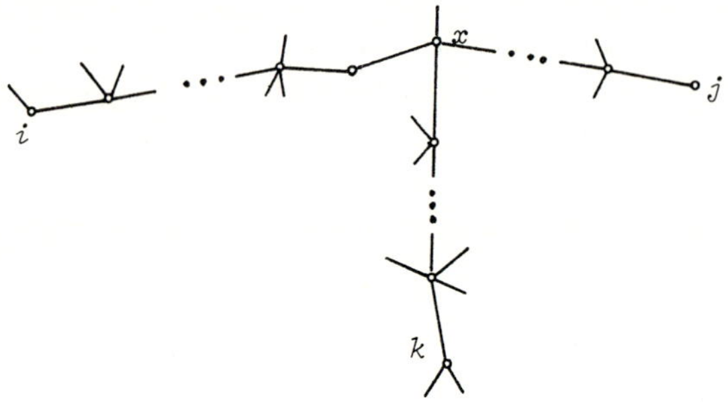

Figure 33

This is evident from examination of Fig. 33, which shows the mutual disposition of any three vertices of a tree. Here x is the first common vertex of the chains $[i,j]$ and $[k,j]$. Clearly, $d(i,k) + d(k,j) = d(i,j) + 2d(k,x)$, so that $d(i,j) \leq d(i,k) + d(k,j)$, equality being obtained if and only if $d(k,x) = 0$, i.e. $k \varepsilon [i,j]$.

From this it also follows that $d(k,x) = \frac{1}{2}[d(k,i)+d(k,j)-d(i,j)]$. Thus a weighted tree gives rise to the metric $d(i,j)$. In particular the system of distances $\|d(i,j)\|_1^N$ between pendant vertices of a tree is a metric.

A vertex of a tree which is adjacent to only one other vertex is said to be *pendant*, one adjacent to two vertices is *communicating*, and one adjacent to three or more vertices is a *branching* vertex. Sometimes non-pendant vertices are said to be *internal*.

The question arises, how completely is the initial tree characterized by its generated matrix $\|d(i,j)\|$ of distances between pendant vertices?

It is known that the metric $\|d(i,j)\|_1^N$ carries within itself information about all of the branching vertices: For, with each branching vertex x one can associate three pendant points i,j,k such that x is the first common vertex of the chains $[i,k]$ and $[j,k]$, with the distances $d(i,x)$, $d(j,x)$, and $d(k,x)$ determinable on the basis only of the given distances $d(i,j)$, $d(i,k)$, and $d(j,k)$, as was already done for $d(k,x)$. Therefore, in every tree which gives rise to the metric $d(i,j)$ of distances between pendant vertices, for these i,j and k there must exist a vertex y, distant from i,j and k by the same amounts as x, and also being the first common vertex of the chains $[i,k]$ and $[j,k]$, and in particular, a branch point.

At the same time, communicating vertices are not uniquely determined by the system $\|d(i,j)\|_1^N$. The matrix $\|d(i,j)\|$ is unchanged by deletion of a communicating vertex x, if the two adjacent edges ax and xb are replaced by a single edge ab with length $f_{ab} = f_{ax} + f_{xb}$. Similarly, any edge ab can be divided into two successive adjacent edges ay and yb with a new communicating vertex y, by defining f_{ay} and f_{yb} arbitrarily, subject to the restriction that

$$f_{ab} = f_{ay} + f_{yb}.$$

Thus a tree realizing the matrix $\|d(i,j)\|_1^N$ (as its matrix of distances between pendant vertices) is determined to within the existence of communicating vertices. We formulate a more complete and exact statement.

Theorem 1. A symmetric matrix $\|d_{ij}\|_1^N$ with zero elements on the main diagonal is realized by a tree with N pendant vertices such that $d(i,j) = d_{ij}$ ($i,j=1,\ldots,N$) if and only if for any $i,j,k,l = 1,\ldots,N$

(1) $d_{ij} < d_{ik} + d_{kj}$;

(2) among the numbers $d_{ij}+d_{kl}$, $d_{ik}+d_{jl}$, and $d_{il}+d_{kj}$ two are identical, and the third does not exceed them.

A tree with the minimal number of edges, which realizes a matrix $\|d_{ij}\|$, is uniquely determined (to within an isomorphism which preserves the numbering of the pendant vertices).

Proof. If L is a tree for which $d(i,j) = d_{ij}$ then condition (1) means that no pendant point belongs to a chain which joins two other pendant points.

The meaning of condition (2) is clear from Fig. 34, where a system of chains is shown which joins the points i, j, k and l. Here x is the last common point of the chains $[i,j]$ and $[i,k]$, and z is the last common point of the chains $[l,j]$, $[l,i]$ and $[l,k]$. It is possible that $x = z$. From the indexing shown we have

$$d(i,j)+d(k,l)=d(i,l)+d(k,j) \geq$$
$$\geq d(i,k)+d(l,j).$$

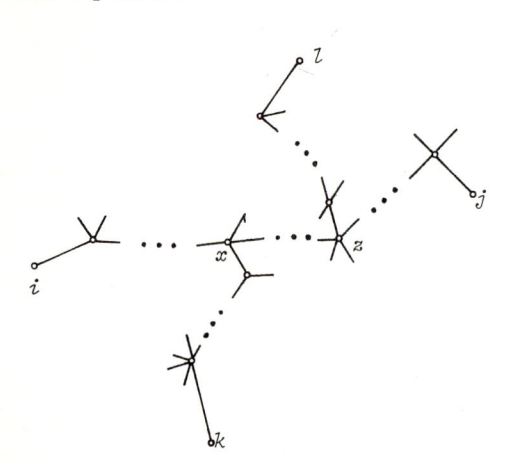

Figure 34

Equality of all three quantities is obtained when $x = z$.

We now prove the reverse assertion: if (1) and (2) hold then there exists a tree having the distances $\|d_{ij}\|$, with its minimal (with respect to the number of its edges) realization being determined to within an isomorphism.

The proof will proceed by induction on N. For $N=2$ conditions (1) and (2) are trivially satisfied, and the matrix $\|d_{ij}\|$ reduces to the single number d_{12}, so that the tree sought is formed by the edge 12 with length d_{12}, with its minimal realization uniquely determined.

Suppose that the statement of the theorem is true for N pendant vertices. We will show that it is true for $N+1$ pendant vertices also.

We construct the tree L_N corresponding to the first N vertices. Consider now the numbers $d_{N+1,i} + d_{N+1,j} - d_{ij}$ ($i,j=1,\ldots,N$), which are positive, from (1). Now we fix the pair (i,j) for which $d_{N+1,i} + d_{N+1,j} - d_{ij}$ is minimal, and consider the three numbers

$$d'_{N+1} = \frac{1}{2}\left(d_{N+1,i} + d_{N+1,j} - d_{ij}\right),$$

$$d'_i = d_{N+1,i} - d'_{N+1} = \frac{1}{2}\left(d_{ij} + d_{N+1,i} - d_{N+1,j}\right),$$

$$d'_j = d_{N+1,j} - d'_{N+1} = \frac{1}{2}\left(d_{ij} + d_{N+1,j} - d_{N+1,i}\right),$$

which are positive, from (1).

In the chain $[i,j]$ of the tree L_N we introduce a new vertex x (by dividing one of the edges into two), in such a way that $d(x,i) = d'_i$ and $d(x,j) = d'_j$ (If in chain $[i,j]$ there is already a vertex x with $d(x,i) = d'_i$ and $d(x,j) = d'_j$ then it is not necessary to introduce a new one). Clearly, such an introduction is correct, since $d(i,x) + d(x,j) = d'_i + d'_j = d_{ij}$ by definition. The tree L_{N+1} is now obtained by adjoining a new pendant vertex $N+1$ to the vertex x by means of an edge whose length is d'_{N+1}.

We will show that in the tree thus constructed $d_{N+1,l} = d(N+1,l)$ for all $l = 1, \ldots, N$. It is clear that

$$d(N+1,i) = d(N+1,x) + d(x,i) = d'_{N+1} + d'_i = d_{N+1,i}$$

$$d(N+1,j) = d(N+1,x) + d(x,j) = d'_{N+1} + d'_j = d_{N+1,j}.$$

Now suppose $l \neq i,j$. By the selection of i and j we have:

$$d_{i,N+1} + d_{j,N+1} - d_{ij} \leq d_{l,N+1} + d_{j,N+1} - d_{lj},$$

so that $d_{lj} + d_{i,N+1} \leq d_{l,N+1} + d_{ij}$.

We now consider the chains $[l,i]$, $[l,j]$ and $[l,N+1]$. Let k be the last common vertex of these chains. By the construction it belongs to either $[i,x]$ or $[j,x]$, since $[N+1,x]$ consists of a single edge. Suppose, say, that k belongs to $[i,x]$, so that

$$d(l,N+1) + d(i,j) = d(i,N+1) + d(l,j).$$

Then $d(i,k) \leq d(i,x)$, so that

$$2d(i,k) = d_{ij} + d_{il} - d_{jl} \leq d_{ij} + d_{i,N+1} - d_{j,N+1} = 2d(i,x).$$

Hence

$$d_{il} + d_{j,N+1} \leq d_{jl} + d_{i,N+1}.$$

Considering the preceding inequality, we obtain

$$d_{il} + d_{j,N+1} \leq d_{jl} + d_{i,N+1} \leq d_{l,N+1} + d_{ij},$$

from which, by condition (2) of the theorem

$$d_{jl} + d_{i,N+1} = d_{l,N+1} + d_{ij}.$$

Hence it follows that

$$d(l,N+1) = d(i,N+1) + d(l,j) - d(i,j) = d_{i,N+1} + d_{lj} - d_{ij} = d_{l,N+1}.$$
Q.E.D.

It remains to show that the tree thus obtained is minimal and is determined to within an isomorphism, which preserves the numbering of pendant vertices. We obtained L_{N+1} from L_N by the addition of no more than two vertices: one of these was the vertex $N+1$, and the other, x, was added to the chain $[i,j]$ with $d(x, N+1) = d_{N+1}$, which had to be contained in L_{N+1} as the last common point of the chains $[N+1, i]$, $[N+1, j]$, in order that the combination $d(N+1, i) + d(N+1, j) - d(i,j)$ have a "fixed" meaning. This proves both the minimality and the uniqueness of L_{N+1}. Q.E.D.

Theorem 1 is a modification of a theorem of Smolenski and Zaretski [7] in the case when the tree is weighted. For results on the representation of metrics by minimal graphs (with respect to the number of edges) of a common form see [8].

We give several corollaries of Theorem 1.

Corollary 1 [93]. A metric $\|d_{ij}\|_1^N$ is generated by a tree if and only if to each subset of four objects their corresponds a submetric which characterizes the tree.

Actually, conditions (1) and (2) of the theorem are formulated for four-element subsets of the initial set, so that it is necessary and sufficient to check their validity for just such subsets.

The next corollary is connected with a notion of considerable importance to us: that of a dendrogram. A *dendrogram* (*tree*) is a rooted tree in which each internal vertex has exactly two edges leading from it to the next level (and one edge leading to it from the preceding level). Thus in a dendrogram all internal vertices besides the root have three "neighbors", while the root is a communicating vertex. The importance of this notion is connected with one of the basic heuristic rules of the theory of macromolecule evolution: in the process of evolution each "antecedent" biopolymer species *diverges* (divides) into two species of "descendants". Therefore the desired phylogenetic tree is in the form of a dendrogram. In fact, divergent evolution need not be dichotomous (See page 160).

Corollary 2. The metric $\|d_{ij}\|_1^N$ is realized by a dendrogram if and only if conditions (1) and (2) of Theorem 1 are satisfied, as well as the following condition (3): for any i,j,k,l, among the numbers $d_{ij}+d_{kl}$, $d_{ik}+d_{jl}$, and $d_{il}+d_{jk}$ there must be differences.

Actually, analysis of Fig. 34 shows that if some vertex of the tree is incident on more than three vertices, then taking as the four vertices i,j,k,l those for which the paths $[x,i]$, $[x,j]$, $[x,k]$ and $[x,l]$ pass through different edges incident on x, we obtain the equality of the mentioned numbers.

We now consider a metric of a special form, for which the construction of a tree is an essentially trivial procedure. The metric $\|d_{ij}\|$ is an *ultrametric* (sometimes called a *Berovski metric* [47]), if it satisfies the stronger inequality: for any i,j,k

$$d_{ij} \leq \max(d_{ik}, d_{kj}).$$

This inequality means that out of the three numbers d_{ij}, d_{ik} and d_{kj}, two coincide, and the third does not exceed their common value. From this it follows that an ultrametric automatically satisfies condition (2) of Theorem 1. We will not prove this fact immediately, rather, we will first make clear the structure of the matrix $\|d_{ij}\|$ for an ultrametric.

Consider the indices i, j for which the value d_{ij} is minimal ($i \neq j$). For any $k \neq i,j$, $d_{ik}=d_{kj}$, since all three numbers d_{ij}, d_{ik} and d_{jk} cannot be pair-wise distinct. Consider further the *a-similarity graph* for $a = d_{ij}$, i.e. that graph G_a, the vertices of which are objects, and which has edges (k,l) only if $d_{kl} \leq a$. From what has been said the relation of d_{ij}-similarity is an equivalence: if $(k,l) \varepsilon G_a$ and $(l,m) \varepsilon G_a$, then $d_{kl} = d_{ij} = d_{lm}$ because of the minimality of d_{ij}. But then $d_{km} = d_{ij}$ by definition of the ultrametric, i.e. $(k,m) \varepsilon G_a$.

The classes of the corresponding partition $R = \{R_1, \ldots, R_m\}$ are "similar" not only in their internal makeup but also externally, i.e. if $k,k' \varepsilon R_s$ and $l,l' \varepsilon R_t$ then $d_{kl}=d_{k'l'}$. If $l=l'$ or $k=k'$ then this equality was proved in the preceding discussion. Now suppose $l \neq l'$, $k \neq k'$; then since $d_{kk'}$ is minimal, $d_{kl}=d_{k'l}$. Similarly, $d_{lk'}=d_{l'k'}$ from the minimality of $d_{ll'}$. Comparing these relationships we obtain the equality $d_{kl}=d_{k'l'}$.

Thus the minimal distance d_{ij} along with the partition $R = \{R_1, \ldots, R_m\}$ for the d_{ij}-similarity graph with distances $\|D_{st}\|_1^m$ between classes, where $D_{st}=D_{kl}$ for $k \varepsilon R_s$, $l \varepsilon R_t$, completely defines the initial ultrametric. This measure $\|D_{st}\|_1^m$ is also clearly an ultrametric, and the preceding construction is applicable to it.

If we designate the various values taken on by d_{ij} ($i,j=1,\ldots,N$) by $d_1 < d_2 < \ldots < d_n$, the d_i-similarity graphs ($i=1,\ldots,n$) characterize the imbedded sequence of partitions R^i ($i=1,\ldots,n$), $R^1 \subseteq R^2 \subseteq \ldots \subseteq R^n$, where R^n is the universal partition consisting of a single class. This system of partitions completely characterizes the rooted tree giving rise to the ultrametric $\|d_{ij}\|$.

The bottom-most (level 0) layer of the tree contains N pendant vertices, corresponding to the initial objects; vertices of layer k ($k=1,\ldots,n$) correspond to classes of the partition R^k; the vertex at layer n is the root. Any vertex i of layer k is directly connected with that vertex j of layer $k+1$ which is associated with the class of the partition R^{k+1} containing the class of R^k to which the given vertex belongs. These classes may coincide, i.e. the vertices may be communicating vertices, and not necessarily branching ones. The length f_{ij} of an edge is defined by the formula

$$f_{ij} = \frac{d_{k+1} - d_k}{2}$$

(with $d_0=0$), so that the lengths of all the edges joining two neighboring layers are identical.

We will now show that this tree gives rise to the initial ultrametric. Suppose $d_{ij} = d_k$. This means that the pendant vertices i and j are "subordinate" to the common vertex x of layer k, which is the last common vertex of the chains joining the root with i and j. Each of the chains $[i,x]$ and $[j,x]$ contains k edges, with total length

$$\frac{d_1}{2} + \frac{d_2-d_1}{2} + \ldots + \frac{d_k-d_{k-1}}{2} = \frac{d_k}{2},$$

so that $d(i,j) = d_k = d_{ij}$, Q.E.D.

Generally speaking, the tree we have obtained is not minimal. The minimization of the number of vertices can be carried out by means of the procedure already described for deleting communicating vertices.

The following theorem has been proved in the process of this construction.

Theorem 2. A metric $\|d_{ij}\|_1^N$ is an ultrametric if and only if it is realized by a rooted tree (possibly having communicating vertices), all pendant vertices of which are at the bottom layer, and in which the lengths of the edges joining vertices of a given layer with those of another are determined only by the two layers and do not depend on specific vertices.

Loosely speaking, an ultrametric corresponds to a variant of protein evolution in which the divergent species accumulate essentially the same number of amino acid replacements in the same period of time, i.e. the speed of evolution of the protein sequences is constant for the various sections of the tree.

Corollary 3. An ultrametric d_{ij} corresponds to a dendrogram if and only if for any i,j and k ($i \neq j \neq k$) at least two of the numbers d_{ij}, d_{jk} and d_{ik} differ from one another.

Actually, the fact that in a rooted tree described by Theorem 2 some vertex is adjacent to three edges joining it with vertices of the next lower level, implies there are at least three pendant vertices i, j,k, "dominated" by this vertex, so that the distances among i, j and k are equal: $d_{ij} = d_{jk} = d_{ik}$.

3. **The construction of dendrograms.** Though Corollaries 2 and 3 characterize dendrograms, it is very seldom that real matrices satisfy their criteria. To construct the dendrogram for an arbitrary matrix in the literature, as a rule the following modified form of the unification algorithm is used.

At step $N-m$ the partition $R^m = \{R_1, \ldots, R_m\}$ and the matrix of average distances

$$d(R_s, R_t) = \frac{1}{N_s N_t} \sum_{i \in R_s} \sum_{j \in R_t} d_{ij}.$$

is examined. At the first step ($m=0$) R^0 is the trivial partition into single-element classes with $d(R_s, R_t) = d_{st}$.

Sometimes, in place of the average distance between objects of classes, the median distance or maximum or minimum distance is used as the distance between classes.

An R_s and R_t are selected for which $d(R_s, R_t)$ is minimal ($s,t=1,\ldots,m$) and then R_s and R_t are unified by the passage from an $m \times m$ matrix $\|d_{ij}\|$ to a new $(m-1) \times (m-1)$ matrix, replacing the two rows (and two columns) corresponding to R_s and R_t by one row (and column) corresponding to $R_s \cup R_t$ according to the formula:

$$d(R_s \cup R_t, R_v) = \frac{N_s d(R_s, R_v) + N_t d(R_t, R_v)}{N_s + N_t} \quad (v \neq s, t).$$

The dendrogram is determined by unifications of classes as branch points.

Consider for example the matrix

	1	2	3	4
1	–	21	8	26
2		–	27	45
3			–	20
4				–

At the first step the closest objects 1 and 3, with $d_{13}=8$ are unified, producing the 3×3 matrix of average distances:

	1,3	2	4
1,3	–	24	23
2		–	45
4			–

Then the classes {1,3} and {4}, and finally there remains only the unification of the two classes {1,3,4} and {2}. These successive unifications generate the following dendrogram (Fig. 35).

Now to determine edge lengths in this dendrogram we use the previously noted fact that in a tree the distance from any pendant vertex i to any "internal" vertex a can be uniquely determined by the indication of those vertices j and k, for which a is the last common vertex of the chains $[i,j]$ and $[i,k]$, according to the formula

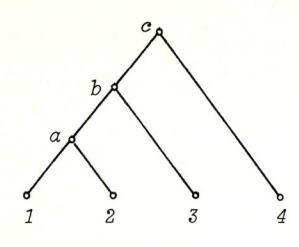

Figure 35

$$d(i,a) = \frac{d(i,j)+d(i,k)-d(j,k)}{2},$$

Now to define the distance between arbitrary internal vertices a and b connected by the edge ab it is sufficient to take any pendant vertex i for which a is contained in the chain $[i,b]$, and define $d(a,b) = d(i,b)-d(i,a)$. However, the result generally depends on i, and therefore it is necessary to average the distance over all pendant i, bearing in mind that some i are connected with b through a and others with a through b. The following procedure is usually used [65,84].

To each internal vertex a (except the root) of the dendrogram there correspond three edges incident on it, and correspondingly, three sets of dependent vertices, according to which of these edges belongs to the chain joining the pendant vertex with a. These sets form a partition $R^a = \{R_1^a, R_2^a, R_3^a\}$ of the set of pendant vertices, which characterizes the vertex a. The root corresponds to a partition into two classes. For example, for the dendrogram of Fig. 35 $R^a = \{\{1\},\{3\},\{2,4\}\}$, $R^b = \{\{1,3\},\{4\},\{2\}\}$, and with the root is associated the partition into two classes: $R^c = \{\{1,3,4\},\{2\}\}$.

For a given vertex a let D_{st} ($s,t=1,2,3$) designate the distance between the classes R_s^a and R_t^a of the partition R^a. The average distances from a to R_1^a, R_2^a, and R_3^a are, respectively $D_1^a = \frac{1}{2}(D_{12}+D_{13}-D_{23})$, $D_2^a = \frac{1}{2}(D_{21}+D_{23}-D_{13})$, $D_3^a = \frac{1}{2}(D_{31}+D_{32}-D_{12})$. If a is the root, then $D_1^a = D_2^a = \frac{D_{12}}{2}$.

The determination of edge lengths is now carried out in successive stages, beginning with some vertex of the lowest level.

For example, for the dendrogram of Fig. 35 we have

$$D_1^a = \frac{8+(21+26)/2-(20+27)/2}{2} = 4,$$

$$D_2^a = \frac{1}{2}(8+(20+27)/2-(21+26)/2 = 4,$$

$$D_3^a = \frac{1}{2}((20+27)/2+(21+26)/2-8) = 19.5.$$

This fixes the lengths of edges $1a$ and $2a$ at 4. Similarly we obtain

$$D_1^b = 1, \qquad D_2^b = 22, \qquad D_3^b = 23.$$

This leads to $f_{ab} = D_a^b - \frac{1}{2}(D_1^a + D_2^a) = -3$, if we calculate from "below", and to the same result $f_{ab} = D_3^a - \frac{1}{2}(D_2^b + D_3^b) = -3$, if the calculation is done from "above." As is evident, this procedure may also lead to negative lines, if, as in the given case, distances are not in agreement with the tree topology. Similarly $f_{4b} = D_2^b = 22$.

For the root c

$$D_1^c = D_2^c = \frac{1}{2} \cdot 31 = 15.5.$$

Therefore from "below" $f_{bc} = D_1^c - \frac{1}{2}(D_1^b + D_2^b) = 4$, and from "above" $f_{bc} = D_1^c - D_2^c = 7.5$, so that "on the average" $f_{bc} = 5.7$. Now the correction

$$f_{2c} = D_3^b - f_{bc} = 23 - 5.7 = 17.3.$$

can be made.

Thus the heuristic procedure described leads to the edge lengths shown in Fig. 36, which give distances between pendant vertices only remotely resembling the initial distances.

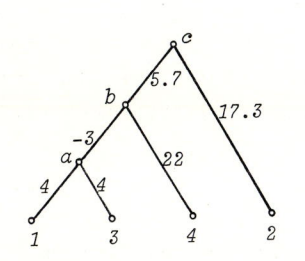

Figure 36

It must be added that the initial data are exactly realized by the tree of Fig. 37,a, which can be transformed into a dendrogram in many ways by the introduction of a new vertex-root between any two neighboring vertices (one of these being shown in Fig. 37,b).

This means that unification may give incorrect results even for a matrix realized by a tree.

We note that negative lengths may appear for edges of a tree not only from inadequacies of the unification algorithm being used, but also for the objective reason that at certain evolutionary levels different, divergent protein sequences may independently register identical amino acid replacements (parallel or *convergent* evolution).

To "correct" the results of the unification algorithm a number of indicators of correspondence between the initial tree and the dendrogram constructed from it have been proposed.

For example, the *Fitch-Margoliash coefficient* [65] measures the average relative squared error of the distances calculated for the dendrogram and the initial distances d_{ij} between objects:

$$\phi = \sqrt{\frac{1}{[N(N-1)/2]-1} \sum_{i<j} \left(\frac{d_{ij}-d(i,j)}{d_{ij}}\right)^2} \, . \qquad (1)$$

Restructuring of a dendrogram for any tree is done in such a way as to minimize ϕ. The more simple, weighted sum of squared differences

$$\chi^2 = \sum_{i<j} \left(\frac{d_{ij}-d(i,j)}{d_{ij}}\right)^2 , \qquad (2)$$

can be used for this since χ^2 has the same monotonic character as ϕ.

The meaning of χ^2 is a squared error estimate of the representation of the numbers d_{ij} by the distances $d(i,j)$, in which the (i,j)th term is smaller, the larger d_{ij} is, other things being equal. In other

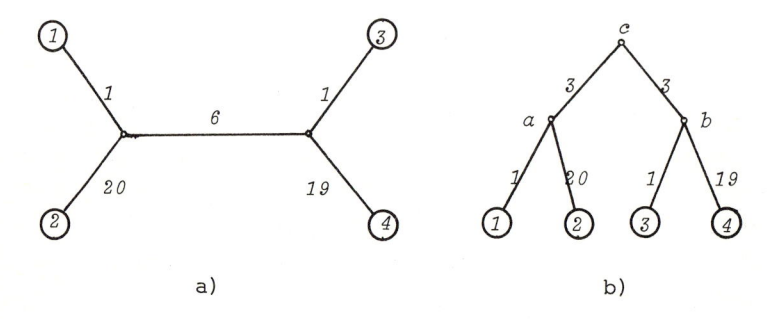

Figure 37

words, the error of approximation of small values of d_{ij} receives greater weight in (2) than that for larger values of d_{ij}. This seems very reasonable, since it corresponds to the accumulation of errors in the course of evolution.

Sums of squares such as (2) play the additional role of aids in constructing a tree for a matrix $\|d_{ij}\|$, and also in determining edge lengths in accord with the idea that the smaller the value of χ^2, the better is the approximation of the matrix $\|d_{ij}\|$. This idea, the method of least squares, was developed in the undergraduate work of A. Zharkiy (Novosibirsk State University, 1977).

The corresponding construction in terms of vectors is as follows.

Suppose a tree has N pendant vertices and M edges numbered 1 to M. The tree obtained from a dendrogram by the removal of the root (the only communicating vertex) has $M = 2N-3$ edges.

The topology of a tree is uniquely characterized by the $\frac{N(N-1)}{2} \times M$ matrices $B = \|b_{ij,k}\|$ (i j, $k=1,\ldots,M$), where $b_{ij,k}$ equals $1/d_{ij}$, if the kth edge belongs to the unique chain connecting the ith and jth pendant vertices. Otherwise $b_{ij,k} = 0$. We designate by f_k the length of the $k\underline{th}$ edge, and the vector of edge lengths $(f_1,\ldots,f_M)$ by f. Then Then the product Bf is the $\frac{N(N-1)}{2}$-element vector of quantities $\frac{d(i,j)}{d_{ij}}$ ($i<j$). Let $\mathbf{1}$ be an $\frac{N(N-1)}{2}$-element vector consisting of all ones. Then the indicator (2) may be expressed in the form of the scalar square

$$\chi^2 = (\mathbf{1}-Bf) \cdot (\mathbf{1}-Bf) \tag{3}$$

of the vector $\mathbf{1}-Bf$. Here, as usual, $x \cdot y$ is the scalar product $\sum_k x_k y_k$ of the vectors $x = (x_k)$ and $y = (y_k)$, such that $x \cdot x = \sum_k x_k^2$.

For a given topology of a tree, i.e. a matrix B, to find the vector $f = (f_1, \ldots, f_M)$ of edge lengths which minimizes the sum of squares (3), one must equate to zero the gradient of (3) as a function of f. In this case, as is usual in the method of least squares, f must be a solution of the system of equations

$$B^T \mathbf{1} = B^T B f, \tag{4}$$

where B^T designates the transpose of B. Written in terms of vector elements (4) becomes

$$\sum_{i<j} b_{ij,k} = \sum_{l=1}^{M} \left(\sum_{i<j} b_{ij,k} b_{ij,l} \right) f_l \quad (k=1,\ldots,M). \tag{5}$$

Another indicator has been proposed by Moore [84]. In the case when a metric is realized in a tree the *Moore coefficient* for each internal vertex a of the derived dendrogram estimates the degree of deviation of the partition R^a from the collection of partitions corresponding to vertices of the "correct" tree.

For arbitrary objects i,j and k let

$$F(i,j,k) = \tfrac{1}{2}(d_{ij}+d_{ik}-d_{jk})$$

be the distance from i to the last common vertex of the chains $[k,i]$ and $[k,j]$, i.e. to the last common ancestor of species i and j. Then:

$$M(a) = \sum_{(i_1,i_2,i_3)} \sum_{i \in R_{i_1}} \sum_{j \in R_{i_2}} \left[\sum_{k \in R_{i_3}} F^2(i,j,k) - \frac{\left(\sum_{k \in R_{i_3}} F(i,j,k)\right)^2}{N_{i_3}} \right]. \tag{6}$$

Here the outer summation is made for all permutations of the classes of the partition R^a, and the term in square brackets measures the dispersion of the distribution of distances $F(i,j,k)$ from i to its last common ancestor with j (depending on k). Clearly, if $\|d_{ij}\|$ is realized by a dendrogram then $M(a)=0$ if and only if R^a is one of the partitions corresponding to the dendrogram [84]. This permits the use of $M(a)$ for directed improvement of the dendrogram for $M(a)>0$.

Consider the smallest neighborhood of a vertex a of the dendrogram given in Fig. 38,a. Here a_1 and a_2 are "offspring" of a, while a and b

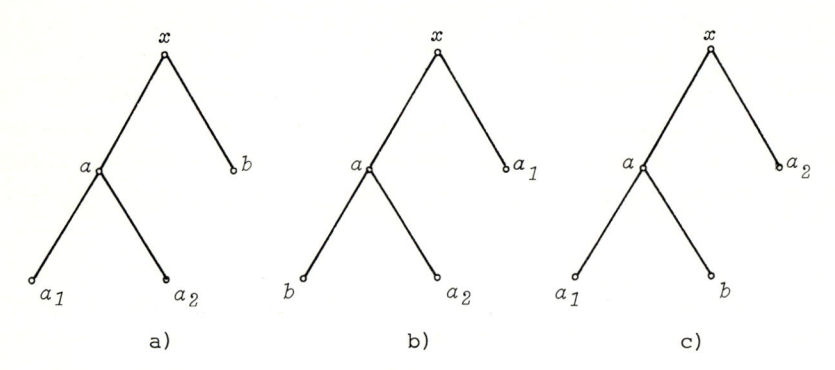

Figure 38

are "offspring" of x. Two methods can be proposed (Fig. 38,b,c) for modifying this local neighborhood while maintaining proper relationships of those parts of the dendrogram attached to a_1, a_2 and b.

An algorithm for successive improvement consists in changing one node at a time (if it decreases $M(a)$) for those nodes with $M(a)>0$. The sum $M = \sum_a M(a)$ acts as an overall indicator of the quality of a restructuring.

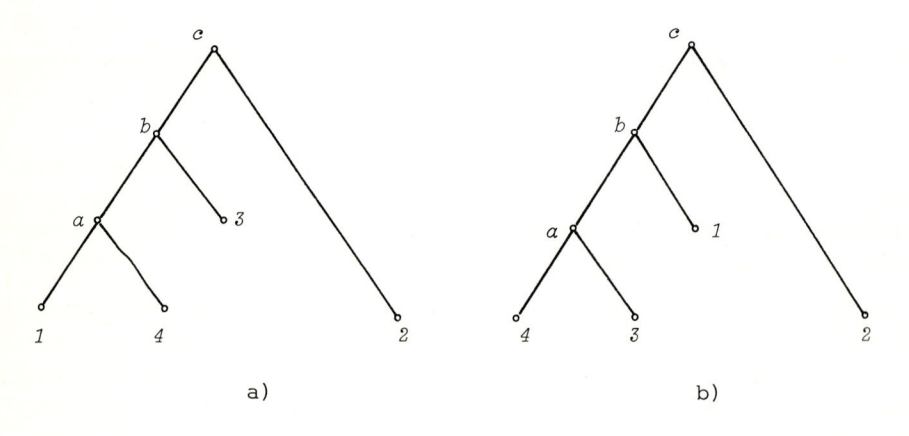

Figure 39

For the tree of Fig. 36 $M(a) = 36$, $M(b) = 18.5$. Consider the two restructurings of the neighborhood of vertex a (Fig. 39), where the Moore coefficient is maximized.

For the variant in Fig. 39,a we have $M(a) = 18$, $M(b) = 36$, so that the total error has not changed. For the dendrogram of Fig. 39,b $M(a) = M(b) = 0$, showing that this dendrogram matches the initial matrix exactly.

Unfortunately the convergence of this algorithm to an exact solution has not been investigated.

Thus the question of the best approximation of a given metric by means of a tree still awaits solution. However in those cases when the speed of fixation of mutations is constant, so that one can expect $\|d_{ij}\|$ to be nearly an ultrametric, the described procedures provide completely "reasonable" solutions (See §2).

4. **Reconstructing the probable structure of ancestral successions.**

The detection of the facts of divergence of species in their order of occurrence by time is not the only use of such trees. For a given dendrogram one can also try to reconstruct the probable structure of the ancestral proteins, on the basis of information about the structure of contemporary protein species corresponding to the pendant vertices of the dendrogram.

For this purpose we make simplifying assumptions, corresponding, to one degree or another, with the real evolutionary process.

First, we assume that for all proteins under consideration one can establish the homologous positions in the primary structure which are the results of mutations having taken place in the same position of the ancestral protein. Second, these homologous positions will be considered independent from one another with respect to evolution, so that successive replacements in a single position (and, correspondingly, in the coding DNA triplet) are not functionally connected with those of other positions.

Thus the problem may be formulated for each separate position of the primary structure of a protein as follows: From information about the codons corresponding to pendant vertices of a dendrogram, reconstruct the codons associated with its internal vertices. In this process, "codon descendants" must resemble as much as possible their "ancestors."

We consider a more precise statment of the problem. Let X be the set of all vertices of a given dendrogram, with $A \subset X$ being the set of pendant vertices. Let Y designate the set of 61 semantically meaningful codons (without the "punctuation marks": UGA, UAA and UAG). With each pendant vertex $i \varepsilon A$ associate a set $F_i \subseteq Y$ of triplets, which code for the amino acid corresponding (in the position being considered) to vertex i.

A single-valued mapping $f: X \to Y$ is *admissible* if for each $i \varepsilon A$ $f(i) \varepsilon F_i$.

The distance $d(y,y')$ between two codons y and y' is the number of corresponding positions in which they differ. The *length of the mapping* $f: X \to Y$ is the total of the distances between $f(a)$ and $f(b)$ for all edges ab in the given dendrogram.

A *characteristic* mapping is a minimal length admissible mapping. The problem of constructing a characteristic mapping is a formalization of the problem of reconstructing ancestral protein sequences. We emphasize that this is a problem of data approximation, not of data representation.

It is evident that if two pendant vertices i and j, adjacent to a "parent" vertex a, have identical amino acids, i.e. $F_i = F_j$, then for a characteristic mapping f $f(i) = f(j) = f(a)$. In fact, if $f(a)$ does not coincide with $f(i) = f(j)$, setting $f(a)$ equal to $f(i) = f(j)$ we obtain a reduction in the length of the dendrogram, independent of the extent to which the initial $f(a)$ differs from the codon of the third vertex adjacent to it. This property permits the replacement of two neighboring pendant vertices by their common ancestor, reducing in the process the tree being considered.

Still another possibility for reducing the calculations lies in the properties of synonymity of the genetic code. In many cases synonymous codons are obtained from one another by the replacement of one of their components. For example, the exchange in the third position of a codon of the component (nucleotide) U for C is always a synonymous transformation. From Table 1 (page 3) it is evident that under further restriction of the amino acids considered, still other possibilities appear for synonymous transformation by a reduction of the number of symbols in the jth component.

It is easy to see that if f is an admissible mapping, then a synonymous transformation in the jth component can only reduce the distance between any neighboring vertices. This means that we can produce at the beginning the synonymous transformations of the sets F_i, reducing their diversity, and then still seek a characteristic mapping.

We now describe a method for building a characteristic mapping. This method associates each internal vertex a with the set of those and only those codons which can be images of a under characteristic mappings. The method was proposed by Fitch [63] for the case when each pendant vertex is characterized by a single codon (all the F_i are single-element sets); and later in [85] it was shown that it is also suitable in the general case.

The construction is carried out in terms of the tree corresponding to the given dendrogram (i.e. obtained by excluding the root and uniting both its adjacent edges into one with their combined length). Fig. 40 shows the tree with given F_i, corresponding to the dendrogram of Fig. 35.

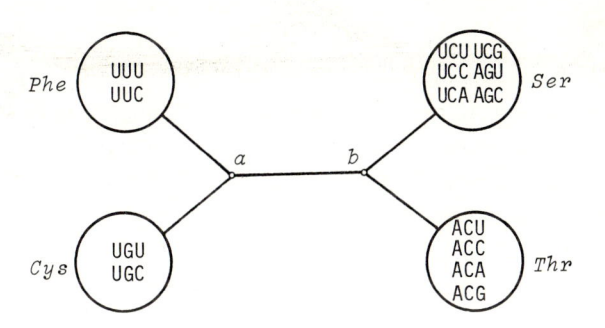

Figure 40

We first produce synonymous transformations of this tree by translating the final characters of all codons to U (in the given case this is possible); obtaining the tree of Fig. 41.

Now each pair a, b of adjacent vertices of the tree determines the mentioned sets $F(a,b)$ of codons, indexed by the numbers 0, 1, 2. This construction is performed inductively, beginning with the pendant vertices. Each pendant vertex i is adjacent to a single internal vertex a; so we set $F(i,a) = F_i$, with the indices of all codons being set equal to zero. Each internal vertex a is adjacent to exactly three vertices b, c and d. If the indexed sets $F(b,a)$ and $F(c,a)$ are already defined then the set $F(a,d)$ is defined as follows. For each pair of codons $b_1 b_2 b_3 \varepsilon F(b,a)$ and $c_1 c_2 c_3 \varepsilon F(c,a)$ all possible codons $a_1 a_2 a_3$ are formed, where a_i is equal to either b_i or c_i ($i=1,2,3$). The index of codon $a_1 a_2 a_3$ is set equal to the sum of the indices of codons $b_1 b_2 b_3$ and $c_1 c_2 c_3$ plus the distance between them. Then in the set thus obtained

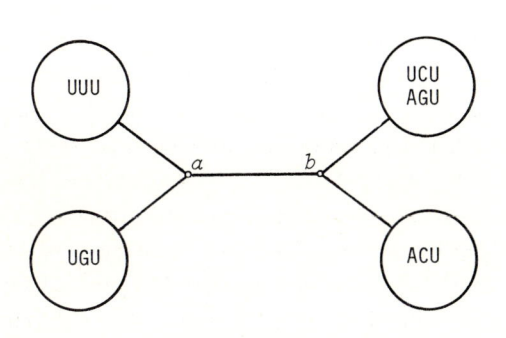

Figure 41

all indices are transformed by subtracting the value of the minimal index, with the subsequent removal of all codons which have a modified index greater than two. If there remain several instances of the same codon in a set, then only the one whose index is minimal is kept. In this way the set $F(a,d)$ is completely determined.

It is known that in general $F(a,b) \neq F(b,a)$. In the example we are considering there are in all only two internal vertices. Clearly $F(a,b) = \{UUU^0, UGU^0\}$, while $F(b,a) = \{ACU^0, AGU^0, UCU^0\}$.

To form the set $F(a)$, the sets $F(b,a)$, $F(c,a)$ and $F(d,a)$ of the three vertices b, c, and d, adjacent to a, are used. For this we consider the set of all possible codons of the form $a_1 a_2 a_3$, where $a_i \in \{b_i, c_i, d_i\}$ for some $b_1 b_2 b_3 \in F(b,a)$, $c_1 c_2 c_3 \in F(c,a)$ and $d_1 d_2 d_3 \in F(d,a)$, with the codon $a_1 a_2 a_3$ being assigned an index equal to the sum of the indices of the initial codons plus the sum of the distances from them to $a_1 a_2 a_3$. The set $F(a)$ is formed from those codons with minimal indices in this set.

In our example $F(a) = \{UUU, UGU, UCU\}$, $F(b) = \{UCU, AGU\}$. Now the meaning of the codons in $F(a)$ becomes clear: a characteristic mapping can be realized only on them. Every other codon "accumulates" a larger distance from pendant vertices. At the same time, for any codon in $F(a)$ one can find a system of codons from the remaining sets $F(b)$, which realizes a characteristic mapping. One can conduct this process in such a way as to remember the "minimal" systems of codons during construction of $F(a,b)$ and $F(a)$. In particular, the further bounding of the excess leads to the following rule: A pair of codons from the sets $F(a)$ and $F(b)$ for neighboring vertices a and b cannot appear in the solution, if the distance between them exceeds the sum of the index of some codon from $F(a,b)$ and the distance between it and a codon taken from $F(b)$.

In our example consider the codon $UUU \in F(a)$. In $F(b,a)$ there are three indexed codons: ACU^0, AGU^0 and UCU^0. Calculating their distances from UUU (plus the indices, which are zero), we obtain 2, 2 and 1, respectively. Consequently together with UUU only UCU from $F(b)$ can be used, since the distance from AGU to UUU exceeds 1. Similarly, only $UCU \in F(b)$ can appear with UCU from $F(a)$, and for $UGU \in F(a)$ the formulated rule does not prohibit any codon from $F(b)$. Thus we have limited the set of possible variants of mappings to four (Fig. 42). Examining the lengths of these mappings shows that they are all equal to three, and consequently Figure 42 shows the complete set of characteristic mappings for the tree of Figure 40.

It is somewhat easier to reconstruct the probable ancestral structure for nucleotide sequences (keeping in mind the possibility of independent replacements at separate positions). A method of solution has been given by Fitch [63] and Hartigan [70a]. This method also consists of two stages.

In the first stage the number of candidates $F(x)$ in a given position is calculated for each vertex of the tree $x \in X$. The values $F(a)$ of a given position of the nucleotide sequence are given for all $a \in A$, where A is the set of pendant vertices. If, for a given $x \in X$, the values $F(y_i)$ for all the direct offspring $y_1, \ldots, y_k$ are already calculated, then $F(x)$ is set equal to the set of all those values (of nucleotides) encountered in the sequence $F(y_1), \ldots, F(y_k)$ with maximal frequency. In the case of a dendrogram $k=2$, so that either $F(x) = F(y_1) \cap F(y_2)$, if $F(y_1) \cap F(y_2) \neq \phi$, or $F(x) = F(y_1) \cup F(y_2)$ for $F(y_1) \cap F(y_2) = \phi$. Thus all the $F(x)$ for $x \in X$ are determined.

The next stage fixes the final form of the characteristic mapping. To do this the set $F(x)$ is modified, beginning at the root. For an arbitrary vertex x and its direct offspring y set $F'(x) = F(y)$ if $F(y) \subseteq F(x)$; otherwise retain in $F(x)$ any one of the elements of the sequence $F(y_1), \ldots, F(y_k)$, $F(y)$, which occurs with maximal frequency, where $y_1, \ldots, y_k$ are the direct offspring of x.

It should be noted that this very algorithm is applicable also to sequences of polypeptides, if one keeps in mind the possibility of independent and equi-probable replacements of individual amino acids by others.

For more realistic situations, when the initial sequences have different lengths and due to deletions and insertions the corresponding positions have not been established beforehand, the problem of reconstructing ancestral structure has been examined in [90, 90a].

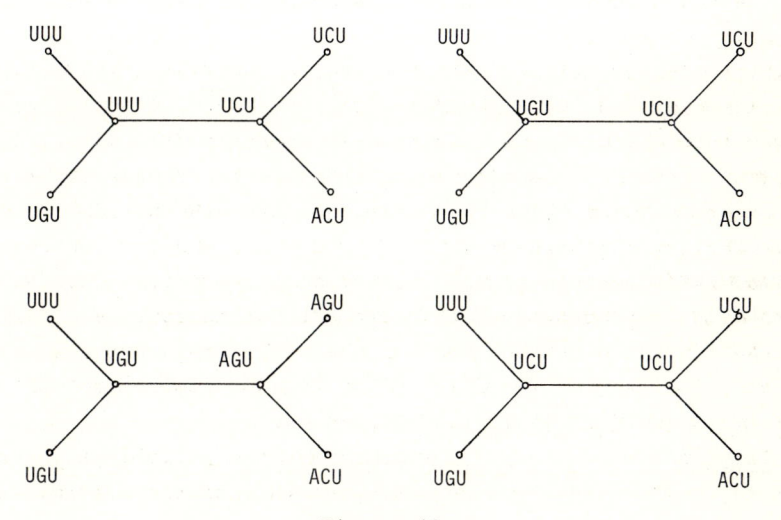

Figure 42

5. **Calculating the internal structure of sequences during tree construction.** The procedure we have described for constructing phylogenetic trees consists of sequential stages: calculation of the matrix of distances between the available genetic texts, formation of the tree topology, estimation of edge lengths, and the reconstruction of ancestral sequences, i.e. sequences corresponding to the internal vertices of the tree. However, considering that the original information available concerns sequences of amino acids or nucleotides, it is natural to try to concomitantly reconstruct both the dendrogram topology and the internal structures of ancestral sequences.

As a measure of the quality of the reconstructed evolution from a set of genetic texts one can use the total number of replacements necessary to generate these texts from a single root sequence. Minimizing this quantity we obtain an evolutionary tree most simply, i.e. by the most parsimonious method which explains the contemporary picture. The corresponding problem is called the problem of maximum parsimony. The principle of obtaining maximal parsimony, which we followed in the preceding section, has long been used in the investigation of the evolution of macromolecules [85,64a,6a].

The solution of the problem of constructing the most parsimonious tree has many interesting properties. We first note those contained in the very formulation of the problem, taking into account that chains of the tree obtained must indicate not only the fact, but the method of transition from one sequence to another, with edge lengths expressing the real differences between corresponding sequences. Consequently, in this problem edges with negative lengths are not allowed. On the other hand, some edges may connect identical sequences, thus having zero length. This possibility of zero-length edges permits the transformation of any tree into a binary tree, i.e. a dendrogram, by the addition of such edges. The optimal tree, then, may be constructed from the beginning as a dendrogram.

We note that for arbitrary contemporary species (i.e. pendant vertices of the tree) i and j the total number of replacements in the chain $[i,j]$ joining them may not be less than the distance d_{ij} between them, since they must be obtained in the tree from a common ancestral sequence by corresponding replacements, and must necessarily accumulate a distance d_{ij} in the process. In these circumstances, then, we always have

$$d_{ij} \leq d(i,j).$$

If in the constructed tree $d(i,j) - d_{ij} > 0$ then in the formation of the sequences i and j there must have appeared not only divergent, but

but also convergent or parallel processes in the evolutionary model represented by the tree. It is these processes that produce an increase in the total number of replacements in comparison with the value d_{ij}, which must have been obtained from only the influence of divergent processes.

We illustrate several, more subtle properties of the task of constructing the most parsimonious tree [6a,64a] in the following pre-arranged example of four nucleotide sequences:

	1	2	3	4	5	6	7	8
1	A	A	A	A	A	G	G	G
2	A	G	G	A	G	A	G	G
3	G	A	G	G	G	A	G	G
4	G	G	A	G	U	U	U	G

The matrix of distances which estimate the number of position-by-position disagreements between the given sequences is:

	1	2	3	4
1	0			
2	4	0		
3	5	3	0	
4	6	6	5	0

At first glance it seems that the sequences 2 and 3 must have a direct common ancestor, as they are the nearest to one another. However, this is not actually so: Their nearness is connected with redundant information, with the presence of those positions in the sequences which are non-essential, i.e. not informative for the construction of the evolutionary tree.

Notice first of all position 8, in which all the sequences contain the nucleotide G: Aside from dependence on tree topology no replacements of this nucleotide are necessary. In the optimal tree it is the same in all sequences, contemporary and ancestral. Clearly, in general, positions in which all the sequences being considered have the same symbol provide no information for construction of the tree.

Positions 5, 6 and 7 present another example of non-informativeness. They are distinguished by the property that in each of them only one symbol appears more than once: in positions 5 and 7 it is G, and in position 6 it is A. For such positions there must be no fewer replacements in the evolutionary tree than there are symbols which appear only once. But these replacements, independently of the topology of the optimal tree, are the most economically organized as follows: In all internal vertices of a given position that element must appear which occurred more than once in the corresponding sequence, unique symbols at pendant vertices being obtained from it by single replacements. Thus these positions provide no information for the formation of the tree.

It is clear that for more than four sequences at least one of the nucleotides will not be unique. With amino acids uniqueness of the symbols in a position is possible only for twenty or fewer sequences. Such a position, containing only uniquely occurring symbols also fails to have any influence on the topology of the tree, since any one of the symbols may be placed at the internal vertices.

The only positions, then, that are helpful in the construction of the optimal tree are those in which two or more symbols occur with a frequency greater than one [6a,64a]. The first four positions in our example are of this type. Removing the four non-informative positions, we have following distance matrix:

	1	2	3	4
1	0			
2	2	0		
3	3	3	0	
4	3	3	2	0

In agreement with this matrix, sequences 1 and 2 have a direct common ancestor, as well as sequences 3 and 4. The optimal tree is shown in Fig. 42a. From the figure it can be seen that use of the distance

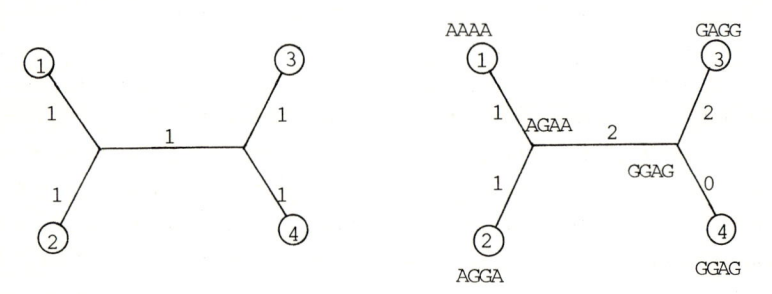

Figure 42a

matrix alone indicates only 5 replacements are necessary, while calculation of symbol content of the sequences shows that fewer than 6 replacements will not suffice.

We see that information about the sequences eases the construction of the optimal tree, but at the same time imposes a lower bound on the number of possible replacements.

A. A. Zharkikh established several additional properties associated with the construction of the most parsimonious trees, which can be recognized from the initial information [6a]. We will formulate several of these.

Identical sequences in an optimal tree have a common direct ancestor with exactly the same sequences attached to it, since this gives the minimal number (zero) of replacements. This property allows the deletion of identical rows, which may lead to an increase in the number of positions having only unique elements, i.e. in the number of non-informative positions. Repeated deletion of identical rows and of non-informative columns in individual cases may significantly reduce the size of the matrix.

The following conditions have a more subtle character, sufficient for joining or, on the other hand, separating individual vertices of a tree.

Let C be a subset of sequences. Then $n(C)$ will designate the number of positions, in each of which all the members of C are identical, but different from any sequence not in C. It can be shown that C forms a weighted subtree, if

$$n(C) > d_{ij}$$

for any $i, j \in C$. In particular, if $C = \{i, j\}$, then i and j have a direct common ancestor in the optimal tree, if they satisfy the inequality

$$d_{ij} > d_i + d_j,$$

where $d_i = \min_{k \neq i,j} d_{ik}$, $d_j = \min_{k \neq i,j} d_{jk}$.

The proof of these properties is based on the fact that no edge of an optimal tree can be longer than the minimum of the distances between pendant vertices separated by that edge. Otherwise one could decrease the total length of the tree by throwing out that edge and joining, by a chain of shorter length, those pendant vertices (pairs) which attain the minimum d_{ij}.

The use of all these properties in the analysis of real data permits a significant reduction in the dimensionality and range of selection of possible variants [6a]. In the case of polypeptide sequences, non-unique optimal reconstructions of the tree and of ancestral proteins can be partially reduced by minimizing differences in the corresponding polynucleotide texts.

§2. The evolution of families of synonymous proteins

1. **The dendrogram of the globins and its analysis**. The methods described above have been used recently for the analysis of a whole series of protein families (See for example [58]). By *protein family* is meant a group of synonymous (iso-functional) polypeptides [31,98].

The *globin family* has been studied the most thoroughly. This family of proteins is responsible at the molecular level for the binding, transfer and release of molecules of oxygen and carbon dioxide. Up to the present more than 200 representatives of this family have been deciphered (from the globins of the mosquito to those of man). Moreover, in several cases (myoglobin of the sperm whale, hemoglobin of horse and man, globin of the worm and others) the spatial organization and chemical functions of these macromolecules is known in detail down to the atomic coordinates. In Section 2.1.2 we already dealt with the possibilities of describing the semantics of hemoglobin molecules.

A few words about this family. Within it there is notable a group of myoglobins, proteins active in muscle tissue. These are connected with the storage and consumption of energy in muscles, the functioning of which requires internal energy reserves. It is the myoglobin molecules that serve as this distinctive oxygen depot. Unlike the hemoglobins, myoglobin chains do not form quaternary structures, but function in the form of separate monomers.

Aside from myoglobin cistrons, in higher mammals there are several cistrons coding for various forms of hemoglobin chains, designated, for example, in man as $\alpha, \beta, \gamma, \delta$ and ε. These chains function in erythrocytes of the blood, in different periods of an individual's growth (*ontogenesis*). The hemoglobin of an adult person consists of 2α- and 2β-chains, as was already noted in Section 2.1.2. However in certain periods of ontogenesis in place of the β-chains in the tetrameric molecule, chains of other hemoglobin cistrons are used: ε-chains are encountered only in the earliest stages of embryonic growth, γ-chains are included in hemoglobin used during most of the interuterine growth, but soon after birth it disappears completely and with this change the embryonic tetramer $2\alpha 2\gamma$ is replaced by adult hemoglobin $2\alpha 2\beta$. Finally, in some postembryonic molecules the δ-chain replaces the β-chain; though the concentration of such proteins is normally very small. Thus one commonality of observed hemoglobin fractions is the presence of products of the α-cistrons. The products of the remaining hemoglobin genes (β-, γ-, and δ-chains) possess remarkable similarity to the α-chain as well

---

†The primary structure of ε-chains has not been completely deciphered, and therefore we will omit them from further discussion.

as among themselves. This gives reason to believe that all known contemporary globin chains arise from a common ancestor.

In recent years impressive success has been obtained in deciphering the structure and organization of the genes which code for globin sequences (basically in mammals: man, rabbits, mice and others), see for example the survey [82a]. It turns out that at the chromosome level the globin genes are concentrated in compact clusters, in which their order of appearance reflects the order of the functioning in the course of ontogenesis. The cluster of β-like cistrons for synthesis of hemoglobins is shown in Fig. 43' (the scale is given in terms of bases and kilobases). First note that the cluster consists of the translatable genes $\varepsilon, G_\gamma, A_\gamma, \delta, \beta$ and the nontranslatable genes $\psi\beta 1, \psi\beta 2$, called pseudogenes. Remarkably, these very pseudogenes divide the fragment into "embryonic" and "postembryonic" sections, including genes active at the embryonic stage, and genes functional only after birth of the organism. In passing, we note that pseudogenes differ from "normal" genes in structure: they have no introns, while usually intron portions of genes constitute a very significant fraction of the whole. For an example see Fig. 43',a, where the intron segments are unshaded. This serves as still another indirect argument in support of the conception of the regulatory role of introns in the reading-out processes.

A special group is constituted by the "respiratory" molecules in insects, lampreys and fish. Limited understanding of the respiratory

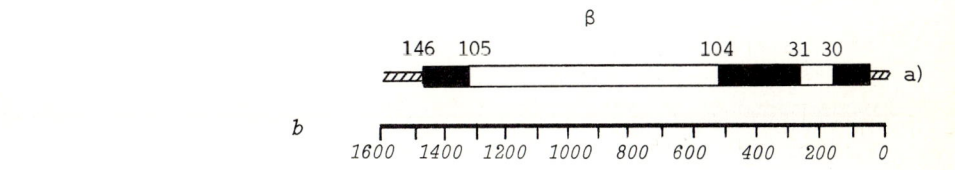

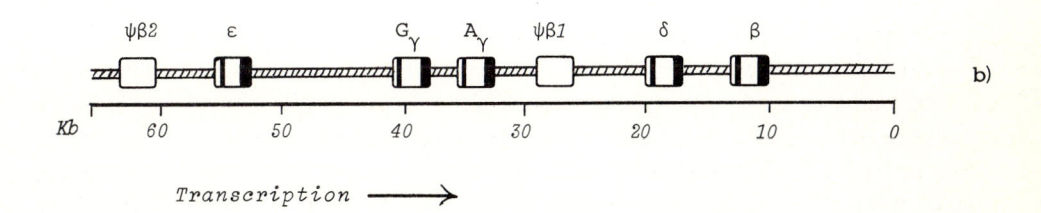

Figure 43'. Structure of human β-hemoglobin gene (a) and linkage arrangement of all human β-like genes in a chromosome. The black and white boxes represent the coding (exon) and non-coding (intron) sequences, respectively. The length of introns is equal to approximately 125-150 and 800-900 base pairs (b), located between codons 30 and 31 and 104 and 105, respectively. The scheme b) shows the positions of the embryonic (ε), fetal (G_γ, A_γ), and adult (δ, β) β-like globin genes and two β-like pseudogenes ($\psi\beta 1, \psi\beta 2$). The length of the entire β-like globin cluster is of 50-60 thousands of base pairs (i.e., 50-60 Kb in the figure).

function of these animals does not permit comparison of the molecules with those groups enumerated above, and so we will call them simply globins.

The collection of primary structures of the globin family which have been studied is sufficiently rich that one can suppose the corresponding dendrogram to reflect all the fundamental stages in the history of this protein.

In Fig. 43 is shown a dendrogram constructed for the matrix of minimal mutational distances of 68 different globin sequences using the unification algorithm [69]. The parts of the tree corresponding to the evolution of α-chains and β-like chains were locally reconstructed using the Moore coefficient [69]. For our purposes the numeric values of edge lengths were non-essential, and they are not included. However, the figure does reflect the relative lengths. Negative lengths were obtained for some edges, and these are shown in the figure with "reverse" directions: upward from the point of branching. Negative edges are few, and their lengths are not great, as one should expect from our previous understanding of the reasons for their appearance.

We first note that the dendrogram of Fig. 43 is in close agreement with classical representations of evolutionary nearness of the species under consideration, which are based on the analysis of external, morphological traits.

We now turn to a more detailed analysis of the branching structure of this tree, in accordance with [30,31]. The symbols d_1, d_2, d_3 and d_4 designate the major stages in the evolution of the globin family. At d_1 the family is divided into two groups: a group of insect globins (1,2) and a group of vertebrate globins (3-68). At d_2 the vertebrates divide into the myoglobins (3,4) and lamprey globins (5), and the group of hemoglobins proper (6-68). At d_3 the latter group is divided into the α-chains and the β-like chains. The β-like chains are divided at d_4 into the group of γ-chains and group of β- and δ-chains of mammals. These branch points can be naturally interpreted as stages, leading to the appearance of new genes coding for new variants of hemoglobin chains. This is not true of d_1, since we still are not sure it represents the appearance of globin genes. The remaining branch points of the tree correspond to divergences of the same type of chain, but in different species.

To clarify the foregoing we note that the basis of cistron divergence is the independent accumulation of mutational replacements during

---

†The alert reader will easily notice the non-uniformity of representation of the various species in the tree. This is due to the fact that the α- and β-chains of mammalian hemoglobin are the most heavily studied.

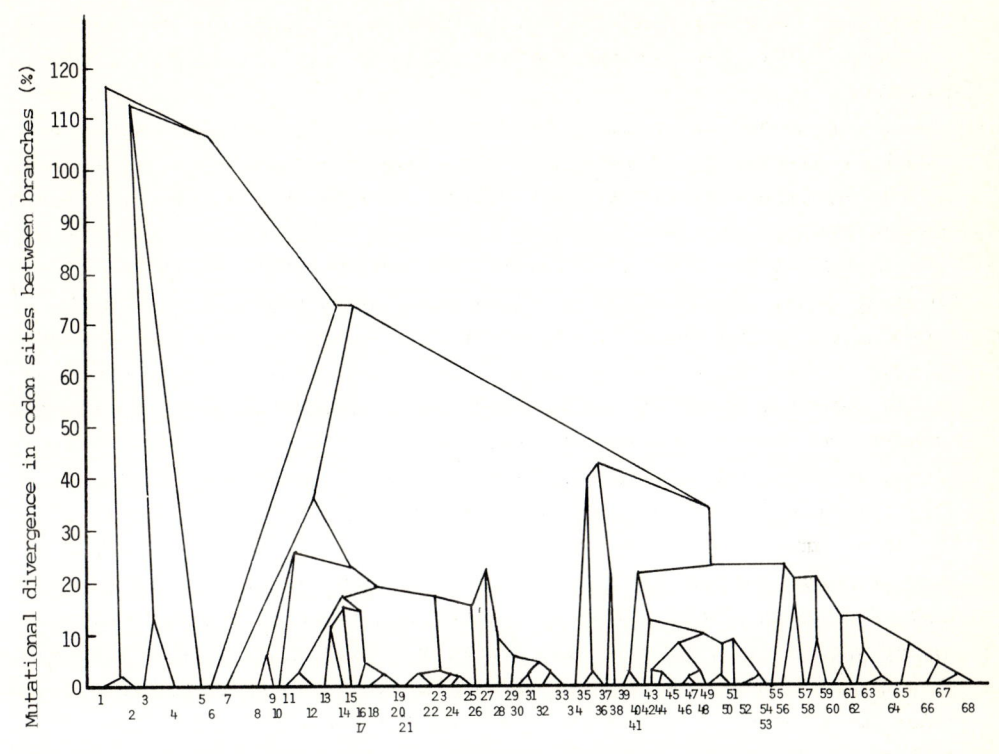

Figure 43. The dichotomous globin tree [69]

1 - globin *Chironomus thummi*; 2 - allelic variant, globin *Chironomus thummi*; 3 - myoglobin sperm whale; 4 - myoglobin horse; 5 - globin lamprey; 6 - globin carp; 7 - α chicken; 8 - α rabbit; 9 - allelic variant α rabbit; 10 - α tree shrew; 11 - α mouse NB; 12 - α mouse C-57 Bl; 13 - α sifaka; 14 - α lemur; 15 - α *Galago crassicandatus*; 16 - α *Macaca mulatta*; 17 - α *Macaca fuscata*; 18 - α gorilla; 19 - α chimpanzee; 20 - α human; 21 - α donkey; 22 - α horse slow; 23 - α horse fast; 24 - α horse fast, allelic variant; 25 - allelic variant α horse slow; 26 - α pig; 27 - α llama; 28 - α bovine; 29 - α goat; 30 - allelic variant α goat; 31 - α goat, fraction A; 32 - α sheep, fraction D; 33 - α sheep, fraction A; 34 - γ human; 35 - β kangaroo; 36 - allelic variant β kangaroo; 37 - β sifaka; 38 - β lemur; 39 - β mouse AKR; 40 - β mouse Sec; 41 - β mouse C-57 Bl; 42 - β rabbit; 43 - β squirrel monkey; 44 - β tamarin; 45 - β *Ateles geoffroyi*; 46 - δ squirrel monkey; 47 - δ tamarin; 48 - δ spider monkey; 49 - β *Macaca fuscata*; 50 - β *Macaca mulatta*; 51 - δ human; 52 - β gorilla; 53 - β chimpanzee; 54 - β human; 55 - β horse, slow fraction; 56 - β pig; 57 - β llama; 58 - β bovine foetal; 59 - β sheep foetal; 60 - β bovine, fraction A; 61 - β bovine, fraction B; 62 - β sheep barbary, fraction C; 63 - β goat, fraction C; 64 - β sheep, fraction C; 65 - β sheep, fraction B; 66 - β sheep, fraction A; 67 - β goat, fraction A; 68 - β goat, fraction A, allelic variant.

evolution. We first examine the so-called *inter-species divergence*, in which hereditary differences are accumulated in descendants of the same gene, separated by an isolating barrier. In this situation the ancestral population is clearly divided into a number of subpopulations, isolated from one another in a genetic sense. Since the subpopulations do not mix there is no exchange of genetic material among them. The isofunctional cistrons of each population independently accumulate various mutations. As a result of this process in contemporary species we find amino acid differences among proteins which perform identical functions. It is known that the number of subpopulations is not necessarily two, so that the notion of a dendrogram (corresponding to dichotomous divergence) is not adequate for the process of interspecies divergence.

Divergence associated with the appearance of new genes (points d_2, d_3 and d_4 in Fig. 43) is called intra-species divergence. A possible mechanism for intraspecies divergence in connection with hemoglobin evolution was first described clearly in the work of V. Ingram [74]. From his study of globin chain similarities in humans he proposed that the appearance of new globin cistrons in organisms is explained by the duplication of old copies obtained by successive, independent evolutionary processes [74, 30, 31]. In this case genetic evolution results from the relative inability of the duplicated genes to recombine, since they are non-homologous.

If the evolutionary process were limited only by interspecies divergence, the appearance of new functions (while retaining the old), which leads to more complex organization, i.e. to *progressive* evolution [27], would be impossible: New functions must be represented in the genetic memory by new genes (a determining, though lesser, role may be played by other mechanisms as well [98]). Duplicate genes may accumulate different mutations, permitting a gradual transition to new functions.

Though gene duplication is an extremely rare phenomenon, rarer still is the appearance of more than two copies of a gene. This is the reason that intra-species divergence is adequate for the description of the dendrogram.

The points d_2, d_3 and d_4 in Fig. 43 correspond to major duplications of pre-cistronic globins. First (d_1) the globins of vertebrates and nonvertebrates diverged (evidently without duplication), then through duplication the myoglobins were separated out (d_2), then the pre-cistron for the hemoglobins diverged, with subsequent divergence of α- and β-chains (d_3); d_4 corresponds to a duplication for the ancestral β-cistron, as a result of which the γ-chains of mammals were isolated.

The question arises, what is the evolutionary meaning of these divergences? Do these hypothetical duplications represent real ages in the history of living forms?

To answer these questions our dendrogram, in which time is measured indirectly by the number of fixed amino-acid replacements, must be connected with real geologic time. An initial premise for such a connection is the postulate that amino acid differences in comparable contemporary proteins are greater, the earlier they diverged. The most simple thing would be to assume that the number of differences is proportional to the time of the independent evolutionary processes of these proteins. Then to estimate the times for all the branch points of the tree it would only be necessary to know the rate of fixation of any single replacement. To determine this rate one can use reliable independent paleontological data concerning the divergence times of some species, i.e. the times since their common ancestor.

However such a calculation does not take into account the important fact that replacements in a single position of a chain can occur (and have occurred) repeatedly. To include this circumstance in our calculations consider the following reasoning [99,30,31]. Let λ be the average number of replacements in a given position per unit of time. Since the appearance of a mutation is an event of small probability, it is natural to assume that the occurrence of repeated replacements in a given position is described by a Poisson distribution, so that the probability of n replacements in time T is given by

$$p_n(T) = \frac{(\lambda T)^n e^{-\lambda T}}{n!}.$$

The probability of no replacements in time T is then

$$p_0(T) = e^{-\lambda T}.$$

The number of coinciding positions in two sufficiently long polypeptide chains which have diverged in time T gives an estimate of this quantity. We designate the fraction of differing positions by Q, so that

$$Q \simeq 1 - p_0(T) = 1 - e^{-\lambda T}.$$

For $\lambda T \ll 1$ the approximating equality

$$Q \simeq \lambda T$$

holds. This formula permits the estimation of λ by means of those chains which have separated comparatively recently (in geologic terms). This is the method used by E. Zuckerkandl and L. Pauling [99]. They found that between the α- and β-chains of hemoglobin in swine, horses, oxen and rabbits, on the one hand, and the corresponding chains in humans, on the other, there are an average of 22 amino acid differences. From paleontologic data it is known that the common ancestor of these mammals lived approximately 80 million years ago, so that $Q = \frac{22}{145} = 0.15$ (145 is the average length of the chains being compared), $T = 8 \cdot 10^7$ years. From this $\lambda \simeq 2 \cdot 10^{-9}$ mutations per year.

Substituting this value of λ in the formula for $p_0(T)$ gives the *Pauling scale* for the translation of mutational differences into units of real time (Fig. 44). The function $Q(T)$ is graphed in this figure, and several points of interest are marked on it [30,31]. The point d_2 ($Q = 0.75$) corresponds to 650 million years, the end of the Pre-Cambrian, when invertebrate sea aminals appeared. Consequently in this period specification of the energetic process in muscles had already occurred, as well as in other tissues (tissue respiration as distinguished from pulmonate). Point d_3 (the average fraction of differences between α- and β-like chains, $Q = 0.52$) corresponds to the Devonian period, about 380 million years ago, when vertebrates moved onto dry land. It was then, apparently, that the cleavage from the former respiratory mechanism occurred -- the transition to oxygen respiration from air by means of lungs. The point d_4 ($Q = 0.26$) corresponds to $T = 150$ million years, the Jurassic period, and the appearance of marsupials. The separation of γ-chains is associated with the singling-out of the embryonic stage in the growth of mammals. The substitution of γ-chains for β-chains in embryonic hemoglobin is evidently connected with the fact that oxygen enters the embryo through the placenta from the mother's blood, and the function of hemoglobin $2α2γ$ is to take oxygen away from the $2α2β$ hemoglobin of the mother.

Thus the evolution of globin structures as a whole, in accord with the tree of Fig. 43, correlates well with phenotypic modifications and restructurings of the respiratory function. Moreover, by the method of constructing such trees one can investigate the real, molecular state of affairs in evolution.

Having satisfied ourselves that the construction of the dendrogram as a whole is adequate, we now turn to a more detailed analysis of the globins in various sections of the dendrogram.

It has been found that the rate of fixation of mutations, calculated for various sections of the protein phylogenetic tree, is nearly constant, even though these sections are associated with strongly dis-

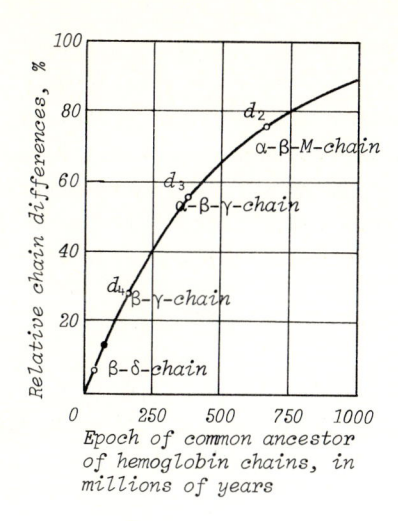

Figure 44

Relative chain differences, % vs *Epoch of common ancestor of hemoglobin chains, in millions of years*

Curve labels: d_2 α-β-M-chain; d_3 α-β-γ-chain; d_4 β-γ-chain; β-δ-chain

tinguished environmental periods and intense morphological evolution. For example the average mutational distance from carp hemoglobin to the α- and β-chains of humans is nearly the same as the distance between these α- and β-chains [76,77,32], so that the fixation rates of mutations in the lines leading to carp hemoglobin and the α- and β-chains of humans from their common ancestor are practically the same. However, as is well known, contemporary carp differs little from this ancient ancestor morphologically (its dwelling conditions have changed little in 400 million years), while in those lines leading to man there have been violent bursts in the generation of forms under the impetus of major variations and sharp changes in environmental conditions. This leads one to the conclusion that molecular changes in globin genes are weakly connected with the progressive evolutionary complexity of organization in living systems: The average rate of fixation of replacements in proteins does not depend on other conditions, such as ecological ones.

This conclusion appears paradoxical, since it contradicts the idea of genetic change as the basis of all evolutionary transformations. The paradox was destroyed by several scientists who hypothesized non-Darwinian gene evolution, according to which, the overwhelming majority of mutations fixed in a population during the course of evolution are non-adaptive, i.e. neutral (they make little change in the fitness of the organism for its surroundings), from which, in the framework of the genetic theory of populations, it is concluded that the average rate of fixation of mutations is constant [76,77,32].

2. **Analyzing the evolution of globin sequences from their internal structure.** The constancy of the average fixation rate of protein replacements, discovered in the analysis of phylogenetic trees constructed on the basis of distance matrices, may be considered an artifact. Within the approach itself lie tendencies toward smoothing of differences in the rate of the evolutionary divergence of species: first, from the transition to summed characteristics in measuring the nearness of sequences in the distance matrix, and second, from the use of the agglomerative unification procedure, which is tightly bound up with

properties of the ultrametric associated with the uniform flow of the evolutionary process.

To clear up this important question in the theory of evolution the alternate approach of Fitch and Zharkikh can be used. In this approach the reconstruction of a phylogenetic tree is based on the direct position-by-position comparison of protein primary structures (see Section 3.1.5). We will describe the results obtained in the paper [34a] by the use of this approach.

Figure 45 shows the phylogenetic tree for 159 globin sequences: 49 α-like, 72 β-like, 36 myoglobin chains and 2 lamprey globins. This collection is much more complete than the one described in the preceding section, but unfortunately it does not encompass the other one completely (missing in particular are the globins of invertebrate animals).

We first compare the structures of the trees in Figs. 43 and 45. Against the background of general agreement a number of differences appear, perhaps the most notable of which is the following: In the new tree the line for the α-like chain of carp is singled out later than the duplication of the ancestor cistron for α- and β-like chains, while in the tree of Fig. 43 this event precedes the duplication point. The variant of Fig. 45 appears better grounded, since apart from all the others the α-like carp chain has a deletion in common with all the other α-like chains of animals, at the time when there is no such deletion in the β-cistrons. The importance of this consideration flows from the fact that the fixation of such large-scale mutations as deletions is a very infrequent event.

A second source of differences stems from the fact that unique replacements in the latter approach are located on pendant edges of the tree, which is also more valid. Thus the two variant α-chains of goat hemoglobin (Fig. 45) differ by 5 unique replacements, all of them located on one branch of the tree, which originates at the direct common ancestor (which therefore coincides with the second of these sequences). Its nearest neighbors, the α-chains of sheep, differ by two or three replacements in all, and, by the logic of algorithms based only on the distance matrix, must be wedged in between the indicated chains of goat hemoglobin, which is patent nonsense. By similar reasoning one can explain several curiosities in the taxonomic arrangement of individual species in trees constructed for other protein families on the basis only of distance matrices. Thus, in the cytochrome tree from [65] the tortoise is closer to birds than to snakes. But if unique replacements are excluded, the sequences of all these species fall together. Most likely, these species diverged from a common ancestor in a non-dichotomous group, but with different velocities: The more rapid accumulation

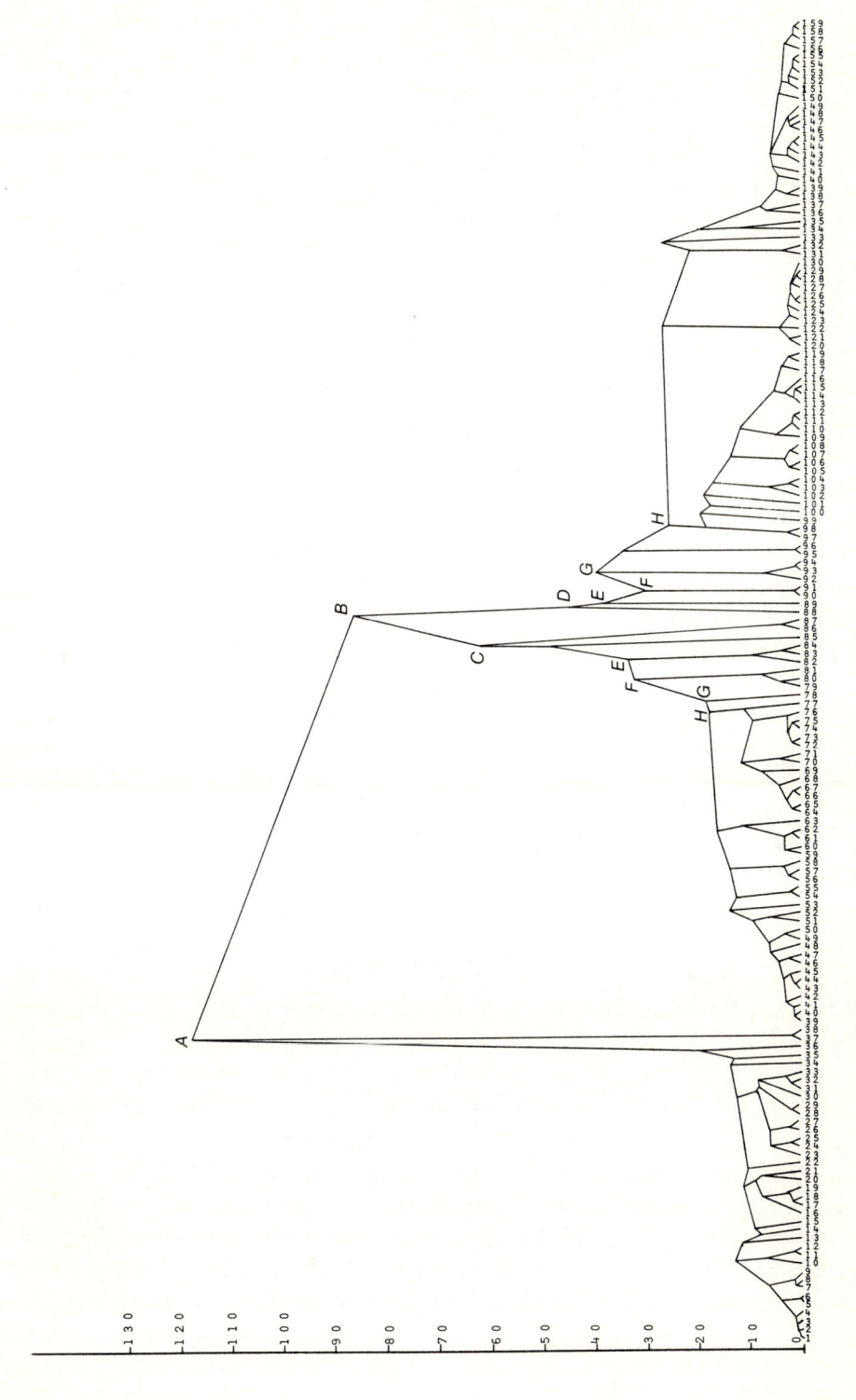

Figure 45. The phylogenetic tree of the globin superfamily constructed by the method of A. A. Zharkikh.

Figure 45 (legend): Designations: A,B,C...- some key points of divergence. The values for rates of substitution were calculated by adding an independent paleontological time scale to the following points: A – the root of the tree (ancestor of vertebrates), C – teleost ancestor, D – tetrapod ancestor (first amphibia), E – amniote ancestor (the point of divergence of birds and mammals), F – mammalian ancestor, H – common ancestor of contemporary mammalian orders; B corresponds to duplication of pre(α – β) cistron, i.e. to divergence of α- and β-hemoglobin chains. Other points mark some important divergences in primate phyletic line.

I. Myoglobins. 1– Human; 2– Chimpanzee, *Pan troglodytes*; 3– Gorilla, *Gorilla gorilla*; 4– Gibbon, *Hylobates agilis*; 5– Olive baboon, *Papio anubis*; 6– Irus macaque, *Macaca fascicularis*; 7– Woolly monkey, *Lagotrix lagotricha*; 8– Squirrel monkey, *Saimiri sciureus*; 9– Common marmoset, *Callithrix jacchus*; 10– Thick tailed galago, *Galago crassicaudatus*; 11– Potto, *Perodicticus potto*; 12– Slow loris, *Nycticebus coucang*; 13– Sportive lemur, *Lepilemur mustelinus*; 14– Common treeshrew, *Tupaia glis*; 15– Hedgehog, *Erinaceus europeaus*; 16– Badger, *Meles meles*; 17– Fox, *Vulpes vulpes*; 18– Domestic dog, *Canis familiaris*; 19– Wild dog, *Lycaon pictus*; 20– California sea lion, *Zalophus californianus*; 21– Harbor seal, *Phoca vitulina*; 22– Rabbit; 23– Sperm whale, *Physeter catodon*; 24– Common dolphin, *Delphinus delphis*; 25– Black Sea dolphin, *Tursiops truncatus*; 26– Amazon River dolphin, *Inia geoffrensis*; 27– Common porpoise, *Phocoena phocoena*; 28– Horse; 29– Zebra; 30– Pig; 31– Sheep; 32– Red deer, *Cervus elaphus*; 33– Bovine; 34– Red kangaroo, *Megaleia rufa*; 35– Opossum, *Didelphis marsupialis*; 36– Chicken, *Gallus gallus*; 37– Lamprey globin, *Lampetra fluviatilis*; 38– Sea lamprey globin, *Petromyzon marinus*.

II. Hemoglobin, alpha-chain. 39– Human; 40– Chimpanzee, *Pan troglodytes*; 41– Gorilla, *Gorilla gorilla*; 42– Hanuman langur, *Presbytis entellus*; 43– Rhesus monkey, *Macaca mulata*; 44– Japanese monkey, *Macaca fuscata fuscata*; 45– Savannah monkey, *Cercopithecus aephips*; 46– Capuchin monkey, *Cebus apella*; 47– Spider monkey, *Ateles geoffroyi*; 48– Tarsius, *Tarsius bancanus*; 49– Bush baby, *Galago crassicaudatus*; 50– Slow loris, *Nycticebus coucang*; 51– Sifaka, *Propithecus verreauxi*; 52– Brown lemur, *Lemur fulvus*; 53– Tree shrew, *Tupaia glis*; 54– Dog, I; 55– Dog, II; 56– Rabbit, I; 57– Rabbit, II; 58– Rabbit, III; 59– Mouse, I; 60– Mouse, II; 61– Mouse, III; 62– Mouse, IV; 63– Rat, *Rattus norvegicus*; 64– Goat, A; 65– Barbary sheep, *Ammotragus lervia*; 66– Sheep, A; 67– Sheep, D; 68– Goat, B; 69– Bovine; 70– Pig, I; 71– Pig, II; 72– Donkey; 73– Horse, I; 74– Horse, II; 75– Horse, III; 76– Horse, IV; 77– Llama, *Lama peruana*; 78– Gray kangaroo, *Macropus giganteus*; 79– Echidna, I, *Tachyglossus aculeatus aculeatus*; 80– Echidna, II; 81– Echidna, III; 82– Goose, *Anser anser*; 83– Chicken, I; 84– Chicken, II; 85– Viper, *Vipera aspis*; 86– Carp, *Cyprinus carpio*; 87– Catostomus clarkii.

III. Hemoglobin, beta-chain. 88– Frog, *Rana esculenta*; 89– Chicken; 90– Echidna, *Tachyglossum aculeatum*; 91 Echidna, II; 92– Potoroo, *Potprous Tridactylus*; 93– Red kangaroo, *Megaleia rufa*; 94– Gray kangaroo, *Macropus giganteus*; 95– Human gamma-chain, I; 96– Human gamma-chain, II; 97– Rabbit, I; 98– Rabbit, II; 99– Horse; 100– Llama, *Lama peruana*; 101– Pig; 102– Bovine, F; 103– Goat, F; 104– Sheep, F; 105– Bovine, D; 106– Bovine, C; 107– Bovine, A; 108– Bovine, B; 109– Barbary sheep, C(NA); 110– Goat, C; 111– Barbary sheep, C; 112– Sheep, C; 113– Goat, E; 114– Goat, D; 115– Goat, A; 116– Sheep, B; 117– Barbary sheep, B; 118– Sheep, A (Soay breed); 119– Sheep, A (Clan breed); 12– Mouse, I; 121– Mouse, II; 122– Mouse, III; 123– Mouse, IV; 124– Mouse, V; 125– Mouse, VI; 126– Mouse, VII; 127– Mouse, VIII; 128– Mouse, IX; 129– Mouse, X; 130– Mouse, XI; 131–

Badger, *Meles meles;* 132- Dog; 133- Treeshrew, *Tupaia glis;* 134- Brown lemur, *Lemur fulvus;* 135- Sifaca, *Propithecus verreauxi;* 136- Bush baby, *Galago crassicaudatus;* 137- Slow loris, *Nyctycebus coucang;* 138- Ramarin, *Saguinus mystax, S.nigricollis;* 139- Capuchin monkey, *Cebus apella;* 140- Night monkey, *Aotus trivirgatus;* 141- Squirrel monkey, *Saimiri sciureus;* 142- Spider monkey, delta-chain, *Ateles geoffroyi;* 143- Night monkey, delta-chain, *Aotus Trivirgatus;* 145- Squirrel monkey, delta-chain, *Saimiri sciureus;* 146- Gibbon, delta-chain, *Hylobates lar;* 147- Gorilla, delta-chain, *Gorilla gorilla;* 148- Chimpanzee, delta-chain, *Pan troglodytes;* 149- Human, delta-chain; 150- Spider monkey, *Ateles geoffroyi;* 151- Hanuman langur, *Presbytis entellus;* 152- Savannah monkey, *Cercopithecus aethiops;* 153- Rhesus monkey, *Macaca mulata;* 154- Japanese monkey, *Macaca fuscata fuscata;* 155- Irus monkey, *Macaca irus;* 156- Gibbon, *Hylobates lar;* 157- Gorilla, *Gorilla gorilla;* 158- Chimpanzee, *Pan troglodytes;* 159 Human.

of unique mutations in snakes led to the mistake in the taxonomy.

Since in Fig. 45 ancestral sequences were constructed simultaneously, so that edge lengths accurately reflect the number of replacements among them, one can try to estimate the rate of the evolutionary process at individual stages of evolution, connected with those points of divergence for which more or less accurate paleontologic data are known. These points are: A - the root of the tree, the common ancestor of vertebrates (about 500 million years ago), C - the common ancestor of fish and land vertebrates (about 400 million years back), D - the common ancestor of tetrapods (the first amphibian, existing approximately 340 million years ago), E - the common ancestor of amniotes, the point of divergence of birds and mammals (about 300 million years ago), F - the common ancestor of mammals (about 150 million years ago), G - the common ancestor of placental mammals (about 120 million years ago) and H - the common ancestor of the principle contemporary order of mammals (about 75 million years back).

On the basis of these data in the paper [34a] the average rates of evolution of the globins were estimated in sections of the tree situated between neighboring pairs of these distinguished points, using the Pauling scale. These rates were calculated not only for the molecules as wholes, but also for the major substructures: the semantic and non-semantic sections, with additional estimates being made for the individual functional centers: the heme-specific center, the centers of contact and the center of binding of 2,3-diphosphoglycerate.

The calculated speeds are far from uniform, some values being as much as 20 times larger than others. The greatest speed was found in the period between 500 and 300 million years ago, when the quaternary structure of hemoglobin was being formed: the contact centers and allosteric centers. The rates of fixation of replacements in the functional centers were found to be several times larger than those in the non-semantic sections. The rates of change were particularly high in the centers which ensure cooperative effects in the functioning of the tetramer. From paleontologic data, it is this period that saw the emergence of land vertebrates and the transition to oxygen respiration from air.

In the interval from 300 to 150 million years ago the evolutionary process of hemoglobins slowed sharply, with the most conservative sections being the centers that ensure cooperative properties (the heme-specific center, the contact α_1-β_2 and the center of binding of 2,3-diphosphoglycerate). Not one mutation became fixed in these centers throughout this period. Finally, in the last 150 million years the fixation rates in various branches of the tree fluctuated less strongly, averaging considerably lower than in the period of emergence of land vertebrates. On the average β-chains evolved more rapidly than α-chains in this later period. A similar result was obtained in [79].

We will consider these results in the light of evolutionary and genetic ideas.

First of all our attention is called to the concurrence of three different processes: the increased rate of accumulation of replacements in the earliest stage of the tree shown in Fig. 45, the formation of quaternary structure in hemoglobin, and the coping with the fundamentally new ecological conditions associated with land dwelling -- the transition to atmospheric respiration of oxygen. The point B marks the duplication of the pre-cistron of α- and β-chains, with their subsequent mutational divergence (the point d_3 of Fig. 43 corresponds to it). This point corresponds to the formation of developed quaternary structure in hemoglobins, since all the earlier divergent globins normally operated in the form of protomers, on the level of tertiary structure. The transformation of globins to quaternary structure at this stage is completely "logical", since it is such structures that bring about cooperative functional effects necessary for living in a high-oxygen environment [32]. It should be remembered that the multimeric structure of hemoglobin provides not only for the cooperative binding of oxygen molecules, but also for their release, under imposed conditions, more easily than with protomers. This feature "rescues" animals from the effects of large changes in partial oxygen pressure [49].

Strictly speaking, the appearance of quaternary structure in globins should be dated somewhat earlier than the point d_3, when identical globin chains were able to aggregate, forming multimers with weak cooperative functional effects. However, the corresponding co-adaptive aggregating variants of globin chains lost one another as a result of meiosis [43]. Duplication of the pre-cistron of the α- and β-chains secured this initial co-adaptivity, which up to then had been incomplete. The duplicated genes were then able to evolve independently. And in fact for more than 300 million years the α- and β-chains of, for example, man have accumulated a significant number of amino-acid differences. But are these differences really independent? It is clear that the contacts $α_1-β_2$ and $α_1-β_1$ which determine, in essence, the quaternary hemoglobin structure, were under the particularly strict "supervision" of natural selection: If serious mutations arose in the centers of contact of the protomers then they became fixed as quickly as possible according to the principle that "a defect corrects a defect" (see Section 2.1.3), which was supported in the paper [36] by the analysis of the structures of contact centers at key branch points of the tree.

The higher rate of evolution in this earlier period can be understood if one considers that the rate of fixation of an arbitrary mutation in a population depends on the degree of fitness of its carriers for environmental conditions. The emergence of land vertebrates represented a sharp change in the ecological environment, and the protomeric globins, evolutionally fitted for a water invironment, lost their value in these fundamentally new circumstances. Among the spontaneously-occurring mutations at this time there must have arisen some fraction of alleles more suited to the new surroundings, i.e. there would be a sharp reduction in the time expected for the appearance of new variants, more preferred than the others. Thus the quickly evolving forms survive, their evolution having a clearly adaptive, "Darwinian" character. In a certain sense this evolution is progressively directed, since it is aimed at adapting to such a long-acting environmental factor as high partial pressure of oxygen. At the molecular level this is manifest in the improvement of quaternary structure of globins and the evolutionary adjustments in their cooperative functions.

The situation is quite different when the quaternary structure is already formed and is coping adequately with its functions. Under these conditions the chances of meeting more preferred variants arising from new mutations is lowered, and a slowing of hemoglobin evolution occurs. From this point the "optimized" quaternary structure of the globins is under the strict control of natural selection, which is now of a stabilizing nature. In this situation neutral or nearly neutral synonymous mutations may be fixed (by means of random genetic drift [31]), being manifest very little in protein structure and functions, and therefore not amenable to natural selection. Such evolution, generally speaking, has no expressed direction and can be characterized by quite large fractions of neutral replacements. For this reason it is called neutral evolution.

The relatively higher rate of evolution of β-like chains in comparison with α-like chains is easily explained by their different functional loads in ontogenesis: α-chains are involved in all hemoglobin fractions, i.e. they are active at all stages of ontogenesis, while the β-like chains $\beta, \gamma, \delta, \epsilon$ interact only with α-chains, not among themselves. In general, the increased diversity of the β-like chains through the associated duplication of ancestral genes (Fig. 43',a) must be accompanied by a quickening of their evolution. But the differences they have accumulated as well as the resolving ability of our methods are not sufficient for a thorough-going analysis of this question. It can be conjectured, however, that beginning with the point of divergence of β- and δ-chains (the formation of the β and δ cistrons), replacements in the α cistron became more strictly evaluated by natural selection (the difference in the rate of entrainment of the positions of the β- and δ-chains grew precisely following this duplication [34a]).

The primary structures of α-, β-, γ- and δ-globins of mammals are sufficiently similar that one can assume homologous positions are involved in their centers of mutual recognition (contact) [30,31]. But then "serious" replacements in the contact centers of the α-chains could hardly become fixed, since for this to happen "simultaneous" co-adaptive mutations would have to occur in nearby positions of β-, γ- and δ-cistrons. The probability of such a coincidence of replacements in several cistrons at once is practically zero.

Thus this evolutionary feature of α-cistrons in mammalian hemoglobin is one more example of global restrictions on the semantic evolution of proteins.

Since Darwin's time a basic problem of the theory of evolution has been the mechanism for the formation of species. Are the changes in globin protein structures connected with the emergence and divergence of species on the morphological level? None of the foregoing bears directly on this question, since it is not connected with the formation of species, as such, but with the increasing complexity of organisms. The fundamental, tetrameric organization of hemoglobins with developed cooperative properties is characteristic, for example, for all contemporary species of mammals from the shrew to man. Can it be that, the "finest hour" of hemoglobin (the adaptive, quickened formation of quaternary structure) having passed, during the next two to three hundred million years it actually evolved under a neutral regimen? One might answer affirmatively, in view of the relatively constant average rates of evolution in various phyletic lines during this period (see page 165).

The real situation, however, is more complex. It may be that the process of formation of species is reflected in molecular texts in ways which are not taken into account in the modeling of intrapopulation allele dynamics [76,77].

If constancy is observed in the fixation rate of mutations, it can be attributed to the fact that for tens and hundreds of millions of years, in the absence of serious functional changes from mutations, all the potentially possible positions of the ancestral proteins were included. But the phylogenetic tree is constructed only from observed positional differences, without consideration of the real number of mutations. Even for the most parsimonious tree with respect to the total number of mutations, shown in Fig. 45, in [34a] it is shown that in those globin positions which have varied during the past 300 million years each amino-acid variant has been fixed on the order of two or three times. This is evidence that, for a given protein, the possibilities for the evolutionary selection of amino-acid replacements have to a significant degree been exhausted [34a].

The explanation of the "neutralist paradox" should be sought in the real contents of those positions which have changed, i.e. in the character of the semantic evolution of genes.

In the process of comparing the accumulated mutations in any two lines leading to contemporary species "a" and "b", all those positions which have changed in comparison with the probable ancestral structure can be divided into three groups: 1) positions in which a replacement occurred in line "a" but not in line "b"; 2) positions in which "b" changed, but not "a", and 3) positions in which replacements were fixed in both lines.

It turns out that for the globins of carp and the α-chain of humans, among the positions of the third group there is not one that is involved in a functional center, while in the first two groups such "semantic" positions constitute more than half. Similar results are obtained for any two taxa (at the level of orders or above) which were distributed sufficiently long ago in the evolutionary process. For species which were distributed more recently, as for example, primates, this regularity is not observable, probably due to the too short period of their independent evolution.

Thus the rates of fixation of mutations in various sections of the tree may, on the average, be nearly identical, although the history of each globin chain has its own, frequently unique features, determined by the evolution of functional, semantic positions. Neutral mutations, known to be synonomous, are fixed in "external" noncentral positions.

Thus even in connection with a given gene which is surely not a key one in processes of species formation we clearly fix the specific nature of the species' differences, even on the molecular level, in the semantic section of the molecule. This indicates that the real arena of neutral evolution is much narrower than one would suppose on the basis of the usual constructions.

As a whole, the example of hemoglobins shows that the process of progressive evolution of forms in the direction of more complicated organization is reflected even on the molecular level: The evolution of globins clearly was directed toward ever greater specification of sections of the primary structure (the appearance of new functional centers), i.e. the distinctive enrichment of the semantic content of genetic information within the bounds of the elementary semantic units of genetic systems, the cistrons.

Of course one should bear in mind that the evolution of land organisms is an extremely complex process, depending on many factors; so that it is doubtful whether it took the same course as the hemoglobin system (even on the molecular level). Nevertheless, changes in the hemoglobin genes themselves indisputably played a notable role in it.

Thus we see that the construction of dendrograms for specific families of proteins provides a number of non-trivial pieces of information about the course of macromolecular evolution.

At the present time there is in progress intensive accumulation of factual material, which will permit the consideration of the phylogenetic nearness of various types of proteins. For example, Yčas [100], having established the probable ancestral structures for the most diverse families of proteins (globins, nucleases, lysozymes, cytochromes, etc.), compared them, and was able to identify related families of proteins. This permitted him to conjecture that primordial ancestral amino-acid sequences evolved in different tertiary spatial structures.

Besides the constantly increasing list of deciphered primary structures of the most diverse proteins, information about the structure of polynucleotide texts, tRNA in particular, is multiplying rapidly. The phylogenetic analysis of tRNA is of considerable interest, the goal being to describe the very earliest stages of evolution of genetic systems, since tRNA is a most important part of the translation apparatus, common to all living things. There is ample basis to consider that the noted features of similarity in the structure of tRNA of various fractions from different species shows the commonality of their origin. Therefore, the construction of the tree for tRNA may provide interesting information about the evolution of the genetic code and the translation apparatus [31,52].

In recent years significant progress has been made in deciphering the spatial semantic structure of macromolecules. It can be conjectured that quantitative estimation of the similarity of tertiary structures will permit significantly more complete and specific descriptions of the semantic evolution of genetic texts in the future. It is clear that, as in the analysis of primary structures, the methods of graph theory will play a leading role in this.

Epilogue: Cryptographic Problems in Genetics

From the contents of this book it is clear that the genetic method of investigation is the reconstruction of the structure and properties of the whole by studying its fragments, i.e. it is by its very nature cryptographic.

Cryptographic problems arise in studying any layer of the genetic language. We will recall several of these, both ones which have been solved and some that still await solution.

The first, and most impressive, was the deciphering of the genetic code. This problem was finally solved by direct biochemical experiments. However a problem arose at one stage of the deciphering process: Given the composition of coded words (nucleotide triplets), to find the order of the characters (nucleotides) in them.

One solution was proposed by V A. Ratner. Comparing mutations involving nucleotide replacements in codons and the results of intracodon recombinations, he was able to establish the order of the nucleotides within codons [29].

At the level of cistrons, the units of translation, at least two cryptographic problems present themselves.

The first of these is the proper reconstruction of the primary structure of cistrons and of the proteins coded by them. There are subtle chemical methods of establishing the nucleotide and amino-acid composition of short fragments of genetic text (5 - 10 nucleotides or amino acids). To establish the structure of longer texts the following method is used. By means of specific proteins (i.e. enzymes such as nucleases and proteinases) the chain being studied is broken down into short, separate fragments. The basic "cryptographic" difficulty is that these fragments are not broken off one after another in sequence, so that to find their ordering it is necessary to repeat the fragmentation process of the initial text, using an enzyme with different specificity. From the resulting large collection of overlapping fragments, it is not difficult to find the desired ordering by hand.

Using this method, up to the present time the primary structures of several hundreds of proteins have been deciphered [58]. However, this is an insignificant fraction of the total number in existence. Moreover the selection of the proteins to be analyzed is usually made on the basis of their amenability to biochemical study, which has little connection with their "value" in a genetic sense.

Frequently genetically interesting texts are not amenable to direct biochemical investigation. However geneticists have available an indirect method for establishing the primary structure of genes and the proteins coded by them, through the construction of recombination maps. In this method individual positions of the primary structure are represented by mutations, and correspondingly the recombination ordering of mutations characterizes the order of these positions in the primary protein structure. Recombination mapping of genes is fundamental to the study of their structure.

With scriptons, we have shown in §1.5 that both recombination and complementation mapping can be used in the reconstruction of their structure.

The problems we have mentioned are associated with the linear organization of genetic systems. However, the most important macromolecular objects of cells are in their very nature three dimensional. Even with individual cistrons, to actualize the information they contain, the linearity of the primary structures is transformed into complex spatial protein structures which are responsible for the functioning of the system. The study of these three-dimensional structures may also be considered as a cryptographic problem, since here also the methods of analysis (both biochemical and genetic) come down to the establishment of the structure and properties of the "whole" by comparing various of its "fragments." The classic example of this is the reconstruction of three-dimensional molecular protein structure from two-dimensional X-ray data (by a map of the electron density of the protein crystals).

Unfortunately, this method is very cumbersome, its basic difficulty being, perhaps, at the first stage, in the obtaining of protein crystals which still have their "active" structure. Therefore the deciphering of the "*physical code*" itself, i.e. the rules of automatic transition from primary to tertiary protein structure, is extremely important[18,26].

Another possible method in the study of proteins is the very significant complementation analysis, the resolving power of which is considered by geneticists to be much weaker than that of recombination mapping. This is understandable, since the complementation test was applied primarily as an express method for determining functional genetic units, and, as it turned out, little could be said about the internal organization of these units. In fact, as we have tried to show in the second chapter, the possibilities associated with this test are considerable. The data obtained from it can be used for quite detailed analysis of functional (semantic) protein organization.

It is important to realize that when a geneticist begins an experiment, as a rule he does not know with what level of complexity of the genetic system he is dealing: individual cistrons, scriptons, operons or more extensive fragments of the genome. Moreover, the estimation of the number of cistrons in a locus is even one of the central problems he must solve.

With this in mind, it is advisable to begin the analysis with the investigation of the macrostructure of the complementation matrix. If it has a clearly discernable linear component, then this is evidence of structural linearity in the system being studied.

It must be said that this approach reflects, in essence, the conjecture already proposed within the framework of classical genetics, that these linear, invariant components of our constructions are images of the real structural topography of genetic loci. The striking ability of genetic systems to retain the linear features of their structural organization throughout a long and tangled history, set the stage for the remarkable achievement of classical genetics in deciphering the structure of the genetic material. Perhaps the most impressive example of this stability of the primary action of genes is the *scute* locus in *Drosophila* (see Section 1.5.3).

The cryptographic problems we have listed are related to the very lowest levels of organization of genetic materials in cells. Of no less interest (and probably of no less complexity) are problems of this type connected with the "top levels", the structural and functional organization of whole blocks of the hereditary program, their grammar and the rules for developing the program in the process of ontogenesis. In this regard, the functional connections of blocks as well as their scheme of coordination clearly have a complex, non-linear character.

For example, by means of "cryptographic" methods of graph theory one can study the "peripheral" portions of intracellular processes, those ramified and sometimes very complex networks of chemical transformation known as the *metabolic processes*. It is useful to mark individual stages in the metabolism process in biochemical genetics with mutations (through changes in the activity of the associated enzymes). Special tests have been worked out (for example, the test for syntrophism or cross-feeding [30,48]), which permit the characterization of paired interactions of mutants. With the matrix of such local connections in hand one can try to reconstruct the whole chain of metabolic reactions.

In addition, already clearly visible is the tempting possibility of applying the language and methods of graph theory in deciphering the genetic control system for ontogenesis. The genetic control mechanisms for ontogenesis are well understood at present only for some viruses (the phages λ, T4, etc.). From graph theory one can expect to construct original at present still "non-structural", but "temporary" portraits characterizing the interactions of the various genes during growth. But the statement of this fundamental problem is a cryptographic one: to construct, from "fragments" (mutational changes in certain stages of a process), the "whole", i.e. the whole of ontogenesis or the individual sequences of ontogenetic events.

Finally, one of the important cryptographic tasks of genetics, for a while yet, evidently, will remain the problem of determining the functional and spatial organization of the entire genome, to which we gave attention in §2.2.

These examples argue persuasively that the methods of graph theory can and must play a notable role in the solution of the most diverse problems of genetics.

Appendix: Some Notions About Graphs

SOME NOTIONS ABOUT GRAPHS

A collection $G = (A, V, \Pi)$ is a *graph* if A is a set of *vertices*, V is a set of links and Π is a (not necessarily everywhere defined) mapping $\Pi: A \times A \rightarrow V$, which associates with each pair (a,b), for which it is defined, an element $v = \Pi(a,b)$ of the set V — a *link* which connects the vertices a and b in such a way that v does not connect any other pair of vertices, i.e. $v = \Pi(x,y) \rightarrow \{x,y\} = \{a,b\}$. The mapping Π divides V in a natural way into three parts: V_e — the set of those links $v = \Pi(a,b)$ which join both a with b and b with a, i.e. $v = \Pi(a,b)$ and $v = \Pi(b,a)$; V_a — the set of links $v = \Pi(a,b)$ for which $v \neq \Pi(b,a)$; and V_l — the set of links $v = \Pi(a,b)$ for which $a = b$.

V_a is called the set of *arcs*, V_e the set of *edges*, and V_l the set of *loops*. Each arc, edge or loop is uniquely characterized by the vertices a and b which it connects. In the sequel we will designate an arc (edge) connecting a with b by (a,b) (or more briefly ab or ba, respectively), and a loop at the vertex a by (a,a).

On paper a graph is usually represented as a *diagram*: with *vertices* (elements of A) represented by points, and links $\Pi(a,b)$ by lines connecting the corresponding points a and b. In such diagrams an arc $\Pi(a,b)$ is represented as an arrow from a to b.

For example, the diagram of Fig. 46 corresponds to the graph with $A = \{1, 2, 3, 4\}$, $V = \{v_1, v_2, v_3, v_4, v_5, v_6\}$, in which $\Pi(1,2) = \Pi(2,1) = v_1$, $\Pi(2,3) = v_2$, $\Pi(1,4) = v_3$, $\Pi(4,1) = v_6$, $\Pi(3,4) = \Pi(4,3) = v_4$, and $\Pi(4,4) = v_5$.

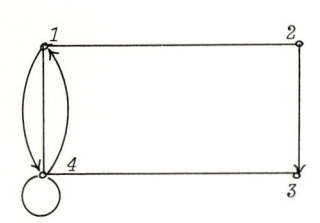

Figure 46

Here v_1 and v_4 are edges, v_2, v_3, v_6 are arcs and v_5 is a loop. If $\Pi(a,b)$ is a link, then it is *incident on* the vertices a and b, and such vertices are said to be *adjacent*.

A sequence $i_1, i_1 v_1 i_2, i_2 v_2 i_3, \ldots, i_k v_k i_{k+1}$ such that $v_j = \Pi(i_j, i_{j+1})$ for all $j = 1, \ldots, k$ is called a *route* of length k from vertex i_1 to vertex i_{k+1}. In the case when each link of a route is an arc the route is frequently called a *path*, and if all the links are edges it is called a *chain*. Sometimes these conventions are extended to include the case when some of the links are loops, in order to allow the possible repetition of one or more vertices. Agreeable with the above definitions, an individual vertex forms a path whether or not there is a loop at that vertex.

A route (chain, path) is *non-repeating* if all its links are distinct.

The length of the shortest route joining vertices i and j is called the *distance* $d(i,j)$ between i and j. The *radius* of a graph is designated by the number $\min_i \max_j d(i,j)$.

A route $i_1, i_1v_1i_2, \ldots, i_kv_ki_{k+1}$ is called *cyclic* if $i_1 = i_{k+1}$; a cyclic path is called a *contour*; a cyclic non-repeating chain is called a *cycle*.

An *oriented graph* has no edges.

In order to specify an oriented graph it is sufficient to designate those ordered pairs of vertices (a,b) for which Π is defined; an arc $\Pi(a,b)$ being completely specified by the pair (a,b). Consequently an oriented graph is specified by the collection of ordered pairs $(a,b) \in A \times A$ which correspond to its arcs, loops being specified by pairs of the form $(a,a) \in A \times A$. In other words an oriented graph is completely defined by a subset (*binary relation*) $R \subseteq A \times A$ of pairs corresponding to its arcs. Thus in dealing with oriented graphs it is possible and frequently suitable to use the set-theoretic language of binary relations.

Another way to specify oriented graphs is by using *Boolean matrices* — matrices containing only zeros and ones. With a given relation (oriented graph) R is associated a Boolean $N \times N$ matrix $r = \|r_{ij}\|$ of the following form:

$$r_{ij} = \begin{cases} 1, & \text{if } (i,j) \in R, \\ 0, & \text{if } (i,j) \notin R. \end{cases}$$

Clearly such a correspondence of relations and matrices is one-to-one: The relation $R_r = \{(i,j) \mid r_{ij} = 1\}$, on the generated matrix r coinciding with the initial relation R. The matrix r is called the *relation matrix* or *adjacency matrix* of R.

There are, then, three means of specifying oriented graphs (diagrams, relations, Boolean matrices) with extremely simple rules for translating from one set of terms to another. A set of facts can be made more clear by comparing them in these differing terminologies.

The relation $\{(1,2), (3,4), (2,2), (2,3), (4,1)\}$ is equivalent to the diagram of Fig. 47 and the matrix

$$\begin{vmatrix} 0 & 1 & 0 & 0 \\ 0 & 1 & 1 & 0 \\ 0 & 0 & 0 & 1 \\ 1 & 0 & 0 & 0 \end{vmatrix}.$$

Also of interest besides oriented graphs are *regular graphs*: graphs whose links are all edges. Sometimes this term is used for non-oriented graphs without loops. Every regular graph can be represented as an oriented graph with the replacement of each edge by two arcs (a,b) and (b,a). Therefore the study of regular graphs may be replaced by the study of corresponding oriented graphs, the "translation" from the language of oriented graphs to that of regular graphs and vice versa being a trivial exercise.

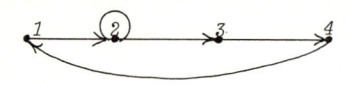

Figure

Thus regular graphs may be represented in terms of relations or Boolean matrices. It is evident that a Boolean matrix corresponds to a regular graph if and only if it is symmetric.

In this book, in the absence of particular stipulations and when it will not be misleading, we will use whichever of these notations (diagrammatic, relational, Boolean-matric) seems appropriate.

For an arbitrary relation $R \subseteq A \times A$, $R{<}a{>}$ will represent a set, frequently called the *image of the object a with respect to the relation R*, or *the cut (section) of the relation R with respect to the object a*, and defined by

$$R{<}a{>} = \{b \mid (a,b) \in R\}.$$

In terms of diagrams the set $R{<}a{>}$ consists of those and only those vertices which are the terminal points of arcs originating at a.

The graph obtained from R by reversing the orientation of its arcs (replacing pairs (a,b) with pairs (b,a)) is said to be *inverse to R* and is designated by $R^{-1} = \{(a,b) \mid (b,a) \in R\}$. Clearly the matrix of an inverted graph is the transpose of the matrix of the initial graph, i.e. rows are interchanged with columns, the elements r_{ij} being interchanged with r_{ji}.

A *hypergraph* on the set of vertices A is a pair $\Gamma = (A, \{S_i\}_{i \in I})$, where $\{S_i\}_{i \in I}$ is a collection of nonempty subsets $S_i \subseteq A$, called *edges* of the hypergraph (*hyperedges*). A hypergraph Γ may be specified by the relation $\gamma \subseteq I \times A$, where $\gamma{<}i{>} = S_i$, so that $(i,a) \in \gamma \leftrightarrow a \in S_i$. Such a relation γ (hypergraph Γ) may also be represented by a Boolean $n \times N$ matrix $g = \|g_{ia}\|$, where $n = |I|$ (the cardinality of I) and $N = |A|$ by the rule

$$g_{ia} = \begin{cases} 1, & \text{if } a \in S_i, \\ 0, & \text{if } a \notin S_i. \end{cases}$$

The transposed matrix g^T is associated with the inverse relation γ^{-1} and with the inverse hypergraph $\Gamma^{-1} = (I, \{T_a\}_{a \in A})$ where $i \in T_a \leftrightarrow a \in S_i$ as well.

A graph with loops at each vertex corresponds to a relation containing all pairs of the form (i,i) ($i \in A$) (such a relation is called *reflexive*) and to a matrix with ones along its principal diagonal. Analogously, a graph with no loops corresponds to a matrix with zeros on its principal diagonal, and to a relation containing no pairs of the form (i,i), called an *anti-reflexive* graph.

An oriented graph is *symmetric* if for each arc (a,b) it contains the reverse arc (b,a), in other words, if $R = R^{-1}$. Matrices of symmetric graphs are symmetric.

Oriented graphs corresponding to regular graphs are symmetric.

An oriented graph without loops is *asymmetric* if each link "goes" only one way, i.e. if for each arc (a,b) in the graph, the arc (b,a) is not in the graph. In terms of inverse relations the asymmetry of R means that $R \cap R^{-1} = \phi$. An analogous graph, but with loops, such that $(a,b) \in R$ and $(b,a) \in R \leftrightarrow a = b$, is *antisymmetric*.

Clearly, for a given relation R there exists a unique relation, minimal among those symmetric relations containing R: $R \cup R^{-1}$, obtained by adding all arcs of the form (b,a) for which $(a,b) \in R$. This relation $\tilde{R} = R \cup R^{-1}$ is called the *symmetric closure* of the relation R. In terms of diagrams this operation corresponds to *deorientation* — transforming a graph to a regular graph by the replacement of arcs with edges.

The maximal asymmetric relations contained in a symmetric relation R (there may be many) are obtained by removing from each "edge" $\{(a,b), (b,a)\}$ one of its arcs. For a graph the corresponding operation consists of orienting all edges (replacing them with arcs), and is called the *orientation* of the graph.

A relation (graph) R is *transitive* if, along with the pairs (a,b), (b,c), it contains the pair (a,c). A regular graph is transitive if, along with each pair of edges ab and bc, it contains also the edge ac. In contrast to the preceding situations, these definitions are not entirely consistent: A symmetric oriented graph corresponding to a regular transitive graph satisfies the transitivity condition for $c \neq a$. If c were equal to a then by the definition of transitivity, together with (a,b) and (b,a), the graph would have to contain the loop (a,a), which is not in the original regular graph. For complete transitivity of this graph it is necessary to add loops at all vertices from which an edge extends (i.e. set the diagonal element to one for each nonempty row of the matrix).

This inconsistency between a regular graph and its corresponding graph appears only with loops. In many applications the occurrence or non-occurrence of loops is completely unimportant (we are unconcerned about the diagonal elements of the matrices), so that this inconsistency may be ignored.

Information about the transitivity of a graph can be reformulated as follows. We say that a vertex j is *accessible* from a vertex i if there exitst a path from i to j. A relation is transitive if and only if it coincides with its relation of *accessibilities*. Actually, the definition of trnasitivity means that whatever is accessible in two steps is accessible in one step, and consequently whatever is accessible in three steps is accessible in two steps, etc.

In the general case a graph of accessibilities R does not coincide with the initial graph R. It can be shown that this graph is uniquely minimal among the transitive graphs which contain R; and hence it is called the *transitive closure* of the graph R.

A set S of vertices (objects) is a *clique* in the graph R if (a,b) ε R for any a,b ε S, i.e. $S \times S \subseteq R$, or equivalently, if any two vertices of S are joined by arcs in both directions.

A transitive graph has the following important property: The collection of vertices of any contour form a clique, since all vertices lying on the same contour are accessible from one another. In a symmetric graph all the vertices lying on a particular path are mutually accessible, so that through transitivity they also form a clique.

A symmetric transitive reflexive relation is called an *equivalence relation*. In accord with the above, the vertices of an equivalence graph[†] can be divided into a system of maximal cliques, no one of which is connected with another. If there were an edge joining a vertex of one clique to a vertex of another than all vertices of these two cliques would be mutually accessible, and consequently their union would also form a clique. An equivalence relation R may thus be written in the form

$$R = S_1 \times S_1 \quad \ldots \quad S_m \times S_m, \qquad (1)$$

where $S_1, \ldots, S_m$ are maximal cliques which form a *partition of the set* A (they are also called *equivalence classes* of the relation R). The matrix corresponding to an indexed set of objects in the order $S_1, S_2, \ldots, S_m$ reduces to one of block-diagonal form, so that for any $r,t = 1, \ldots, m$, if i,j ε S_r then $r_{ij} = 1$, while $r_{ij} = 0$ for i ε S_r, j ε S_t $(r \neq t)$.

The associated partition $S = \{S_1, \ldots, S_m\}$ of the set A is connected in a one-to-one way with an equivalence. To every partition S (i.e. collection of disjoint non-empty subsets covering the set A) there corresponds a relation of the form (1) which is an equivalence relation,

---

[†] As was remarked earlier, under the transition to a graph, loops (reflexivity) may be disregarded.

in which the classes $S_1, \ldots, S_m$ of the partition are maximal cliques of the associated graph. This means that yet another language may be used for equivalence relations (graphs) — the language of partitions $S = \{S_1, \ldots, S_m\}$ which, in many applications, has greater clarity than the languages of relations, matrices and diagrams.

In particular, the intersection $P \cap R$ of equivalence relations P and R is itself an equivalence relation with a corresponding partition, the classes of which are non-empty intersections of the classes of the original partitions. The resulting partition is designated by $P \cap R$ and is called the *intersection of the partitions* P and R, where, for convenience we designate partitions by the same letters as their corresponding relations.

Similarly, corresponding to inclusion of equivalence relations $R \subseteq P$ there is inclusion of partitions $R \subseteq P$, meaning that the partition R is more fractional than P, in the sense that each class of P is the union of several classes of R.

Now let R be an arbitrary relation. The following equivalence relations are naturally associated with R. The *connectivity* relation in R, which is the accessibility relation in $R \cup R^{-1}$. Vertices i and j are connected if there is a chain of the non-oriented graph (i.e. where no attention is paid to the orientation of arcs) between them. The transitivity, symmetry and reflexivity of the connectivity relation are evident. The classes of a connectivity relation are called *components* (of connectedness) of the original graph. In other words, components are maximal sets of mutually connected vertices with no connections between different components. A *connected* graph consists of a single component.

The *biconnection relation* is the symmetric part of the accessibility relation. Vertices i and j are *biconnected* if and only if there is a path from i to j or from j to i. That this is an equivalence relation, is evident. The classes of a biconnection relation are called *bicomponents* (*components of strong connectivity*) of the graph. In other words, a bicomponent is a maximal subset of mutually connected vertices.

Vertices i and j are biconnected if and only if they lie on the same contour (characterized by a path from i to j or from j to i), and therefore it is evident that vertices in different bicomponents do not lie on a common contour.

We form a new graph, having as its vertices the bicomponents $S_1, \ldots, S_m$ in which an arc goes from S_r to S_t if and only if there exists an arc in the original graph from some vertex $i \in S_r$ to some vertex $j \in S_t$. The graph of bicomponents is antisymmetric and contains no contours, and thus, of course, no contours connecting vertices of different bicomponents.

A transitive antisymmetric graph is called a *partial order*. From the foregoing, a graph of the bicomponents of a transitive graph is a partial order.

A *complete (linear, perfect) graph* has, for every pair of vertices a,b an arc connecting them, i.e. $(a,b) \in R$ or $(b,a) \in R$. A complete partial order is called an *order*: The relation linearly orders the objects so that a path leads from an initial vertex through all vertices in succession. Clearly the graph of bicomponents of a complete transitive relation (often called a *linear quasiorder*) is an order, so that such a graph divides the entire set A into classes (bicomponents), which are its maximal cliques. These classes are linearly ordered, in the sense that from any vertex of a class arcs lead to all those (and only those) vertices which belong to the class or to its successor. It can also be shown that for any ordered partition of the set A there is a corresponding linear quasiorder, the arcs of which are directed from all vertices of preceding classes to all vertices of succeeding classes (and also "internally"). Every partial order is contained in some linear order. For an ordered partition P, the notation aPb means that the class of P which contains a precedes the class containing b.

The relation $R \subseteq A \times A$ is said to be *homomorphic* to the relation $P \subseteq B \times B$ if there exists a single-valued mapping $f: A \rightarrow B$ such that

$$(a,b) \in R \leftrightarrow (f(a), f(b)) \in P.$$

Such a mapping f is called a *homomorphism* from R to P.

Let $a, c \in A$ be such that $f(a) = f(c)$ (for some homomorphism from R to P). Then $R<a> = R<c>$, for $b \in R<a> \leftrightarrow (a,b) \in R \leftrightarrow (f(a), f(b)) \in P \leftrightarrow (f(c), f(b)) \in P \leftrightarrow (c,b) \in R \leftrightarrow b \in R<c>$.

Similarly, it can be shown that $R^{-1}<a> = R^{-1}<c>$.

This leads to the following definition. For a given $R \subseteq A \times A$, a pair of objects a,c are considered *indistinguishable* if $R<a> = R<c>$ and $R^{-1}<a> = R^{-1}<c>$. This indistinguishability relation is an equivalence, and the set A is partitioned into non-intersecting classes $R_1, \ldots, R_m$ of indistinguishable objects. On this collection (set) of classes we consider the relation P, consisting of those and only those pairs (R_s, R_t) for which there exist vertices $a \in R_s$, $b \in R_t$ such that $(a,b) \in R$. Clearly, if $(R_s, R_t) \in P$ then for any $c \in R_s$ and $d \in R_t$ the containment $(c,d) \in R$ holds, i.e. $R_s \times R_t \subseteq R$, so that $(a,b) \in R \rightarrow (c,b) \in R \rightarrow (c,a) \in R$. This means that the mapping $f: A \rightarrow B$, which associates with each object $a \in A$ the equivalence class R_s which contains it, is a homomorphism. This homomorphism and the partition of B are called *canonical*.

It is evident that a relation is an equivalence if and only if it is homomorphic to a graph with loops at each vertex and no other links. Linear quasiorders are similarly characterized by the fact that they are homorphic to a linear order. Indistinguishability relations of such graphs give partitions corresponding to them.

An ordinary graph is called a *tree* if it is connected and has no cycles. There are a number of other characterizations for trees: a) a connected graph with N vertices and $N-1$ edges, b) a graph in which any two vertices are connected by one and only one chain, c) a connected graph which loses that property with the elision of any edge [7].

A type of tree of frequent interest is one in which a particular vertex, called the *root*, is singled out. Such a *rooted tree* is considered to be a *hierarchy*, with the top level containing the root, the second level those vertices directly connected to the root, the third level those vertices connected to vertices in the second level, etc.

A *spanning tree* of a graph R is a tree whose vertex set coincides with the set of vertices of R, and in which each edge is an edge of the graph R. To construct a spanning tree we use the following procedure. Beginning with an arbitrary vertex, include in the tree all edges incident on it which lead to "new," as yet unexamined vertices. If, in this process, no new adjacent vertices are produced, it means that the generated tree is the spanning tree of the selected component of the graph. Now it is necessary to repeat the process, using some, as yet unprocessed vertex. In this way all components of the graph will be selected and their spanning trees generated.

References

1. Andreeva, N. S.: Trehmernaya struktura fermentov. *Žurnal Vsesojuznogo himičeskogo obščestva. im. D. I. Mendeleeva*,16,No.4,1971. (Three-dimensional protein structure.)
1a. Aleskerov, F. T.: *Intervaľnyi vybor i evo primenenie v mnogokriteriaľnyh zadačah prinjatija rešenii*. Avtoref. dissertacii. M., Institut problem upravlenija,1980,c.17. (*Interval selection and its application in multi-criteria decision making*.)
2. Bačinskii, A. G., Ratner, V. A.: Optimaľnosť i pomehoustoičivosť genetičeskih tekstov. V kn. *Voprosy matematičeskoi genetiki*, ICiG, Novosibirsk,1974,242-261. (Optimality and robustness of genetic texts.)
3. Benzer, S.: Elementarnye ediničy nasledstvennosti. V kn. *Himičeskie osnovy nasledstvennosti*, IL,M.,1960,56. (The elementary units of heredity.)
4. Geršenzon, S. M., Aleksandrov, Ju. N., Maljuta, S. S.: Mutagennoe deistvie DNK i virusov u drozofily. *IMBiG AN USSR*, Naukova dumka, Kiev,1975. (mutagenic action of DNA and viruses in *Drosophila*.)
5. Dubinin, N. P.: Teorija gena, istorija i sovremennye problemy. *Bjull. Mosk. ob-va ispyt. prirody*,69,No.1,1964. (The theory of genes; history and contemporary problems.)
6. Dubinin, N. P., Hvostova, V. V.: Mehanizm obrazovanija složnyh hromosomnyh reorganizacii. *Bio. žurnal*,1935,4,935. (The mechanism of formation of complex chromosomal reorganizations.)
6a. Žarkih, A. A.: Algoritm postroenija filogenetičeskih drev po aminokislotnym posledovateľnostjam. V sbornike *Matematičeskie modeli evoljucii i selekčii*. (V. A. Ratner, Red.). Novosibirsk, Izd-vo Instituta citologii i genetiki SO AN SSSR,1977,c.5-52. (An algorithm for constructing phylogenetic trees for amino-acid sequences.)
7. Zykov, A. A.: *Teorija konečnyh grafov*. Nauka, Novosibirsk,1969. (*The theory of finite graphs*.)
8. Imrih, V., Stockii, E. D.: Ob optimaľnyh vloženijah metrik v grafy. *DAN SSSR*,1971,200,No.2,279-281. (On the optimal imbeddings of metrics in graphs.)
9. Inge-Večtomov, S. G., Popova, I. A., Gukovskii, D. I., Krivov, V. N.: Rekombinačija i komplementačija v lokuse ad-2 u drozzei *Saccharomyces cerevisiae*. V kn. *Issledovanija po genetike*, Izd-vo LGU,5,1974,35-48. (Recombination and complementation in the locus ad-2 of the yeasts *Saccharomyces cerevisiae*.)
10. Kemeny, G., Snell, G.: *Kibernetičeskoe modelirovanie*. Sov. radio, M.,1972, (*Cybernetic modeling*.)
11. Kuličkov, V. A., Žimulev, I. F.: Analiz prostranstvennoi ogranizačii genoma *Drosophila Melanogaster* na osnove dannyh po ektopičeskoi konjugačii politennyh hromosom. *Genetika*,1976,12,No.5,81-89. (Analysis of the spatial organization in the genome of *Drosophila melanogaster* on the basis of data about the ectopic conuugation of polynemic chromosomes.)
12. Kuperštoh, V. L.: O roli poroga suščestvennosti individuaľnyh svjazei v zadače klassifikačii. V sb. 25 . (On the role of an existence threshhold for individual connections in classification problems.)
13. Kuperštoh, V. L., Mirkin, B. G.: Yporjadočenie bzaimosvjazannyh ob"ektov. *Avtomatika i telemehanika*,1971,No.6,77-83; No.7,91-97. (The ordering of interconnected objects.)
14. Kuperštoh, V. L., Trofimov, V. A.: Algoritm analiza struktury matričy sfjazi. *Avtomatika i telemehanika*,1975,No.11,170-180. (An analysis algorithm of the structure of connection matrices.)
15. Kuperštoh, V. L., Mirkin, B. G., Trofimov, V. A.: Summa vnutrennih sfjazei kak pokazatel kačestva klassifikačii. *Avtomatika i telemehanika*,1976,No.3. (The sum of internal connections as an indicator of the quality of a classification.)
16. Kurganov, B. I., Poljanovskii, O. L.: Cetverticnaja struktura i

16. Kurganov, B. I., Poljanovskiĭ, O. L.: Cetvertičnaja struktura i allosteričeskaja reguljačija adtivnosti fermentov. Ž. Vsesojuznogo him. ob-va im. D. I. Mendeleeva,1971,16,No.4,421-431. (Quaternary structure and the allosteric regulation of enzyme activity.)
17. Kušev, V. V.: *Mehanizmy genetičeskoĭ rekombinačii*. Nauka, Leningrad,1971. *(Genetic recombination mechanisms.)*
18. Lim, V. I.: Strukturnye prevraščenija belkovoĭ čepi pri formirovanii nativnoĭ globuly. Gipoteza "izbytočnyh" spiralei. *DAN SSSR*, 1975,222,No.6,1467-1469. (Structural transformations of protein chains in the formation of native globules. The hypothesis of redundant spirals.)
19. Mirkin, B. G.: Ob odnom klasse otnošeniĭ predpočtenija. V sb. *Matem. voprosy formir. èkonom. modelei*. IÈiOPP, Novosibirsk,1970,90-102. (On a class of preference relations.)
20. Mirkin, B. G.: Zadači approksimačii v prostranstve otnošeniĭ i analiz nečislovyh priznakov. *Avtomatika i telemehanika*,1974,No.9, (Approximation problems in a space of relations and the analysis of non-numeric indicators.)
21. Mirkin, B. G.: *Problema gruppovogo vybora*. Nauka,M.,1974. (The problem of group selection.)
22. Mirkin, B. G.: *Analiz kačestvennyh priznakov*. Statistika,M.,1976. *(The analysis of qualitative indicators.)*
23. Mirkin, B. G., Rodin, S. N.: K analizu bulevskih matrič, Svjazannyh s rešeniem nekotoryh genetičeskih zadač. *Kibernetika*,1974,No.2, 108-114. (Toward the analysis of Boolean matrices connected with some genetic problems.)
24. *Problemy analiza diskretnoĭ informačii. C. I*, IÈiOPP,Novosibirsk, 1975. *(Problems in the analysis of discrete information, Part I.)*
25. *Problemy analiza diskretnoĭ informačii. C. II*, IÈiOPP,Novosibirsk, 1976. *(Problems in the analysis of discrete information, Part II.)*
26. Ptičyn, O. B.: Fizičeskie prinčipy samoorganizačii belkovyh šepei. *Yspehi sovrem. biol.*,1970,69,vyp.1,26-48. (physical principles in the self-organization of protein chains.)
27. Ono, S.: *Genetičeskie mehanizmy pregressivnoĭ evoljučii*. Mir,M., 1973. *(The genetic mechanisms of progressive evolution.)*
28. Rao, S. R.: *Lineĭnye statističeskie metody i ih primenenija*.Nauka, M.,1968. *(Linear statistical methods and their applications.)*
29. Ratner, V. A.: *Genetičeskie upravljajuščie sistemy*. Nauka,Novosibirsk,1966. *(Genetic control systems.)*
30. Ratner, V. A.: *Prinčipy organizačii i mehanizmy molekuljarnogenetičeskih pročessov*, Nauka,Novosibirsk,1972. *(Organizational principles and mechanisms of molecular-genetic processes.)*
31. Ratner, V. A.: *Molekuljarno-genetičeskie sistemy apravlenija*. Nauka, Novosibirsk,1975. *(Molecular-genetic control systems.)*
32. Ratner, V. A.: O nekotoryh teoretičeskih problemah molekuljarnoĭ evoljučii. *Zyrn. obščei biol.*,1976,37,No.1,18-29. (On some theoretical problems of molecular evolution.)
33. Ratner, V. A., Furman, D. P., Nikoro, Z. S.: Issledovanie genetičeskoĭ topografii lokusa *scute* y *Drosophila Melanogaster*. *Genetika*, 1969,5,No.6,72-85. (The investigation of the genetic topography in the *scute* locus of *Drosophila melanogaster*.)
34. Ratner, V. A., Rodin, S. N.: O matematičeskoĭ obrabotke matrič allelizma. II: Postroenie i interpretačija drev mežalleľnoi komplementačii. V kn. *Voprosy matematičeskoĭ genetiki*, ICiG,Novosibirsk, 1974,214-241. (On the mathematical treatment of allelism matrices, II: The construction and interpretation of interallelic complementation trees.)
34a. Ratner, V.A., Rodin, S. N., Žarkih, A. A.: Issledovanie molekuljarnoi filogenii globinov utočnennym metodom. B sbornike *Matematičeskie modeli evoljučii i selekčii* (V. A. Ratner, Red.). Novosibirsk,Ezd-bo Instituta čitologii i genetiki SO AN SSSR,1977,c.53-96. (Investigation of the molecular phylogeny of globins by a more precise method.)
35. Ratner, V. A., Rodin, S. N., Šenderov, A. N.: Problema mežalleľnoĭ komplementačii. *Uspehi sovr. biol.*,1975,3,399-419. (The interallelic complementation problem.)

36. Ratner, V. A., Kananjan, G. H.: Postroenie filogenetičeskih drev dlja funkcionalnyh centrov globinov. V kn. *Issledovanija po matematičeskoi genetike*,ICiG,Novosibirsk,1975,125-168. (The construction of phylogenetic trees for the functional centers of globins.)
37. Rodin, S. N: Analiz allelnyh otnošenii recessivnyh letalei, inducirovannyh u drozofily čyžerodnymi DNK i nekotorymi virusami. *Genetika*,1974,10,NO.9,94-105. (The analysis of allelic relations of recessive lethals, induced in *Drosophila* by foreign DNA and several viruses.)
38. Serebrovskii, A. S., Dubinin, N. P.: Iskusstvennoe polučenie mutacii i problema gena. *Uspehi èksper. biol.*,1929,4,235. (Artificial production of mutations and the gene problem.)
39. Soidla, T. T.: O strukture lokusa ade-2 droždei-saharomičetov. *Genetika*,1972,8,No.6,72-80. (On the structure of the *ad-2* locus in saccharomycetes yeasts.)
40. Soidla, T. T., Inge-Večtomov, S. G., Simarov, B. V.: Mežallelnaja komplementacija v lokuse AD-2 y droždei Saccharomyces cerevisiae. *Issledovannija po genetike*,LGU,1967,3,148-164. (Interallelic complementation in the *ad-2* locus of the yeasts *Saccharomyces cerevisiae*.)
41. Stal, F.: *Mehanizmy nasledstvennosti*. Mir,M.,1966. (*Hereditary mechanisms*.)
42. Trofimov, V. A.: K analizu čikličeskoi struktury matričy svjazi. V sb. *Voprosy analiza složnyh sistem*, Nauka,Novosibirsk,1974,77-83. (Toward the analysis of cyclic structure in connection matrices.)
43. Finčem, Dž.: *Genetičeskaja komplementacija*. Mir,M.,1968. (*Genetic complementation*.)
44. Friedman, G. S.: Nekotorye rezultaty v zadače approksimacii grafov. V kn. [24]. (Some results on the approximation problem for graphs.)
45. Furman, D. P., Rodin, S. N., Ratner, V. A.: Issledovanie genetičeskoi topografii lokusa *scute* y *Drosophila melanogaster*. Soobščenija II-V,*Genetika*,1977,Nos.2,4,6. (Investigation of the genetic topography of the *scute* locus in *Drosophila melanogaster*.)
46. Harris, G.: *Osnovy biohimičeskoi genetiki čeloveka*. Mir,M.-1973. (*The bases of the biochemical genetics of man*.)
47. Hausdorf, F.: *Teorija Množestv*. ONTI,M.-L.,1937. (*Set theory*.)
48. Heis, U.: *Genetika bakterii i bakteriofagov*. Mir,M.,1965. (*The genetics of bacteria and bacteriophages*.)
49. Šaronov, Ju. A., Šaronova, N. A.: Struktura i funkcii gemoglobina. *Molekuljarnaja biologija*,1975,9.No.1,145-172. (The structure and functions of hemoglobin.)
50. Škurba, V. V.: O matematičeskoi obrabotke odnogo klassa biohimičeskih èksperimentov. *Kibernetika*,1965,No.1. (On the mathematical treatment of a class of biochemical experiments.)
51. Šmalgauzen, I. I.: Što takoe nasledstvennaja informacija? *Problemy kibernetike*,1966,16,23-35. (What is hereditary information.)
52. Eigen, M.: *Samoorganizacija materii i evoljucija biologičeskih makromolekul*. Mir,M.,1973. (*Self-organization of matter and the evolution of biological macromolecules*.)
53. Adams, M., Buehner, M., Chandrasekhar, K., Ford, G. C., Hackert, M. L., Liljas, A., Lentz, P., Rao, S. T., Tossmann, M. G., Smiley, I. E., White, J. L., In *Protein-protein interactions* (Ed. by Jaenicke, R., and Helmreich, E.), Springer-Verlag, Berlin,Heidelberg, New York,1972,139.
54. Ahmed, A., Case, M. E., Giles, N. H.: The nature of complementation among mutants in the histidine-3 region of *Neurospora crassa*. *Brookhaven Symp. Biol.*,1964,17,53-65.
55. Benzer, S.: On the topology of the genetic fine structure. *Proc. Nat. Acad. Sci. Wash.*,1959,45,1607.
55a. Blake, S. S. F.: Exons encode protein functional units. *Nature*, 1979,277,No.5698,598.

55b. Carramolino, L., Ruiz-Gomez, M., Guerrero, M., Campuzano, S., Modolell, J.: DNA map of mutations at the *scute* locus of *Drosophila melanogaster*. *EMBO Journal*,1982,1,10,1185-1191.
56. Child, G.: Phenogenetic studies of *scute-1* of *Drosophila melanogaster*. I. The associations between the bristles and effects of genetic modifiers and temperature. *Genetics*,1935,20,N 2,109.
57. Crick, F. H. C.: Orgel, L. E., The theory of inter-allelic complementation. *J. Mol. Biol.*,1964,8,161.
58. Dayhoff, M. O.: *Atlas of protein sequence and structure*, v. 5,Nat. Biom. Res. Foundation, Silver Spring, Maryland, USA,1972.
59. De Serres, F. J.: Carbon dioxide stimulation of the *ad-3* mutants of *Neurospora crassa*. *Mutation Res.*,1966,3,420-425.
60. Fishburn, P. C.: Intransitive individual indifference with unequal indifference intervals. *J. Math. Psychol.*,1970,7,No.1,144-149.
61. Fishburn, P. C.: An interval graph is not a comparability graph. *J. Combinatorial Theory*,1970,8,442-443.
62. Fitch, W. M.: An improved method of testing for evolutionary homology. *J. Mol. Biol.*,16,9-16,1966.
63. Fitch W. M.: Toward defining the course of evolution: minimum change for a specific tree topology. *Syst. Zool.*,1971,20,406-416.
64. Fitch, W. M.: A comparison between evolutionary substitutions and variants in human haemoglobins. *Annals of the New York Academy of Sciences*,1974,241,439-448.
64a. Fitch, W. M.: On the problem of discovering the most parsimonious tree. *Amer. Naturalist*,1977,111,No.978,223-257.
65. Fitch, W. M.: Margoliash, E., Construction of phylogenetic trees, *Science*,1967,155,N 4,279.
65a. Flament, C.: Hypergraphs arborés. *Discrete Math.*,1978,21,223-275.
66. Fulkerson, D. R., Gross, O. A.: Incidence matrices and interval graphs. *Pacif. J. Math.*,1965,15,No.3,835-885.
66a. Gilbert, W.: Why genes in pieces? *Nature*,1978,271,501.
67. Gilmore, P. C., Hoffman, J. J.: A characterisation of comparability graphs and of interval graphs. *Canad. J. Mathem.*,1964,
68. Goldschmidt, R.: Die Entwicklungsphysiologische Erklärung des Falls der sogenannten Treppenallelormorphe des Genes *scute* von *Drosophila*. *Biol. Zbl.*,51,507,1930.
69. Goodman, M., Barnabas, J., Maesuda, G., Moore, G. W.: Molecular evolution in the descent of man. *Nature.*,1971,233,No.5322,604-613.
69a. Green, M. M.: Transposable elements in *Drosophila* and other diptera. *Ann. Rev. Genetics*,1980,14,109-120.
70. Greer, J.: Three-dimensional structure of abnormal mutant human haemoglobin. *Cold Spr. Harb. Symp. Quant. Biol.*,1972,36,315.
70a. Hartigan, J. A.: Minimum mutation fits to a given tree. *Biometrics*,1973,29,53-65.
71. Hartman, P. E., Hartman, Z., Stahl, R. C., Ames, B. N.: Classification and mapping of spontaneous and induced mutations in the histidine operon of *Salmonella*. *Advances in Genetics*,1971,v.16,1-34.
72. Houston, L. L.: Purification and properties of a mutant bifunctional ensyme from the *hisB* gene of *Salmonella typhimurium*. *J. Biol. Chemistry*,1973,248,No.12,4144-4149.
73. Houston, L. L.: Specialized subregions of the bifunctional *hisB* gene of *Salmonella typhimurium*. *J. Bacteriology*,1973,113,No.1,82-87.
74. Ingram, V. M.: *The haemoglobins in genetics and evolution*. Columbia Univ. Press,New York,1963.
74a. Karp, R.: Reducibility among combinatorial problems, *Complexity of Computer Computations* (Ed. by R. Miller and J. Thatcher), Plenum, New York,1972,85-103.
75. Kendall, D. G.: Incidence matrices, interval graphs and seriation in archaelogy. *Pacific J. Math.* ,1969,28,No.3,565-570.

76. Kimura, M.: Evolutionary rate at the molecular level. *Nature*,1968, 217,624-626.
77. Kimura, M., Ohta, T.: On some principles governing molecular evolution. *Proc. Nat. Acad. Sci.*,USA,1974,71,No.7,2848-2852.
78. Kuratowski, G.: Sur le probleme des courbes gaushes en topologie. *Fund. Mathem.*,1930,15,271-283.
79. Langley, C. H., Fitch, W. M.: An estimation of the constancy of the rate of molecular evolution. *J. Mol. Evol.*,1974,3,161-178.
80. Lekkerkerker, C. G., Boland, J. Ch.: Representation of a finite graph by a set of intervals on the real line. *Fundam. Math.*,1962, 51,No.1,45-64.
81. Loper, J. C., Grabnar, M., Stahl, R. C., Hartman, Z., Hartman, P. E.: Genes and proteins involved in histidine biosynthesis in *Salmonella*. *Brookhaven Symp. Biol.*,1964,17,15-52.
82. Luce, R.: Semi-orders and a theory of utility discrimination. *Econometrica*,1956,24,178-191.
82a. Maniatis, T., Fritsch, E. F., Lauer, J., Lawn, R. M.: The molecular genetics of human hemoglobins. *Ann. Rev. Genetics*,1980,14,145-178.
83. Michel, J.: An interval graph is a comparability graph. *J. Combin. Theory*,1969,No.2,189-190.
84. Moore, G. W., Gookman, M., Barnabas, J.: An iterative approach from the standpoint of the additive hypothesis to the dendrogram problem posed by molecular data sets. *J. Theor. Biol.*,1973,38,423-457.
85. Moore, G. W., Barnabas, J., Goodman, M.: A method for constructing maximum parsimony ancestral amino acid sequences on a given network. *J. Theor. Biol.*,1973,38,459-485.
86. Perutz, M. F., Lehmann, H.: Molecular pathology of human haemoglobin. *Nature*,1968,219,No.5157,902-909.
87. Perutz, M. F., Muirhead, H., Cox, J. M., Goaman, L. C. G.: Three-dimensional Fourier synthesis of horse oxyhaemoglobin at 2.8 A resolution: the atomic model. *Nature*,1968,219,No.5150,131-139.
88. Ratner, V. A., Rodin, S. N.: Theoretical aspects of genetic complementation. In *Progress in Theoretical Biology* (Ed. by R. Rosen, Snell, F. M.),1976,Acad. Press, v.4,1-63.
89. Roberts, F. S.: Non-transitive indifference. *J. Math. Psychology*, 1970,7,No.2,243-258.
89a. Sakano, H., Huppi, K., Heinrich, G., Tonegawa, S.: Sequences at the somatic recombination sites of immunoglobulin light-chain genes. *Nature*,1979,280,No.5720,288-294.
90. Sankoff, D.: Minimal mutation trees of sequences. *SIAM J. Appl. Math.*,1975,28,No.1,35-42.
91. Scott, D., Suppes, P.: Foundational aspects of theories of measurement. *J. Symbolic Logic.*,1958,23,113-128.
92. Sellers, P. H.: An algorithm for the distance between two finite sequences. *J. Combin. Theory A*,1974,16,No.2,253-258.
92a. Sellers, P. H.: Pattern recognition in genetic sequences. *Proc. Nat. Acad. Sci. U.S.A.*,1979,76,No.7,3041.
93. Simoes Pereira, J. M. S.: On the tree realisation of a distance matrix. *Theorie des graphes*,Actes Journees Int. Etudes ICC,Rome, 1966; Paris, Dunod,1967,383-388.
94. Stern, C.: Genetic mechanisms of development (with localized initiation of differentiation). *Cold Spring Harbor Symp. Quant. Biol.*, 1956,21,375.
95. Stertevant, A. H., Schultz, J.: The inadequacy of the subgene hypothesis of the nature of the scute allelomorphs of *Drosophila*. *Proc. Nat. Acad. Sci. USA*,1931,17,265.
96. Tucker, A.: A structure theorem for the consecutive 1's property. *J. Combin. Theory*,1972,B12,No.2,153-162.
97. Tucker, A.: Matrix characterisations of circular-arc graphs. *Pacif. J. Mathem.*,1971,39,No.2,535-545.
97a. Tucker, A.: Circular arc graphs: new uses and a new algorithm. In *Theory and applications of graphs* (*Proc. Internat. Conf.*,Western

Mich. Univ., Kalamazoo, Mich.,1976),580-589. *Lecture Notes in Math.*, 642,Springer,Berlin,1978.
98. Zuckerkandl, E.: The appearance of new structures and functions in proteins during evolution. *J. Molec. Evol.*,1975,7,1-57.
99. Zuckerkandl, E., Pauling, L.: Evolutionary divergence and convergence in proteins. In *Evolving genes and proteins* (Ed. by V. Bryson and H. J. Vogel),Acad. Press,1965,97-166.
100. Ycas, M.: On certain homologies between proteins. 1976,7,No.3,215-244

Index of Genetics Terms

Adenine	1	-interspecies	162
Allele	16	-instaspecies	162
Amino acid	1	Domain	7
Anticodon	5	Dominance	15
Antibody	7	Duplication	13
Antigen	7		
		Enzyme	6
Base, hydrogenous	1	-proteolytic	8
		Evolution, neutral	165
Cell		-progressive	162
-diploid	13	Exon	7
-haploid	13		
Center, allosteric	93	Family, protein	158
-catalytic	93		
-contact	93	Gamete	13
-diphosphyoglycerate binding	95	Gene	5,8,12
-functional	8,93	Gene - regulator	6
Chromosomes	4,12	Genes	
-homologous	13	-homologous	16
Cistron	5,70	-cohesive	14
Code		Genome	1
-genetic triplet	2,8	Globins	158
-physical	182	Guanine(G)	1
Codon	1		
Collinearity	10,12-13	Heme	95
Complementation	16	Hemoglobin	94
-interallelic (intracistronic)	19,92	Heterozygote	16
-matrix	19	Homozygote	16
Complon	36		
Conjugation, chromosome	14	Immunoglobulin	7
Convergence	145	Initiator (or translation)	5
Crossing-over	14	Insertosome	80
Cytosine(C)	1	Interaction, allosteric	95
		Interphase	127
Deletion	13,15	Intron	7
Deoxyribonucleic acid(DNA)	1	Inversion	75
Dimer	94		
Divergence	140	Layers (of genetic language)	8
		-semantic	10
		-structural	10

–structural	10	Pauling scale	164
Lethal	14	Polypeptide	1
Locus	16	Promotor (of transcription)	5
		Proteins	1,93
Map, genetic (recombination)	17		
Mapping, complementation	20	Recombination	14
Mapping function	17	Regulator (gene)	6
Meiosis	13	Replication (of DNA)	2
Messenger ribonucleic acid		Repressor	6
(mRNA)	4	Reversion	66
Metabolism	180	Ribosome	4
Mitosis	12		
Molecule of DNA	2	Scripton	5
Multimer	2	Semantics	8
Mutation	13	Structure, protein	2
–frameshift	13,73		
–lethal	14	Terminator	5,6
–missense	13	Test, complementation	19
–nonsense	13	–recombination	16
–point	19,66	Tetramer	94
–recessive	15	Thymine (T)	1
–synonymous	13	Transcription	4,71
		Translation	5,71
Ontogenesis	1	Transposon	80
Operator (of a scripton)	6	Tree, phylogenetic	133
Operon	6	Triplet	
		Uracil (U)	4
		Zygote	12

Index of Mathematical Terms

Accessibility of vertices	184
Adjacency	180
Agglomerative algorithms	113
Algorithm	
-Fulkerson-Gross	51
-"structure"	115
-"transfer"	112
-"unification"	112
Approximation problem	preface
Arc	180
Bicomponent	185
Biconnectivity (of vertices)	185
Chain	180
Class, equivalence	184
Clique	184
Closure	
-symmetric	183
-transitive	184
Complon	42
Component of a graph	185
Concentrate	107
Connectivity (of vertices)	185
Cut (of a relation)	182
Cycle	181
Diagram	180
Disjunction	38
Distance (graphs)	181
-Hamming	104
-minimal mutational	134
Edge, graph	180
-hypergraph	182
Equivalence, interval	25
Farthest neighbor method	113
Graph	180
-a-similarity	115, 141
-arc	50
-asteroidal	33
-asymmetric	183
-complementation	21
-complete	186
-connected	185
-interval	25
-noncovering	34
-oriented	181
-regular	181
-signed	106
-tree	50
-triangulated	31
Hierarchy	187
Homomorphism (relations)	186
-canonical	186
Hyperedge	182
Hypergraph	182
-G-interval	48
-interval	36
-minimal	41
Incidence	180
Inclusion of a partition	185
Intersection (partition)	185
Interval	
-of an ordering	36
-of an ordered partition	63
Length (maps)	20
-(mappings)	150
Link	180
Loop	180
Macrostructure	102

Map	20	Radius (graphs)	181
Mapping, admissible	150	Relation, binary	181
-characteristic	150	-accessibility	184
Mapping function	17	-anti-reflexive	183
Matrix		-biconnectivity	185
-Boolean	181	-connectivity	232
-Complementation	21	-equivalence	184
-linear	35,39,62	-inverse	182
-quasi-diagonal	25	-quasilinear	21
-relation	181	-reflexive	183
Median	104	-semi-order	34
Metric, Berovski	141	-symmetric	183
		-transitive	183
Nearest neighbor method	113	Representation problem	preface
Neighborhood	41	Root of a tree	187
		Route	180
Object(s)			
-complonic	42	Spanning tree	187
-covering of	63		
-indistinguishable	186	Tree	140,187
Order	186	-rooted	187
-interval	21	-weighted	136
-partial	186	Triangulator	28
Ordering, admissible	62	Triplet(s)	
Orientation (graphs)	183	-asteroidal	32
		-overlapping	55
Partition	184		
-canonical	187	Ultrametric	141
-compact	107		
-structured	102	Vertex (graphs)	180
Partition with structure	102	-branching	137
-ordered	22	-communicating	137
Path	180	-internal	137
		-pendant	137
Quasilinearity condition	21	-simplicial	42
Quasiorder, linear	186		

Biomathematics

Managing Editors: K. Krickeberg, S. A. Levin

Editorial Board: H. J. Bremermann, J. Cowan, W. M. Hirsch, S. Karlin, J. Keller, R. C. Lewontin, R. M. May, J. Neyman, S. I. Rubinow, M. Schreiber, L. A. Segel

Volume 10
A. Okubo

Diffusion and Ecological Problems: Mathematical Models

1980. 114 figures, 6 tables. XIII, 254 pages
ISBN 3-540-09620-5

Contents: Introduction: The Mathematics of Ecological Diffusion. – The Basics of Diffusion. – Passive Diffusion in Ecosystems. – Diffusion of "Smell" and "Taste": Chemical Communication. – Mathematical Treatment of Biological Diffusion. – Some Examples of Animal Diffusion. – The Dynamics of Animal Grouping. – Animal Movements in Home Range. – Patchy Distribution and Diffusion. – Population Dynamics in Temporal and Spatial Domains. – References. – Author Index. – Subject Index.

Volume 9
W. J. Ewens

Mathematical Population Genetics

1979. 4 figures, 17 tables. XII, 325 pages
ISBN 3-540-09577-2

Contents: The Golden Age. – Technicalities and Generalizations. – Discrete Stochastic Models. – Diffusion Theory. – Applications of Diffusion Theory. – Two Loci. – Many Loci. – Molecular Population Genetics. – The Neural Theory. – Generalizations and Conclusions. – Appendices. – References. – Author Index. – Subject Index.

Volume 8
A. T. Winfree

The Geometry of Biological Time

1980. 290 figures. XIV, 530 pages
ISBN 3-540-09373-7

Contents: Introduction. – Circular Logic. – Phase Singularities (Screwy Results of Circular Logic). – The Rules of the Ring. – Ring Populations. – Getting Off the Ring. – Attracting Cycles and Isochrons. – Measuring the Trajectories of a Circadian Clock. – Populations of Attractor Cycle Oscillators. – Excitable Kinetics and Excitable Media. – The Varieties of Phaseless Experience: In Which the Geometrical Orderliness of Rhythmic Organization Breaks Down in Diverse Ways. – The Firefly Machine. – Energy Metabolism in Cells. – The Malonic Acid Reagent ("Sodium Geometrate"). – Electrical Rhythmicity and Excitability in Cell Membranes. – The Aggregation of Slime Mold Amoebae. – Growth and Regeneration. – Arthropod Cuticle. – Pattern Formation in the Fungi. – Circadian Rhythms in General. – The Circadian Clocks of Insect Eclosion. – The Flower of Kalanchoe. – The Cell Mitotic Cycle. – The Female Cycle. – References. – Index of Names. – Index of Subjects.

Volume 7
E. R. Lewis

Network Models in Population Biology

1977. 187 figures. XII, 402 pages
ISBN 3-540-08214-X

Contents: Foundations of Modeling Dynamic Systems. – General Concepts of Population Modeling. – A Network Approach to Population Modeling. – Analysis of Network Models. – Appendices: Probability Arrays, Array Manipulation. Bernoulli Trials in the Binomial Distribution.

Volume 6
D. Smith, N. Keyfitz

Mathematical Demography

Selected Papers
1977. 31 figures. XI, 514 pages
ISBN 3-540-07899-1

Contents: The Life Table. – Stable Population Theory. – Attempts at Prediction and the Theory they Stimulated. – Parameterization and Curve Fitting. – Probability Models of Conception and Birth. – Branching Theory and Other Stochastic Processes. – Cohort and Period, Problem of the Sexes, Sampling.

Springer-Verlag
Berlin
Heidelberg
New York
Tokyo

Biomathematics

Managing Editors: K. Krickeberg, S. A. Levin

Editorial Board: H. J. Bremermann, J. Cowan, W. M. Hirsch, S. Karlin, J. Keller, R. C. Lewontin, R. M. May, J. Neyman, S. I. Rubinow, M. Schreiber, L. A. Segel

---

Volume 5
A. Jacquard
The Genetic Structure of Populations

Translators: D. Charlesworth, B. Charlesworth
1974. 92 figures. XVIII, 569 pages
ISBN 3-540-06329-3

Contents: Basic Facts and Concepts: The Foundations of Genetics. Basic Concepts and Notation. Genetic Structure of Populations and of Individuals. – A Reference Model: Absence of Evolutionary Factors: The Hardy-Weinberg Equilibrium for one Locus. The Equilibrium for two Loci. The Inhentance of Quanitative Characters. Genetic Relationships between Relatives. Overlapping Generations. – The Causes of Evolutionary Changes in Populations: Finite Populations. Deviations from Random Mating. Selection. Mutation. Migration. The Combined Effects of Different Evolutionary Forces. – The Study of Human Population Structure: Genetic Distance. I. Basic Concepts and Methods. Genetic Distance. II. The Representation of Sets of Objects. Some Studies of Human Populations. – Appendix A. Linear Difference Equations. – Appendix B. Some Definitions and Results in Matrix Algebra.

Volume 4
M. Iosifescu, P. Tăutu
Stochastic Processes and Applications in Biology and Medicine
Part 2: Models

1973. 337 pages. ISBN 3-540-06271-8

Contents: Preliminary Considerations. – Population Growth Models. – Population Dynamics Processes. – Evolutionary Processes. – Models in Physiology and Pathology.

Volume 3
M. Iosifescu, P. Tăutu
Stochastic Processes and Applications in Biology and Medicine
Part 1: Theory

1973. 331 pages
ISBN 3-540-06270-X

Contents: Discrete Parameter Stochastic Processes: Denumerable Markov Chains. Noteworthy Classes of Denumerable Markov Chains. Markov Chains with Arbitrary State Space. – Continuous Parameter Stochastic Processes: Some General Problems. Processes with Independent Increments. Markov Processes.

Volume 2
E. Batschelet
Introduction to Mathematics for Life Scientists

3rd edition. 1979. 227 figures, 62 tables.
XV, 643 pages
ISBN 3-540-09662-0

Contents: Real Numbers. – Sets and Symbolic Logic. – Relations and Functions. – The Power Function and Related Functions. – Periodic Functions. – Exponential and Logarithmic Functions I. – Graphical Methods. – Limits. – Differential and Integral Calculus. – Exponential and Logarithmic Functions II. – Ordinary Differential Equations. – Functions of Two or More Independent Variables. – Probability. – Matrices and Vectors. – Complex Numbers. – Appendix (Tables A to K). – Solutions to Odd Numbered Problems. – References. – Author and Subject Index.

Springer-Verlag
Berlin
Heidelberg
New York
Tokyo

AIR UNIVERSITY

SCHOOL OF ADVANCED AIR AND SPACE STUDIES

"All the Missiles Work"

Technological Dislocations and Military Innovation

A Case Study in US Air Force Air-to-Air Armament,
Post–World War II through Operation Rolling Thunder

STEVEN A. FINO
Lieutenant Colonel, USAF

Drew Paper No. 12

Air University Press
Air Force Research Institute
Maxwell Air Force Base, Alabama

Project Editor
Belinda L. Bazinet

Copy Editor
Carolyn Burns

Cover Art, Book Design, and Illustrations
Daniel Armstrong

Composition and Prepress Production
Vivian D. O'Neal

Print Preparation and Distribution
Diane Clark

AIR FORCE RESEARCH INSTITUTE

AIR UNIVERSITY PRESS

Director and Publisher
Allen G. Peck

Editor in Chief
Oreste M. Johnson

Managing Editor
Demorah Hayes

Design and Production Manager
Cheryl King

Air University Press
155 N. Twining St., Bldg. 693
Maxwell AFB, AL 36112-6026
afri.aupress@us.af.mil

http://aupress.au.af.mil/
http://afri.au.af.mil/

AFRI AU PRESS
AIR FORCE RESEARCH INSTITUTE

Library of Congress Cataloging-in-Publication Data

Fino, Steven A., 1974–
 "All the missiles work" : technological dislocations and military innovation : a case study in US Air Force air-to-air armament, post-World War II through Operation Rolling Thunder / Steven A. Fino, Lieutenant Colonel, USAF.
   pages cm. — (Drew paper, ISSN 1941-3785 ; no. 12)
   ISBN 978-1-58566-248-7
   1. Air-to-air rockets—United States—History. 2. Guided missiles—United States—History. 3. Airplanes, Military—Armament—United States—History. 4. Airplanes, Military—Armament—Technological innovations. 5. United States. Air Force—Weapons systems—Technological innovations. I. Title. II. Title: Technological dislocations and military innovation, a case study in US Air Force air-to-air armament, post-World War II through Operation Rolling Thunder.
   UG1312.A35F55 2014
   358.4'1825191097309045—dc23
                                                    2014037391

Published by Air University Press in January 2015

Disclaimer

Opinions, conclusions, and recommendations expressed or implied within are solely those of the author and do not necessarily represent the views of the School of Advanced Air and Space Studies, the Air Force Research Institute, Air University, the United States Air Force, the Department of Defense, or any other US government agency. Cleared for public release: distribution unlimited.

This Drew Paper and others in the series are available electronically at the AU Press website: http://aupress.au.af.mil.

The Drew Papers

The Drew Papers are award-winning master's theses selected for publication by the School of Advanced Air and Space Studies (SAASS), Maxwell AFB, Alabama. This series of papers commemorates the distinguished career of Col Dennis "Denny" Drew, USAF, retired. In 30 years at Air University (AU), Colonel Drew served on the Air Command and Staff College faculty, directed the Airpower Research Institute, and served as dean, associate dean, and professor of military strategy at SAASS. Colonel Drew is one of the Air Force's most extensively published authors and an international speaker in high demand. He has lectured to over 100,000 students at AU as well as to foreign military audiences. In 1985 he received the Muir S. Fairchild Award for outstanding contributions to Air University. In 2003 Queen Beatrix of the Netherlands made him a Knight in the Order of Orange-Nassau for his contributions to education in the Royal Netherlands Air Force.

The Drew Papers are dedicated to promoting the understanding of air and space power theory and application. These studies are published by the Air University Press and broadly distributed throughout the US Air Force, the Department of Defense, and other governmental organizations, as well as to leading scholars, selected institutions of higher learning, public-policy institutes, and the media.

Please send inquiries or comments to

Commandant and Dean
School of Advanced Air and Space Studies
125 Chennault Circle
Maxwell AFB, AL 36112
Tel: (334) 953-5155
DSN: 493-5155
saass.admin@us.af.mil

Contents

	List of Illustrations	*vii*
	Foreword	*ix*
	About the Author	*xi*
	Acknowledgments	*xiii*
	Abstract	*xv*
1	Introduction	1
2	Foundations of Technology	13
3	Technological Dislocations	37
4	Rise of the Missile Mafia	51
5	The Gun Resurrected	73
6	An Interim Solution	95
7	Military Innovation	115
8	Conclusion	133
	Abbreviations	137
	Bibliography	139

Illustrations

1	Technological momentum	29
2	The putative tipping point between social constructivism and technological determinism	38
3	Edge dislocation	41
4	Technological dislocations	44
5	Historical analysis and agency	46

Foreword

When Airmen think about technology, it is typically in terms of how best to achieve military effects. This paper explores the deeper and more crucial issues of how organizations and individuals resist, embrace, and shape technological innovation. Steam-powered warships, machine guns, aircraft carriers, intercontinental ballistic missiles, and remotely piloted aircraft all threatened established practice and faced the resistance of entrenched bureaucracies and dogmatic tradition. Yet such rigidity of mind and habit contrasts with the imperative of achieving military advantage and the sparkling allure of the new, the scientific, and the powerful. Lt Col Steve Fino explores the tension between such technological skepticism and technological exuberance—a tension that "weaves itself through the fabric of US military history," with his concept of *technological dislocation*. In this light, he examines how various factors can dislocate the predicted evolutionary pathway of emerging or even established technologies and steer them in new, perhaps surprising, directions.

The development of air-to-air armament in the initial decades of the jet age provides an intriguing case study of technological dislocation. On this stage, various actors such as fighter pilots, senior leaders, institutional preferences, cutting-edge technologies, and enemy combatants played out the first scenes of aerial combat's missile age. The Air Force's inflated rhetoric on missile efficacy, based on misassumptions and technological exuberance, was punctured by the failure of missiles in actual combat. Still, the institution persisted in its belief that "all the missiles worked." Fino examines why senior leaders—some of them renowned combat pilots—clung to this fiction despite dramatic failures spotlighted in Korea and Vietnam. Nevertheless, a handful of relatively junior officers effected a dislocation from the dominant missile technology and installed an external air-to-air cannon on the F-4 Phantom, an aircraft previously exalted as a missile-only guarantor of air superiority. Their clashes with the enemy and the bureaucracy created long-lasting results and demonstrated how the concepts of technological skepticism, exuberance, and dislocation instruct military innovation. Indeed, this study provides today's leaders and strategists much needed insights into how to bring about change and create advantage in the swirling complexity of modern technology and bureaucracy.

Originally written as a master's thesis for Air University's School of Advanced Air and Space Studies (SAASS), Colonel Fino's *Technological Dislocations and Military Innovation* received the Air University Foundation's 2010 award for the best SAASS thesis on the subject of technology, space, or cyberspace. I am pleased to commend this excellent study to all who believe that

FOREWORD

broadly informed research, rigorous argumentation, and clear expression are vital to the advancement of strategic thought and practice.

Timothy P. Schultz
Colonel, USAF, Retired, PhD
Associate Dean of Academic Affairs,
US Naval War College

About the Author

Lt Col Steven A. Fino graduated from the US Air Force Academy as a distinguished graduate in 1996. Following an academic assignment to the University of California–Los Angeles (UCLA), he graduated from Euro-NATO Joint Jet Pilot Training at Sheppard AFB, Texas, in 1998 and was selected to fly the F-15C Eagle. In June 2004 he graduated from the US Air Force Weapons School at Nellis AFB, Nevada. His weapons school research paper, "Achieving Air Superiority in a Gardenia Electronic Attack Environment," won top honors among the 73 graduates. His flying assignments in the F-15C included Langley AFB, Virginia; Kadena AB, Japan; Eglin AFB, Florida; and Nellis.

While at Nellis, Colonel Fino was involved in the operational testing of critical software and hardware upgrades for the F-15C Eagle. In addition to his instrumental role in developing new fourth- and fifth-generation fighter tactics for use against advanced electronic attack-equipped adversaries, Colonel Fino also worked with the F-35 Joint Strike Fighter program staff as a core pilot specializing in air-to-air tactics. Following his assignment at the School of Advanced Air and Space Studies at Maxwell AFB in 2010, Colonel Fino was stationed at the Pentagon in the Air Force chief of staff's strategic studies group, where he provided critical analysis and policy recommendations on a variety of strategic initiatives, including the joint Air Force–Navy Air-Sea Battle initiative.

Colonel Fino has flown combat missions in support of Operations Northern Watch and Southern Watch and homeland defense missions in support of Operation Noble Eagle. He has a bachelor of science degree in materials science from the USAFA, a master of science degree in materials science and engineering from UCLA, a master of science degree in operations analysis from the Air Force Institute of Technology at Wright-Patterson AFB, Ohio, and a master's degree in airpower art and science from SAASS. He is currently pursuing a PhD in engineering systems at the Massachusetts Institute of Technology.

Acknowledgments

Like the technological innovations discussed herein, this paper quickly developed a life of its own, and my attitude toward it fluctuated almost daily between skepticism and exuberance. Col Tim Schultz, SAASS commandant, dean, and trusted advisor, always provided a valuable moderating influence. Two other SAASS professors were particularly instrumental in steering me toward investigating the role of science and technology in the military: Dr. Stephen Chiabotti, who enlightened me to the science and technology historical subfield; and Dr. Michael Pavelec, who helped me transform a jumble of random ideas into a semicoherent research prospectus. Their collective efforts ensured that my foray into the annals of airpower history was satisfying and rewarding.

The superb staff at the Air Force Historical Research Agency (AFHRA) and the wonderful librarians at the Muir S. Fairchild Research Information Center at Maxwell aided my adventure. My SAASS professors—especially Dr. Hal Winton—helped refine my writing skills, while my SAASS colleagues helped hone my argumentation skills. I emerged from this 11-month program better able to conceptualize and articulate the diverse challenges facing the Air Force and our nation.

I extend special thanks to retired colonel Gail "Evil" Peck. Evil, a talented F-4 driver who flew in Vietnam, is also a gifted civilian instructor at the USAF Weapons School at Nellis. I had the pleasure (or agony) of sitting through 20-plus hours of Evil's instruction on the AIM-7 Sparrow air-to-air guided missile during my weapons school training in 2004. When I needed an authoritative source to answer my F-4 systems questions, I knew where to turn, and Evil graciously accommodated me. He also put me in touch with retired major Sam Bakke, an F-4 pilot who worked for Col Frederick "Boots" Blesse in 1967 at Da Nang Air Base, South Vietnam. Mr. Bakke provided a valuable firsthand account of the 366th Tactical Fighter Wing's internal dynamics and critical insight into the innovation process central to this thesis. Mr. Bakke in turn put me in touch with retired lieutenant colonel Darnell "D" Simmonds, an F-4 pilot who served with Col Robin Olds in the 8th Tactical Fighter Wing at Ubon, Thailand, and a member of the only F-4 crew to score two MiGs with the external gun in a single mission. I am proud to serve in the Air Force legacy that Evil, Sam, and D helped cultivate.

Finally, there is no way to adequately say "thank you" to my wife and our wonderful children. I draw my strength and motivation from their untiring patience and support. To them I dedicate this work.

Abstract

History reveals a Janus-faced, nearly schizophrenic military attitude toward technological innovation. Some technologies are stymied by bureaucratic skepticism; others are exuberantly embraced by the organization. The opposing perceptions of skepticism and exuberance that greet military technologies mirror the different interpretations of technology's role in broader society. Thomas Hughes's theory of technological momentum attempted to reconcile two of the disparate perspectives—social constructivism and technological determinism. The theory of technological dislocations advanced by this thesis is a refinement of Hughes's theory and is more reflective of the complex, interdependent relationship that exists between technology and society.

Drawing on a single, detailed historical case study that examines the development of air-to-air armament within the US Air Force, post–World War II through Operation Rolling Thunder, this paper illustrates how an unwavering commitment to existing technologies and a fascination with the promise of new technologies often obfuscate an institution's ability to recognize and adapt to an evolving strategic environment. The importance of a keen marketing strategy in outmaneuvering bureaucratic skepticism, the benefits of adopting a strategy of innovative systems integration vice outright systems acquisition, and the need for credible, innovative individuals and courageous commanders who are willing to act on their subordinates' recommendations are all revealed as being critical to successful technological innovation.

Chapter 1

Introduction

Da Nang was a mess. We shared operational use of the base with the Vietnamese and neither the previous American nor Vietnamese commander appeared to have a handle on the wide variety of problems that faced them. . . . To make matters worse, the senior officers in the wing were doing little or no flying.

—Maj Gen Frederick "Boots" Blesse, USAF

As the new deputy commander for operations at the 366th Tactical Fighter Wing (TFW), Da Nang Air Base, South Vietnam (SVN), then-colonel "Boots" Blesse, a Korean War double ace, was determined to transform his unit into a "respectable combat outfit." He and his assistant, Col Bert Brennan, hammered out new wing directives, established new traffic patterns to minimize aircraft exposure to potential ground attack, and developed new landing procedures to curb the frequent mishaps that occurred on the poorly designed and often wet Vietnamese runway. More importantly, Blesse and Brennan understood that "you can't push a piece of string" and made a pact shortly after their arrival in April 1967 that they "would be two full colonels who flew 100 missions 'Up North.'" Whereas some Air Force colonels in Vietnam tried to limit their exposure to the more dangerous combat missions, merely biding their time before rotating back home to the states after their one-year assignment, Blesse and Brennan were determined to fly "the same missions as the buck pilots."[1]

Thus, when the wing commander, Col Jones Bolt, stopped by to see Blesse on 13 May 1967, the message came as quite a shock. "We have several other missions besides the Hanoi run and I expect you to be active in them all," he informed Blesse. "You can't be going to Pack Six every day, so get back to spreading yourself around."[2] Although heartbroken, Blesse knew his commander was right. He had flown two Pack Six missions the two previous days—on one, even loitering in the target area for an extra 10 minutes "hoping to see enemy aircraft." It wasn't that Blesse was "hogging" the combat missions; he had a personal stake in the outcome of the next aerial engagement with the North Vietnamese MiGs.[3] So it was with some anxiety and much reservation that Blesse watched the next day's two flights of four F-4C Phantoms lumber off the runway at Da Nang. It was Sunday afternoon, 14 May

INTRODUCTION

1967. The F-4s had a mission "Up North," and several of them were loaded with the Air Force fighter's newest air-to-air weapon.[4]

Piloting the lead aircraft—call sign Speedo 1—was Maj James Hargrove, Jr. Because he occupied the front seat of the F-4, he was the aircraft commander. In the backseat sat 1st Lt Stephen H. DeMuth. DeMuth was also a pilot, as were all Air Force F-4 backseaters during the Vietnam War, but he and the other pilots flying in the rear seat had grown accustomed to being referred to, somewhat derogatorily, as the "GIB" (guy in back). Mimicking the previous two days' missions, Hargrove's four-ship of F-4s teamed with an additional flight of four F-4s—callsign Elgin 1—to provide MiG combat air patrol (MiGCAP) cover for 19 388th TFW F-105 Thunderchief fighter-bombers from Korat Royal Thai AFB, Thailand, that were tasked with striking targets near Hanoi. The specific target that Sunday afternoon was the Ha Dong army barracks, located approximately four miles south of the capital city. After the members of Speedo flight completed their prestrike aerial refueling in the skies over Thailand and began their trek north toward Hanoi, Air Force early warning controllers alerted them to the suspected presence of enemy MiGs in the target area. The aircraft of Speedo flight assumed their tactical formation, slightly behind and 2,000 feet above the F-105 strikers, and eagerly searched the area with their state-of-the-art AN/APQ-100 radars. As the strike force neared the target, the Air Force controllers continued to warn the F-4s that MiGs were patrolling the area. Just then, the lead F-105 called, "MiG, 12 o'clock low, coming under."

Flying at 19,000 feet and more than 500 knots airspeed, offset slightly to the right of Hargrove in Speedo 1, Capt James Craig, Jr., in Speedo 3, and his GIB, 1st Lt James Talley, were the first F-4 crew to spot the MiGs, passing head-on, underneath the F-105 strikers just ahead of Speedo flight. A passing glance out the left side of the F-4 and the shimmer of silver wings against the cloudy undercast alerted Craig to two more MiGs at nine o'clock. Hargrove called for the flight to turn left, descend, and engage the enemy aircraft. Midway through the turn, Craig recognized that the "enemy MiGs" he had seen to the left were in fact friendly F-105 strikers. Pausing momentarily in disgust at his misidentification and now wondering where the earlier-spotted MiGs were, Craig resumed his visual scan of the airspace surrounding the F-105s and quickly, and this time correctly, identified four MiG-17s, split into two elements of two aircraft each, chasing down the F-105s. Communicating the observed MiG formation to the other Speedo flight members, Craig and his element mates in Speedo 4 started to maneuver into position against the trailing two MiGs. Hargrove in Speedo 1 jettisoned his cumbersome external fuel tanks and announced that his element would attack the leading two MiGs. Hargrove's wingmen, Capt William Carey, and 1st Lt Ray Dothard in Speedo 2, jettisoned their external fuel tanks and maneuvered into a supporting position slightly aft of Speedo 1.

Speedo 1 and 2 tightened their left turns, the four American pilots straining against the rapidly increasing G-forces, and accelerated downhill toward the MiGs, hoping to position themselves at the MiGs' six o'clock before the enemy fighters could react. It was to no avail. The MiGs may have seen the white vapor trails streaming off the F-4 wingtips in the humid afternoon air, or they may have detected the characteristic black smoke spewing from the Phantom's General Electric J79 engines tracing the F-4s' maneuvers against the blue sky above.[5]

The MiGs started a hard, diving left turn toward Hargrove and his wingman, eventually passing head-on before they disappeared into the clouds behind and below the F-4s; there was no time for Hargrove to mount an attack. Frustrated, Hargrove began a climbing right turn, exchanging kinetic energy for potential energy and maneuvering away from the deadly antiaircraft artillery (AAA) that preyed on fighters caught flying too low to the ground. As the needle on the altimeter spun through 7,000 feet, Hargrove looked outside and surveyed the area. Exuberantly recounting the engagement for Blesse after he landed back at Da Nang, Hargrove described the scene: "Wall to wall MiGs, Colonel. You should have been there!"[6] Indeed, F-4 and F-105 pilot reports submitted after the mission revealed the presence of 16 MiG-17s in the skies facing Speedo flight that afternoon.[7] At this point, Speedo flight had accounted for only four.

Whereas the North Vietnamese MiGs quickly and successfully shook Speedo 1 and 2, Speedo 3 and 4's MiG prey were initially not so lucky. Craig and his wingman were able to dive on the MiGs, achieving the ideal six o'clock position from which to launch their Sparrow radar-guided or Sidewinder heat-seeking missiles. Craig pointed the nose of the F-4 at one of the MiGs and told Talley in the back seat to get a radar lock.[8] While Talley worked the radar, Craig ordered his wingman to jettison the external fuel tanks as Speedo 1 and 2 had done earlier—standard procedure to increase the F-4's performance for an imminent dogfight. Unfortunately, only one of Craig's two wing tanks fell away from the aircraft, leaving one tank partially filled with fuel still attached to the aircraft, seriously handicapping the Phantom's maneuverability and stability.

With Craig in the front seat trying desperately to jettison the remaining fuel tank and Talley in the back seat working feverishly to attain a radar lock, the MiG suddenly initiated a hard, descending 180-degree left turn toward Speedo 3 and 4. Recognizing the fleeting weapons opportunity as the MiG

INTRODUCTION

rapidly approached minimum missile employment range, Craig pointed his F-4 at the turning MiG and launched a Sparrow missile despite lacking the requisite radar lock needed to accurately guide the missile to the target. The aircraft shuddered as the 12-foot missile ejected from its nesting place under the belly of the F-4, but the missile motor never fired, and it fell harmlessly to the ground as the MiG disappeared into the clouds below. Craig and his wingman began a climbing right turn, looking to escape the lethal low-altitude AAA employment zone as Speedo 1 and 2 had done earlier.

Midway through their climb, Craig visually acquired another MiG two-ship off the left side, low, in a left-hand turn. In a maneuver nearly identical to their first, Speedo 3 and 4 entered a tight, descending left turn and arrived undetected just behind the MiGs. Craig again pointed the nose of his F-4 at one of the MiGs as Talley adjusted the radar scan in hopes of achieving a radar lock on the enemy aircraft. Talley was successful this time, and from a mile away, in a left-hand turn, with the radar seemingly locked on to the target, Craig again squeezed the trigger and launched a Sparrow missile. Unfortunately, the result was the same—the missile separated from the aircraft and then promptly fell 4,000 feet to the ground. Now twice frustrated and too close to the MiGs to launch another missile, Craig and his wingman initiated a high-speed "yo-yo" maneuver to gain lateral and vertical separation from the MiGs and started searching for yet another target.[9]

Meanwhile, Speedo 1 and 2 had similarly engaged another two flights of two MiGs each, with unfortunately similar results—both of Hargrove's Sparrow missiles failed to guide, much less score a hit. After more than five minutes of intense ai combat, the F-4s in Speedo flight had launched four Sparrow missiles, and none had worked as advertised—all fell harmlessly to the ground. The F-4s could ill afford to remain in the fight much longer. Well outnumbered by the MiGs, the American aircrews were losing situational awareness while quickly depleting their F-4's precious energy and maneuverability with continued attacks. Their luck was beginning to run out.

Following his last unsuccessful Sparrow missile attack, Hargrove directed his element to pursue another MiG. By turning to pursue the MiG in sight, though, Hargrove inadvertently maneuvered his element directly in front of an attacking MiG. Fixated on the MiG in front of them, Hargrove and his wingman failed to detect the two incoming enemy Atoll heat-seeking missiles launched from the MiG now behind them. Luckily, the North Vietnamese missile performance was comparable to the Americans' that day, and the missiles failed to guide toward the F-4 element. The MiG continued to press the attack, rapidly closing the range between the aircraft. Only a last-second, passing glance alerted Hargrove to the presence of the attacking MiG-17, the front of the enemy aircraft rhythmically sparkling with muzzle flashes as the Vietnamese pilot fired his cannons at the F-4s.

As missile failures continued to frustrate the members of Speedo flight, heir accompanying flight of four F-4s—callsign Elgin—led by Maj Sam Bakke and his GIB, Capt Robert Lambert, approached the target area and quickly joined the melee. Bakke in Elgin 1 selected a MiG and fired two Sidewinder missiles at it. The enemy pilot abruptly initiated a hard defensive turn and successfully outmaneuvered the American heat-seeking missiles. Observing their initial missiles defeated, Elgin 1 and 2 executed a high-speed "yo-yo" maneuver to reposition away from the turning MiG and selected another MiG-17 to attack. That MiG dove into the low clouds before Bakke could maneuver his element into a firing position.

Simultaneously, Elgin 3 and 4, flying in a supporting position slightly above the other two members of Elgin flight, caught a glimpse of another pair of MiGs rapidly closing on and firing at Bakke and his wingman. Hoping to distract the MiG pilots, Elgin 4 fired two Sidewinder missiles in quick succession, but neither missile was launched within proper parameters and both failed to guide toward the target. Elgin 3 also attempted to launch a Sidewinder missile at the attacking MiGs; that missile, despite being launched with the requisite tone and within valid launch parameters, misfired and never left the aircraft. Then, as Elgin 3 and 4 were engaging the MiGs that were attacking Elgin 1 and 2, another set of MiGs appeared and began attacking Elgin 3 and 4. Like Speedo flight, Elgin flight's luck was beginning to wear thin.

Once under attack, both Elgin 3 and 4 immediately initiated individual defensive "jink" maneuvers, but not before the MiGs' bullets passed within 15 feet of Elgin 4's crew. Fortunately, Elgin 4's maneuvers were effective; the F-4 crew successfully shook the MiG attacker and, in a remarkable stroke of good luck, ended up in perfect Sidewinder firing position behind another MiG that inexplicably flew directly in front of them. They tried to take advantage of the precious opportunity, but par for the day, that Sidewinder missile also failed to guide toward the target. The crew of Elgin 3 successfully shook an attacking MiG, and following the last unsuccessful Sidewinder missile attack by Elgin 4, the two aircraft, now both low on fuel, decided to exit the fracas. They turned south out of the target area and joined a flight of F-105s that were heading home after dropping their ordnance on the target.

Elgin 1 and 2 remained in the target area battling the MiGs. After losing sight of the second MiG that dove into the clouds, and as Elgin 3 and 4 were defending themselves from the separate MiG attacks, Bakke and his wingman observed a lone MiG in a left-hand turn a half-mile in front of and 2,000 feet above them. Bakke pointed the F-4 toward the MiG, and Lambert acquired a

INTRODUCTION

radar lock. In his zeal to dispatch the MiG, Bakke squeezed the trigger three times trying to launch a Sparrow missile at the target before he realized that he was too close to the MiG to shoot.[10] Selecting idle power and slowing the F-4 opened the range between the two aircraft. Once outside of minimum missile range, Bakke launched two Sparrow missiles in quick succession at the unsuspecting MiG. The first missile failed to guide, but the second missile " 'homed in' on the target, causing an explosion and fire in the right aft wing root of the MiG-17."[11] The MiG "burst into flame and pitched up about 30 degrees, stalled out, and descended tail first, in a nose high attitude at a rapid rate into the cloud deck" below.[12] Finally, a missile worked; a MiG was destroyed, and Bakke and Lambert had earned a kill.

Bakke and his element mates had no time to celebrate. The North Vietnamese surface-to-air missile (SAM) sites surrounding the target were particularly active that day. The F-105s reported 14 observed SAM launches, one of which claimed an F-105.[13] Fortunately, the SA-2 missile launched toward Bakke's element shortly after it destroyed the MiG missed, detonating almost a mile away. Undeterred, Elgin 1 and 2 continued to attack the MiGs. They engaged another lone MiG with two Sidewinder missiles, but that MiG successfully outmaneuvered both missiles by executing a maximum-G turn.

As they broke off their unsuccessful attack and initiated a climb to higher altitude, the F-4s observed another three MiG-17s flying directly beneath them. Once more, Bakke and Lambert selected a MiG, acquired a radar lock, and fired a Sparrow missile—their last. And once more, the Sparrow missile failed to guide to the target. After separating from the aircraft, the missile veered sharply to the right and rocketed out of sight. Out of missiles, Elgin 1 tried to maneuver the element into position behind the remaining MiGs so that Elgin 2 could engage the enemy aircraft with its missiles, but the last of the remaining MiGs dipped into the clouds below before a stable firing position could be attained. The MiGs never reappeared. Elgin 1 and 2 conducted one last sweep of the target area and then turned south toward the tanker aircraft orbiting over Thailand before continuing home to Da Nang.

Bakke and Lambert's kill was not the only one that day. Immediately before Elgin 3 and 4 defensively reacted to the attacking pair of MiGs, all of the members of Elgin flight observed a "MiG-17 erupt into a ball of flame and dive, at an 80-degree angle, into the cloud shelf." About two minutes later, just prior to Elgin 3 and 4 exiting the target area, Elgin 2 and 3 observed another "MiG-17 in a 60-degree dive, at a high rate of speed, with a thin plume of white smoke trailing the aircraft."[14] Both MiGs were victims of Speedo flight and Blesse's mystery weapon.

Recall that as the members of Elgin flight entered the fight, Hargrove and DeMuth in Speedo 1 were under missile and gun attack by a rapidly closing MiG. Tightening the F-4's turn, Hargrove hoped to both avoid the MiG's bullets and cause the MiG to fly out in front of the Phantom. The tactic worked; the MiG overshot, and Hargrove, slamming the throttles into afterburner, reversed his turn direction to follow the MiG. Unfortunately, the F-4 was too slow, having sacrificed energy and speed executing the tight defensive turn, and the MiG quickly sped away from the lumbering F-4.

Speedo 1 and 2 initiated a climb and searched for other MiG targets. They found two at right, two o'clock, a half-mile away, low. Hargrove started a right turn, selected the trailing MiG in the right-turning formation, and surmised that he was in perfect position to employ the new weapon slung beneath the F-4's belly. Flying between 450 and 500 knots and only 2,000 to 2,500 feet behind the MiG, Hargrove pulled the nose of the F-4 far out in front of the MiG and squeezed the trigger. As the range collapsed inside of 1,000 feet, Hargrove could clearly distinguish the individual aluminum panels that made up the skin of the Russian-built fighter. Hargrove continued to mash down on the trigger. As the range collapsed inside of 500 feet, even more detail on the MiG became apparent. Despite continuing to accelerate toward the MiG on a certain collision course, Hargrove pressed the attack. Watching Hargrove's attack from a supporting position 500 feet behind and 1,000 feet above, slightly offset toward the left, Carey in Speedo 2 began worrying that "Speedo 1 had lost sight of the MiG-17 and would collide with him."[15]

Finally, at 300 feet separation—the point where the image of the MiG completely filled the F-4's windscreen—Hargrove observed the weapon's effectiveness. The weapon was the SUU-16 20-millimeter (mm) gun pod, and at 300 feet the impact of the individual rounds could be observed tearing holes into the MiG's thin aluminum skin right behind the canopy. "At approximately 300 feet, flame erupted from the top of the MiG fuselage. Almost immediately, thereafter, the MiG exploded from the flaming area and the fuselage separated in the area just aft of the canopy."[16]

Desperately trying to avoid the debris from the MiG erupting immediately before him, Hargrove initiated a violent, evasive maneuver to the left, inadvertently toward Speedo 2. Carey and Dothard in Speedo 2, in turn, executed an aggressive climbing turn in their own frantic attempt to avoid hitting both the MiG debris and Speedo 1. In the commotion, Speedo 1 and 2 became separated from each other, and the two fighters never successfully rejoined. Instead, Speedo 2 came upon another set of American fighters, and Hargrove in Speedo 1 directed Carey in Speedo 2 to join with the other fighters and accompany them home.

INTRODUCTION

Speedo 1, now operating alone, attempted to engage an additional MiG with a Sidewinder missile, but Hargrove launched the missile when the F-4 was under too many G-forces, and it missed the target. Hargrove continued to close on the target, intending to employ the gun once again, but passing inside of 2,500 feet he realized he was out of ammunition. Rather than continue to press the attack, the crew of Speedo 1 thought better of using their sole remaining Sidewinder and elected instead to retain the missile for the long trek south to friendly airspace.

Craig and Talley in Speedo 3 also had success with the new SUU-16 20 mm gun pod that afternoon. Frustrated by two unsuccessful Sparrow launches, Craig observed two MiGs at nine o'clock low, in a left-hand turn, and immediately decided to maneuver for a gun attack. As Craig led his element in a diving left turn to engage the pair of MiGs, he noticed another lone MiG trailing the others by 3,000 feet. Craig wisely decided to switch his attack to the trailing MiG. Speedo 3 and 4 executed a barrel roll to gain better position on the trailing MiG, but, like Elgin 3 and 4, they too came under SAM fire. Similarly undeterred, Speedo 3 and 4 continued to prosecute the attack. The MiG tried to shake the chasing F-4s with a sudden reversal in turn direction, but Craig matched the maneuver perfectly and closed to within 1,500 to 2,000 feet before opening fire. Craig later reported, "I followed the MiG through the turn reversal, pulled lead, and fired a two and one-half second burst from my 20-mm cannon." His aim was spot-on. "Flames immediately erupted from his [the MiG's] right wing root and extended past the tailpipe. As I yo-yo'd high, the MiG rolled out to wings level, in a slight descent, and I observed fire coming from the left fuselage area. I initiated a follow up attack. However before I could fire, the MiG burst into flames from the cockpit aft and immediately pitched over and dived vertically into the very low undercast."[17] Shortly thereafter, Craig and his element-mate rejoined with Hargrove in Speedo 1 and together they pressed home, looking forward to the celebration that would take place later that night at the DOOM, the Da Nang officer's open mess.

Because the 366th wing commander was in Hong Kong for a meeting that fateful day in May 1967, Blesse had the pleasure of authoring the wing's daily operational summary report for Gen William Momyer at Seventh Air Force. It read: "SPEEDO Fl[igh]t: Today's success with SUU-16 on the F-4C confirms feasibility of this idea. Wing now has 14 a[ir]c[ra]ft modified and continuing modification at as rapid a pace as possible. We feel certain there will be two pilot meetings tonight. One in Hanoi, the other in the 8th Tac Fighter Wing."[18] Surprisingly, the numerous failures of the air-to-air missiles that afternoon warranted no mention in the summary report; their lackluster performance was not deemed out of the ordinary.[19]

INTRODUCTION

How was it that in the dawning age of solid-state electronic radars paired with advanced air-to-air radar-guided and heat-seeking missiles, the successful combat employment of an antiquated weapons system cumbersomely mounted externally on an F-4 fighter aircraft was heralded so triumphantly by a seasoned combat fighter pilot? Surely, Air Force fighter pilots would have instead preferred, indeed demanded, the latest and most technologically advanced weaponry to help them in the life-or-death struggle that is air combat. If that technology failed to live up to advertised performance requirements, as it did on 14 May 1967 and countless times before that, then one would assume the Air Force pilots would have been up in arms, demanding the technology be quickly improved and refined. Instead, pilots like "Boots" Blesse pursued a decidedly low-tech weapon and fought to get a gun, even in bastardized form, on the F-4C.

The story then of the return of the air-to-air cannon to the F-4 Phantom provides a unique vantage point to peer into the complex interdependent relationship between technology and the US military—a relationship that historically alternates between periods of technological exuberance and technological skepticism. The theoretical lens of *technological dislocations* explains this relationship. To appreciate the theory's utility, a conceptual understanding of the foundational theories of technological change, especially Thomas Hughes's theory of technological momentum, is required and presented in chapter 2. Chapter 3 presents the theory of technological dislocations. Chapters 4, 5, and 6 describe the development of Air Force air-to-air weaponry post–World War II through Operation Rolling Thunder. This historical survey provides a useful case study to evaluate the role of technological dislocations in military history. Armed with this historical knowledge, the concept of technological dislocations can be extended to the larger context of military innovation, which is the subject of chapter 7. Collectively, a thorough understanding of the nature of technological development based on these concepts provides decision makers with the necessary tools to assess technology's influence on strategic decisions.

Notes

All notes appear in shortened form. For full details, see the appropriate entry in the bibliography. Additionally, note that portions of this thesis were used in Fino, "Breaking the Trance."

1. Blesse, *Check Six*, 117–18.
2. Ibid., 124. "Pack Six" refers to Route Package Six. To simplify command arrangements during the Vietnam War, the Navy and the Air Force subdivided North Vietnam into seven geographic regions (Route Packs One through Five, and 6A and 6B). Hanoi and the majority of lucrative North Vietnamese targets were located in Route Pack Six. Michel, *Clashes*, 38. During

INTRODUCTION

a 1984 interview, Maj Gen Jones Bolt described Blesse's enthusiasm: "Boots was a pretty good troup [sic]; he was a little flamboyant sort of fellow; you had to keep your thumb on him. Boots wanted to fly too much." Bolt, oral history interview, 190.

3. Blesse, *Check Six*, 123.

4. PACAF Command Center, Chronological Log, 14–15 May 1967; Bakke, interview. Specifically, the aircraft flying in the #1 and #3 positions were supposed to be loaded with the new weapon. However, one of the flight's aircraft was unable to launch that afternoon due to a malfunction, and an airborne spare aircraft rolled into the Elgin 1 position. Unfortunately, there were not enough of the new weapons to equip the spare aircraft. The narrative of the 14 May 1967 mission that follows in this chapter is based on information in the Air Force Historical Research Agency (AFHRA) Aerial Victory Credit folders: AFHRA, "1967–14 May; Hargrove and DeMuth"; AFHRA, "1967–14 May; Craig and Talley"; and AFHRA, "1967–14 May; Bakke and Lambert." Each AFHRA folder contains a narrative summary and aircrew personal statements and/or memoranda to the "Enemy Aircraft Claims Board" that describe the MiG engagement.

5. Describing the characteristic F-4 smoke trail in subafterburner powers settings, one former combat F-4 pilot noted, "There were times when I could see F-4s 15 or 20 miles away due to the smoke trail—especially at a co-altitude when the F-4s were highlighted against the haze layer." Peck, e-mail.

6. Blesse, *Check Six*, 123.

7. AFHRA, "1967–14 May; Bakke and Lambert." The other F-4 flight, callsign Elgin, encountered another 10 MiG aircraft that afternoon, but based on the proximity of the two fights, there may be some overlap in the reported number of MiGs in Speedo and Elgin flights' accounts.

8. In close combat, F-4 crews generally used their radars in Boresight mode. The 8th Tactical Fighter Wing's (TFW) "Tactical Doctrine" manual described the boresight procedure: "Going to Boresight cages the radar antenna to the dead ahead position. The aircraft commander now steers to place the target within the reticle of the optical sight and places the pipper on the target. The radar target blip will appear in the pilot's radar scope 'B' sweep. The pilot then locks on to the target in the Boresight mode. Once lock-on is acquired, the system is returned to the RADAR mode to provide full system capability with auto tracking. The aircraft commander now begins to pull lead on the target by placing the target tangent to the top of the radome. . . . Upon reaching the 'in range' area, the AIM-7E should be launched." 8th TFW, "Tactical Doctrine," March 1967, 80.

9. Boyd described "the high speed yo-yo" as "an offensive tactic in which the attacker maneuvers through both the vertical and the horizontal planes to prevent an overshoot in the plane of the defender's turn. . . . The purpose of the maneuver is . . . to maintain an offensive advantage by keeping nose-tail separation between the attacker and defender." The offensive maneuver begins with an aggressive pull up into the vertical plane while rolling slightly away from the target. As the distance to the target begins to increase toward an acceptable range, the offender rolls back toward the target and initiates a descent toward the defender's extended six o'clock position. Boyd, *Aerial Attack Study*, 64–73.

10. Bakke, to 366 TFW Enemy Aircraft Claims Board. The F-4 weapons system was equipped with an "interlock" switch that when activated inhibited launching a Sparrow missile unless all of the missile firing parameters were met.

11. Ibid.

12. AFHRA, "1967–14 May; Bakke and Lambert."

13. Message, 388 TFW to NMCC. The message noted that the pilot of the downed F-105, callsign Crab 2, was successfully recovered by rescue forces.
14. AFHRA, "1967–14 May; Bakke and Lambert."
15. Statement from Carey.
16. Hargrove, to 366 TFW (DCO).
17. Craig, to 366 TFW Enemy Aircraft Claims Board.
18. Message, 366 TFW to 7 AF CC. Blesse's reference to the pilot meeting at the 8th TFW reflected his belief that Col Robin Olds, 8th TFW commander, would demand quick implementation of the 366th TFW's innovation within his own F-4 wing at Ubon, Thailand. In his autobiography, *Check Six*, Blesse stated that the summary report read: "We engaged enemy aircraft in the Hanoi area, shooting down three without the loss of any F-4s. One was destroyed with missiles, an AIM-7 that missed and an AIM-9 heat seeker that hit. That kill cost the US government $46,000. The other two aircraft were destroyed using the 20-mm cannon—226 rounds in one case and 110 rounds in the other. Those two kills cost the US government $1,130 and $550, respectively. As a result of today's action, it is my personal opinion there will be two pilot's meetings in the theater tonight—one in Hanoi and the other at the 8th TFW at Ubon." Blesse, *Check Six*, 124. Blesse's recollection of the summary report in *Check Six* is factually incorrect. Rather than firing an AIM-7 Sparrow followed by an AIM-9 Sidewinder that destroyed the MiG as Blesse described, Bakke is clear in his statement following the event: "I fired two Sparrow missiles while pursuing the target in a left turn. One missile did not guide and the other 'homed in' on the target." Bakke, to 366 TFW Enemy Aircraft Claims Board.
19. Coincidentally, the Sparrow missile failures did catch the attention of the Pacific Air Forces (PACAF) commanding general, who four days later demanded "immediate analysis of AIM-7 missile failures during MiG engagements on 12, 13, 14 May 67." Message, PACAF CC to 7 AF and 13 AF, 18.

Chapter 2

Foundations of Technology

But lo! men have become the tools of their tools.

—Henry David Thoreau

On 17 January 1961, Pres. Dwight Eisenhower delivered his farewell address to the nation. Besides extending the customary thanks to Congress and offering best wishes for the next presidential administration, Eisenhower warned of two "threats, new in kind or degree," that loomed over the nation. Both concerned technology. The first admonition is well cited.

> Our military organization today bears little relation to that known of any of my predecessors in peacetime, or indeed by the fighting men of World War II or Korea.
>
> Until the latest of our world conflicts, the United States had no armaments industry. American makers of plowshares could, with time and as required, make swords as well. But we can no longer risk emergency improvisation of national defense; we have been compelled to create a permanent armaments industry of vast proportions. Added to this, three and a half million men and women are directly engaged in the defense establishment. We annually spend on military security more than the net income of all United States corporations.
>
> Now this conjunction of an immense military establishment and a large arms industry is new in the American experience. The total influence—economic, political, even spiritual—is felt in every city, every State house, every office of the Federal government. We recognize the imperative need for this development. Yet we must not fail to comprehend its grave implications. Our toil, resources and livelihood are all involved; so is the very structure of our society.
>
> In the councils of government, we must guard against the acquisition of unwarranted influence, whether sought or unsought, by the military-industrial complex. The potential for the disastrous rise of misplaced power exists and will persist.
>
> We must never let the weight of this combination endanger our liberties or democratic processes. We should take nothing for granted. Only an alert and knowledgeable citizenry can compel the proper meshing of the huge industrial and military machinery of defense with our peaceful methods and goals, so that security and liberty may prosper together.

The second warning is less well known.

> Akin to and largely responsible for the sweeping changes in our industrial-military posture has been the technological revolution during recent decades.

In this revolution, research has become central; it also becomes more formalized, complex, and costly. A steadily increasing share is conducted for, by, or at the direction of, the Federal government.

Today, the solitary inventor, tinkering in his shop, has been overshadowed by task forces of scientists, in laboratories and testing fields. In the same fashion, the free university, historically the fountainhead of free ideas and scientific discovery, has experienced a revolution in the conduct of research. Partly because of the huge costs involved, a government contract becomes virtually a substitute for intellectual curiosity. For every old blackboard there are now hundreds of new electronic computers.

The prospect of domination of the nation's scholars by Federal employment, project allocations, and the power of money is ever present—and is gravely to be regarded.

Yet, in holding scientific research and discovery in respect, as we should, we must also be alert to the equal and opposite danger that public policy could itself become the captive of a scientific-technological elite.[1]

Stephen Ambrose characterized Eisenhower's farewell speech as that of "a soldier-prophet, a general who has given his life to the defense of freedom and the achievement of peace."[2] Not all received the speech so warmly. One Air Force writer questioned Eisenhower's sincerity, commenting, "President Eisenhower . . . had his eye on a place in history as a military hero who revolted against war."[3] Walter McDougall, writing in 1985, described Eisenhower's farewell speech as eerily prescient. "It reads like prophecy now, its phrases sagging with future memories."[4] McDougall lamented that Eisenhower's warnings, regardless of their particular motivation, went unheeded. For McDougall, the burgeoning role of the military-industrial complex and an unhealthy faith in technology's unrelenting march toward "progress" fostered a technocratic ideology that quickly permeated the United States.[5]

Technological Exuberance

Nearly a decade prior to McDougall, Herbert York also called attention to the nation's fascination with technological solutions to international and domestic issues. He suggested that this attitude sprouted from the nation's unique world stature: "The United States is richer and more powerful, and its science and technology are more dynamic and generate more ideals and inventions of all kinds, including ever more powerful and exotic means of mass destruction. In short, the root of the problem has not been maliciousness, but rather a sort of *technological exuberance* that has overwhelmed the other factors that go into the making of overall national policy."[6] While York's unabashed faith in the United States' technological superiority may conjure visions of a social Darwinist argument, the idea that civilian and military leaders can be blinkered by the promise of technology—York's technological exuberance—is consistent with the message in Eisenhower's farewell address and McDougall's observation of a United States slipping toward technocracy.[7]

The link between technology and the military can be especially profound. Merritt Roe Smith observed that the "military enterprise has played a central role in America's rise as an industrial power and . . . since the early days of the republic, industrial might has been intimately connected with military might."[8] Looking toward the future in a decidedly ethnocentric manner reminiscent of York's argument, a US Army War College report written in 2000 claimed that "the ability to accept and capitalize on emerging technology will be a determinant of success in future armed conflict. No military is better at this than the American, in large part because no culture is better at it than the American."[9] Indeed, a cursory review of popular US military history reveals the services' affinity for relying on technological solutions to ensure national security—in York's words, "a sort of technological exuberance."

The trend is particularly evident within the US Air Force. After gaining independence in 1947, the Air Force built upon its World War II image as a technologically advanced fighting force armed with an array of high-speed fighters and massive four-engine bombers. The chief of the fledgling air service, Gen Henry "Hap" Arnold, relished his opportunity to cultivate technology within the service. He described his charge as "get[ting] the best brains available, hav[ing] them use as a background the latest scientific developments in the air arms," and creating instruments "for our airplanes . . . that are too difficult for our Air Force engineers to develop themselves."[10] Having for years been constrained by the world war's unrelenting demands for immediate technological practicality, Arnold was excited to now "look ahead and set free the evangelist of technology that dwelt within him."[11] Neil Sheehan posited that Arnold "intended to leave to his beloved air arm a heritage of science and technology so deeply imbued in the institution that the weapons it would fight with would always be the best the state of the art could provide and those on its drawing boards would be prodigies of futuristic thought."[12]

Arnold had already laid the foundation by war's end. Earlier, the air chief established the Army Air Forces Scientific Advisory Group, a collection of military officers and academics led by scientific whiz Dr. Theodore von Kármán (at Arnold's behest), and tasked it with peering into the future and charting a course for Air Force technological development. The group's 33-volume report, *Toward New Horizons*, was completed in December 1945. The title of the first volume, "Science: The Key to Air Supremacy," was indicative of the report's general conclusions. Von Kármán's executive summary of the volume boldly proclaimed: "The men in charge of the future Air Forces should always remember that problems never have final or universal solutions, and only a constant inquisitive attitude toward science and a *ceaseless and swift adaptation to new developments* can maintain the security of this nation through world air supremacy."[13] Bolstered by the promise of technology, the nascent Air Force of the 1950s marketed itself as *the* military service of the future, proudly ushering in the "Air Age" with visions of gleaming B-36 bombers soaring high across the sky, far above Soviet air defenses, ready to deliver the atomic weapons that American scientific ingenuity had bequeathed to the nation.[14]

A decade later, images of futuristic space rockets and ballistic missiles dominated the public and military consciousness. The Air Force sought to capitalize on the fascination and aggressively lobbied for a manned presence in space independent from that of the newly formed National Aeronautics and Space Administration (NASA).[15] The Air Force's vehicle, the X-20A Dyna-Soar—"a low, delta-winged spaceplane to be launched on a Titan rocket but land like an airplane"—eventually informed NASA's space shuttle designs.[16] The Air Force originally marketed the X-20A as an ideal way to quickly deliver nuclear weapons anywhere in the world. However, as the space antiweaponization movement became more entrenched, the mission of the X-20A to rain down nuclear destruction from space became untenable. The Air Force scrambled to identify a more palatable purpose for the Dyna-Soar. According to McDougall, the subsequent search for a useful application for the impressive but impractical technology was "typical [of a] big project [at the time]: demonstration of technical feasibility, privately funded research and salesmanship leading to military acceptance, extrapolation of existing technology, contrivance of plausible military missions, the savor of 'technological sweetness,' and finally the Sputnik panic."[17] McDougall's lambasting continued: "It [the X-20] was a bastard child of the rocket revolution, an idea too good to pass up, if only because it promised spaceflight without dispensing with wings or a pilot. . . . It was wet-nursed by industry and raised by the military on the vaguest of pretexts."[18] After seven years and $400 million in funding, but with the program still facing "imposing technical challenges, . . . an overly ambitious set of objectives," and an "ill-defined military requirement," Secretary of Defense Robert McNamara cancelled the program in 1964.[19]

By the late 1960s and early 1970s, the military, grasping for technological solutions that would facilitate victory in the jungles of Vietnam and Laos, became entranced with the promise of cybernetic warfare.[20] In 1969 Gen William Westmoreland predicted, "On the battlefield of the future, enemy forces will be located, tracked, and targeted almost instantaneously through the use of data links, computer assisted intelligence evaluation and automated fire control. With first round probabilities approaching certainty, and with surveillance devices that can continually track the enemy, the need for large forces will be less important."[21] Within two years, Westmoreland's vision was largely realized in the jungles of Southeast Asia. Under the auspices of Igloo White, the American military deployed and maintained a system of "acoustic and seismic" sensors along the Ho Chi Minh Trail at an annual cost approaching $1 billion.[22] The sensors' signals were relayed by overhead aircraft "to the heart of the system, an IBM 360/65 computer at Nakhon Phanom Royal Thai Air Force Base." The computer-processed information enabled "real-time tracking of the truck traffic" moving into SVN.[23] Fueled by the tactical intelligence goldmine, the Igloo White system "triggered massive B-52 and fighter strikes aimed at destroying the road structure and the trucks in transit."[24] However, when North Vietnam (NVN) responded in November 1971 by deploying SAMs and fighters to counter the B-52 strikes, they rendered the technologically impressive Igloo White system impotent. The North Vietnamese counter not only curtailed the Americans' ability to act on the high-tech intelligence, but it also capitalized on the shifting "psychology of the [American] war effort" and the new emphasis "on limiting American casualties of all types, and especially avoiding the loss of highly visible assets like the B-52."[25]

In 1983 Pres. Ronald Reagan and the nation again turned to the promise of futuristic technology to provide for the national defense:

> Let us turn to the very strengths in technology that spawned our great industrial base and that have given us the quality of life that we enjoy today.
>
> What if free people could live secure in the knowledge that their security did not rest upon the threat of instant US retaliation to deter a Soviet attack, that we could intercept and destroy strategic ballistic missiles before they reached our own soil or that of our allies?
>
> I know this is a formidable, technical task, one that may not be accomplished before the end of the century. Yet, current technology has attained a level of sophistication where it's reasonable for us to begin this effort. . . .
>
> I call upon the scientific community in our country, those who gave us nuclear weapons, to turn their great talents now to the cause of mankind and world peace, to give us the means of rendering these nuclear weapons impotent and obsolete.[26]

With these words, President Reagan launched the Strategic Defense Initiative (SDI), later derogatorily nicknamed "Star Wars." SDI cultivated visions of space-based lasers and "Brilliant Pebbles" kinetic kill vehicles orbiting high above the earth's atmosphere, always in position and ready to defend the United States and its allies from Soviet ballistic missile attack. Despite the optimistic rhetoric, the SDI technology failed to materialize. However, the failure did not diminish the American military's obsession with technology. Eight years later, the world was offered a front-row seat—via CNN—to witness the impressive state of Reagan-inspired military technology during Operation Desert Storm.

The focus on high-cost and high-tech came to the forefront of the Air Force consciousness again in 2008. Facing a seemingly interminable and daunting counterinsurgency struggle in Iraq and Afghanistan, Secretary of Defense Robert Gates was aghast at the Air Force's preoccupation with acquiring more F-22 stealth fighters. In May 2008, speaking in Colorado Springs, Colorado, Gates suggested that the Air Force, by focusing on future potential "near-peer" competitors at the expense of supporting the current wars, suffered from "next-war-itis."[27] Gates's frustration was also evidenced a month prior. In a speech at Maxwell AFB, Alabama, in April 2008, Gates lamented, "I've been wrestling for months to get more intelligence, surveillance and reconnaissance [ISR] assets into the theater. Because people were stuck in old ways of doing business, it's been like pulling teeth."[28] The secretary demanded that the Air Force quickly field more ISR assets, including low-tech, expendable unmanned aerial vehicles (UAV) (later referred to as remotely piloted aircraft [RPA]). When the Air Force chief of staff and the secretary of the Air Force failed to conform to Gates's wishes, they were relieved of duty.[29]

. . . or Technological Skepticism

Secretary Gates recognized the Air Force's technological skepticism that overshadowed an otherwise blossoming UAV/RPA fleet. Several authors note the Air Force's legacy of shunning development and deployment of UAV/RPAs for a variety of reasons—some technical, but the majority organizational. For example, P. W. Singer cited one individual's assessment, "The Air Force was terrified of unmanned planes; . . . the whole silk scarf mentality."[30] Another former Defense Department analyst joked that "no fighter pilot is ever going to pick up a girl at a bar saying he flies a UAV. . . . Fighter pilots don't want to be replaced."[31] Summarizing these perspectives and characterizing the persistent nature of the Air Force's organizational culture, Singer noted that "being a fighter pilot is . . . in the Air Force leadership's organizational DNA. Given this, it is no surprise then that the Air Force long stymied the development and use of drones, letting DARPA [Defense Advanced Research Projects Agency] and the intelligence agencies take the lead instead."[32] Thomas Mahnken made a similar observation, noting that despite "considerable use" of UAVs such as the Teledyne Ryan BQM-34 Firebee during the Vietnam War, "they did not find a permanent home in the Air Force until decades later.... Favored by neither the bomber nor the fighter communities, unmanned systems lacked an organizational home."[33]

It took the events of 9/11 and the developing counterinsurgency battles in Iraq and Afghanistan to overcome much of the bureaucratic resistance. Singer cited one defense contractor: "Prior to 9/11, the size of the unmanned vehicle market had been growing, but at an almost glacial pace. Thanks to battlefield successes, governments are [now] lavishing money on UAV programs as never before."[34]

The later decision to arm the UAVs also met with considerable skepticism. Mahnken noted that prior to "September 11, [2001], nobody wanted control of (and responsibility for) the armed Predator.... The notion of an unmanned vehicle controlled by an operator located hundreds or thousands of miles away delivering bombs in support of troops in close combat is something that would have previously been inconceivable" to both the Air Force and the Army.[35] Indeed, Singer noted that just prior to 9/11, a senior White House official was needed to resolve the disputes between the Central Intelligence Agency (CIA) and the Air Force over who would be responsible for controlling and, more importantly, funding the paltry $2 million cost of arming the Predator drones with Hellfire missiles.[36]

The story of F-22s and Predator UAV/RPAs is one recent illustration of the complex history of military technology. However, it is not unique. For all of the stories of technological exuberance, an equally rich history of technological skepticism, bolstered by organizational and bureaucratic resistance, also weaves itself through the fabric of US military history.

For example, military bureaucratic resistance stalled development of the Air Force's *raison d'être*—manned flight—for several years. In 1905, less than two years after their historic flight at Kitty Hawk, North Carolina, Wilbur and Orville Wright approached the US War Department seeking a contract to produce airplanes for the US military. Their inquiries merited no response.[37] The Wright brothers then turned to the British War Office at the suggestion of their adviser Octave Chanute, reasoning after the fact that their "invention will make more for peace in the hands of the British than in our own."[38] Those negotiations also languished. The Wright brothers, fearing piracy of their designs, returned to the United States and dismantled their aircraft; they would not fly again until May 1908.[39] In 1907, following renewed European interest in the Wright brothers' Flyer and prodding by Senator Henry Cabot Lodge, the War Department finally solicited bids for an airplane that matched the Wrights' specifications. The Wright brothers' first test flight at Fort Myer on 3 September 1908 easily surpassed the performance requirements, and the US military promptly drafted a contract.[40] It had been almost five years since the first successful flight and three years since Orville and Wilbur first approached the US military.

Similarly, the intercontinental ballistic missile (ICBM) met with considerable skepticism within the Air Force, especially prior to the successful development of the solid-fuel Minuteman missile. Sheehan described how the Air Force ICBM emerged from the inventive imagination of Air Force colonel Bernard Schriever. At a March 1953 meeting of the Air Force Scientific Advisory Board—the latest incarnation of Arnold's earlier von Kármán–led Scientific Advisory Group—Schriever listened to nuclear weapons pioneers Edward Teller and John von Neumann explain how expected improvements in thermonuclear bomb design would, within 10 years, result in a high-yield, low-weight device. Based on the scientists' predictions, Schriever envisioned "the ultimate weapon—nuclear-armed ballistic missiles hurtling across continents at 16,000 miles per hour through the vastness of space."[41] Despite the strategic promise of the ICBM concept, the blue-sky bomber generals of the Air Staff, typified by Gen Curtis LeMay, stymied Air Force development of the missile. Sheehan attributed LeMay's "vociferous" opposition to his concern that ICBM development "would divert funds from aircraft production."[42] Characteristic of the skepticism directed toward ICBMs, LeMay once quipped, These things will never be operational, so you can depend on them, in my lifetime."[43] By 1958 the promise of future ICBM development, embodied in the design of the Air Force's latest Minuteman missile, had surmounted General LeMay's skepticism.[44]

Technological skepticism is not limited to the future-minded, technologically dependent US Air Force. John Ellis described the almost-worldwide resistance to the machine gun that persisted for more than 30 years after its introduction in 1862. He noted that by 1892, "the machine gun [was] well-designed, relatively easy to mass produce and fairly reliable under battlefield conditions."[45] Still, most militaries passed on the technology. Attempting to explain their rationale, Ellis concluded that the majority of the officers in the world's armies were not in tune with the Industrial Revolution and, being groomed within "rigid hierarchical structures," were able to "minimize the impact of the faith in science and the machine."[46] Ellis continued, "When faced with the machine gun and the attendant necessity to rethink all the old orthodoxies about the primacy of the final infantry charge, such soldiers either did not understand the significance of the new weapon at all, or tried to ignore it, dimly aware that it spelled the end of their own conception of war. . . . For them, the machine gun was anathema, and even when their governments bought them out of curiosity, or because their enemies did, they almost totally ignored them."[47]

William McNeil located earlier evidence of technological skepticism in the development, or lack thereof, of English musketry. He noted that the "standard [English] infantry weapon," affectionately nicknamed the "Brown Bess," persisted from 1690 through 1840 "with only minor modifications."[48] McNeil attributed this technological stasis to the military's "choice between the advantages of uniformity and the cost of reequipping an entire army."[49] It chose uniformity over capability. McNeil also observed a similar conservative skepticism in an 1828 English Admiralty memorandum regarding a proposed shift from sail- to steam-powered warships; the Admirals warned, "Their Lordships feel it is their bounden duty to discourage to the utmost of their ability the employment of steam vessels, as they consider that the introduction of steam is calculated to strike a fatal blow at the naval supremacy of the Empire."[50]

As the preceding survey illustrates, instead of exhibiting a pattern of careful, rational decision making, the military's pursuit of technological innovation invites a diagnosis of organizational schizophrenia. Upon further inspection, however, a pattern emerges—revolutionary technological innovations that challenge preconceived notions of warfare such as the airplane, the ballistic missile, the machine gun, or the steamship are usually met with stubborn, bureaucratic paranoia and technological skepticism. If the resistance is overcome and the innovation is allowed to mature, the technology can be embraced by the organization and reinforced with subsequent evolutionary innovation, yielding an image of technological exuberance. This is the case with the evolutionary technologies represented by the B-36 aircraft of the 1950s, the cybernetic warfare systems developed in the 1970s, and the F-22 of the 2000s. However, technological exuberance is not strictly limited to just evolutionary technologies; it can also extend to revolutionary technologies such as the X-20 Dyna-Soar project or Reagan's SDI program.[51] This observed pattern of behavior forms a basis for Hughes's theory of technological momentum.

Technological Momentum

Hughes recognized the "complex and messy" nature of technology: "It is difficult to define and to understand. In its variety, it is full of contradictions, laden with human folly, saved by occasional benign deeds, and rich with unintended consequences. Yet today most people in the industrialized world reduce technology's complexity, ignore its contradictions, and see it as little more than gadgets and as a handmaiden of commercial capitalism and the military."[52] Confounding matters, even the term *technology* is often muddled by differing connotations. As Eisenhower noted in his 1961 farewell address, the notion of technology was relatively new to the post–World War II world. Prior to that, what would be referred to as *technology* today might have been called "applied science," the "practical arts," or simply "engineering."[53] Hughes provided his own definition of technology—"craftsmen, mechanics, inventors, engineers, designers, and scientists using tools, machines, and knowledge to create and control a human-built world consisting of artifacts and systems."[54] There are advantages to Hughes's liberal definition of technology: it avoids the restrictive connotations of artifacts engineered solely for utility and instead recognizes processes as possible manifestations of technology.[55] Based on this understanding of technology and recognizing the historical patterns of technological evolution, Hughes purported that "massive [technological] systems . . . have a characteristic analogous to the inertia of motion in the physical world"—momentum.[56]

Hughes first coined the term "technological momentum" to describe the pattern of technological evolution that he observed in his study of the interwar German chemical industry and the exclusive contract for synthetic gasoline that materialized between the German chemical firm I. G. Farben and the nascent National Socialist regime.[57] For Hughes, the "dynamic force" of technological momentum provided an alternative to the popular "conspiracy thesis" presented at the Nuremberg trials where Farben scientists and engineers were accused of entering into a "conspiratorial alliance [with the Nazis] . . . to prepare [for] wars of aggression."[58] Hughes acknowledged that Farben's research into hydrogenation offered a means to convert Germany's vast deposits of brown coal into a more practical resource, gasoline. Hughes also acknowledged that access to indigenously produced gasoline renewed the "possibility of Germany regaining her economic and political position among the world powers." But he discounted the Nuremberg accusations that Farben directors engaged in Machiavellian-style behavior that sought to stoke a "future military market."[59] Rather, for Hughes, Farben's early commitments to developing the hydrogenation process contributed to a powerful and nearly autonomous "drive to produce and a drive to create."[60] Unfortunately for the engineers and managers at Farben, almost immediately after the investment of significant time and resources yielded a successful process, the Great Depression erased much of the world's demand for gasoline. Farben was left with "a vested interest in a white elephant."[61] Unwilling to cut their losses, the company officers sought industrial protection from Nazi officials. For Hughes, the "commitment of engineers, chemists, and managers experienced in the [hydrogenation] process, and of the corporation heavily invested in it, contributed to the momentum" that led to the arrangement.[62] In short, "the technology, having gathered great force, hung heavily upon the corporation that developed it and thereby contributed to the fateful decision of the vulnerable corporation to cooperate with an extremist political party."[63]

Hughes refined his theory of technological momentum over the next 30 years, continuing to stress the role of technological maturation and organizational acceptance as important components of technological momentum. Describing the influence, Hughes noted, "People and investors in technological systems construct a bulwark of organizational structures, ideological commitments, and political power to protect themselves and the systems. Rarely do we encounter a nascent system, the brainchild of a radical inventor, so reinforced; but rarely do we find a mature system presided over by business corporations and governmental agencies without the reinforcement. This is a major reason that mature systems suffocate nascent ones."[64]

Hughes frequently cited examples of technological momentum within the military-industrial complex. For example, commenting on nuclear weapons, Hughes noted, "The inertia of the system producing explosives for nuclear weapons arises from the involvement of numerous military, industrial, university, and other organizations as well as from the commitment of thousands of persons whose skills and employment are dependent on the system. Furthermore, cold war values reinforce the momentum of the system."[65] According to Hughes, understanding these vested interests helps opposition to nuclear disarmament. "Disarmament offered such formidable obstacles not simply because of the existence of tens of thousands of nuclear weapons, but because of the conservative momentum of the military-industrial-university complex."[66]

Such organizational motivations are not new. An economist might characterize Hughes's technological momentum as simply a manifestation of the principle of sunk cost.[67] However, within the field of the history of technology, Hughes's theory of technological momentum provided a unique and important bridge between two opposing theories of technological change—technological determinism and social constructivism.

Technological Determinism

Henry David Thoreau poetically derided the rise of machines in everyday life: "But lo! men have become the tools of their tools."[68] Historian Lewis Mumford similarly lamented, "Instead of functioning actively as an autonomous personality, man will become a passive, purposeless, machine conditioned animal."[69] Indeed, acknowledging the increasing influence that technology exerts over humankind is, to a certain extent, dehumanizing. Nevertheless, significant historical trends have often been solely attributed to technological development. For example, some blame Eli Whitney's cotton gin for the Civil War. They argue that Whitney's invention restored the profitability of the cotton market, thereby reinvigorating the American slavery system, which consequently caused the Civil War that resulted in the death of more than 620,000 soldiers.[70] Similarly, some suggest that the Protestant Reformation can be traced to Gutenberg's printing press and its capability to provide for the first time "direct, personal access to the word of God" to individuals outside the priesthood.[71] And Jared Diamond traced the demise of Native American cultures to animal domestication in Eurasia.[72]

Merritt Roe Smith and Leo Marx noted that "popular narratives" such as these frequently

> convey a vivid sense of the efficacy of technology as a driving force of history: a technical innovation suddenly appears and causes important things to happen.... The thingness or tangibility of mechanical devices—their accessibility via sense perception—helps to create a sense of causal efficacy made visible. Taken together, these before-and-after narratives give credence to the idea of "technology" as an independent entity, a virtually autonomous agent of change.... It is typified by sentences in which "technology," or a surrogate like "the machine," is made the subject of an active predicate: "The automobile created suburbia." "The atomic bomb divested Congress of its power to declare war."... "The Pill produced a sexual revolution."... These statements carry the further implication that the social consequences of our technical ingenuity are far-reaching, cumulative, mutually reinforcing, and irreversible.[73]

Critics of technological determinism claim that the perspective is too reductionist, marginalizing important societal and environmental influences that affect technological development. However, Nassim Nicholas Taleb suggested it is human nature to be reductionist, preferring "compact stories over raw truths."[74] According to Taleb, we suffer from the "narrative fallacy"—it is difficult for us "to look at sequences of facts without weaving an explanation into them, or, equivalently, forcing a logical link, an *arrow of relationship*, upon them."[75] Nevertheless, he noted that there is value in causal interpretation: "Explanations bind facts together. They make them all the more easily remembered; they help them *make more sense*."[76] Too often, though, the causal relationship is improperly or inadequately constructed. Understanding this human predisposition toward reductionism helps explain why the principles of technological determinism are so seductive.

Despite its reductionist nature, technological determinism appears to possess some historical veracity, as technology sometimes exerts a significant influence over society.[77] For example, it is difficult to discount the societal impact of the automobile, the computer connected to the Internet, or nuclear weapons and ICBMs. Certainly, it would be hard to pry these essential technological systems away from society or the military. In Hughes's parlance, these systems have developed substantial technological momentum. They support the technological determinists' contention that "the advance of technology leads to a situation of inescapable necessity.... Our technologies permit few alternatives to their inherent dictates."[78] Moore's Law, which describes the astonishing growth of the number of transistors on an integrated circuit, is a prime example of technology's "inherent dictates." Thus, according to the determinist perspective, integrated circuit technology adheres to Moore's Law, not because society demands it but because the technology naturally continues to advance at its own exponential pace.[79]

Reinforcing the technological determinist position that society does not significantly influence technological development, there is historical evidence of similar technologies emerging from disparate social environments. The development of ICBMs in both the United States and the Soviet Union is one example.[80] ICBMs emerged from both nations despite their vastly different cultural contexts—Schriever leading the US ICBM effort, and the Soviets benefiting from the technical prowess of their chief rocket scientist, Sergei Korolev.[81] Additionally, technological determinists point to Wernher von Braun's German V-2 ballistic missiles of World War II, suggesting that with the first successful V-2 missile launch, the development of future ICBMs in the United States and the Soviet Union became a foregone conclusion.[82] Accordingly, both nations stumbled into the ICBM race not based on calculated decisions but on the promise of technology. As one historian noted in decidedly deterministic language, "The United States built its missile arsenal without any agreed understanding—even within elite circles, much less among the general population—of why it was doing so."[83]

Giovanni Dosi's theory of a technological trajectory addresses the notion of technological progress's universality.[84] Despite borrowing heavily from Thomas Kuhn's social constructivist interpretation of scientific progress, Dosi's technological trajectory concept has a decidedly deterministic tone.[85] Dosi defined a *technological trajectory* as the "direction of advance within a technological paradigm."[86] He noted that these "technological paradigms have a powerful *exclusion effect*: the efforts and the technological imagination of engineers and of the organizations they are in are focused in rather precise directions while they are, so to speak, 'blind' with respect to other technological possibilities.[87] Donald MacKenzie described Dosi's technological trajectory as a "direction of technical development that is simply natural, not created by social interests but corresponding to the inherent possibilities of the technology."[88] There is also a connection between Dosi's theory of technological trajectories and Hughes's theory of technological momentum. Dosi asserted that "once a path [of technological development] has been selected and established, it shows a *momentum* of its own, which contributes to define the directions toward which the 'problem solving activity' moves."[89]

The Social Construction of Technology

Social constructivists challenge the reductionism associated with the technological determinist interpretation of history. While MacKenzie acknowledged the deterministic connotations associated with Dosi's theory of technological trajectories, he challenged the interpretation, instead suggesting that the trajectory is propagated by social influences as a social "self-fulfilling prophecy"—"those lines of technical development that do not get pursued do not improve; those that get pursued often do."[90] Thus, for MacKenzie, socially constructed forces drive the technological trajectory, not the nature of the technology itself. Drawing on Hughes's discussion of technological momentum, MacKenzie similarly suggested that the trajectory results from people "invest[ing] money, careers, and credibility in being part of 'progress,' and in doing so help[ing] create progress of the predicted form."[91]

Hughes also acknowledged the role of societal influences in furthering a technological system, particularly when technical or organizational problems are encountered during technological development. Describing these obstacles as "reverse salients," Hughes noted, "As technological systems expand, reverse salients develop. Reverse salients are components in the system that have fallen behind or are out of phase with the others."[92] MacKenzie expanded upon Hughes's definition: "A reverse salient is something that holds up technical progress or the growth of a technological system."[93] Emphasizing the social influences implicit in reverse salients, MacKenzie noted, "System builders typically focus inventive effort, much like generals focus their forces, on the elimination of such reverse salients; they identify critical problems whose solution will eliminate them.... But it may not always be clear where progress is being held up, nor what should be done about it. Even with agreement on goals,... the nature of the obstacles to the achievement of these goals and the best means of removing them may be the subject of deep disagreement."[94]

Thus, according to the social constructivists, failure to acknowledge the "economic, political, organizational, cultural, and legal" contexts that surround technology results in an imperfect understanding of technological development.[95] "Technological development [is] a nondetermined, multidirectional flux that involves constant negotiation and renegotiation among and between groups shaping the technology."[96] Within this construct, John Law's "heterogeneous engineer" is an individual well suited to mediate between the opposing social groups while simultaneously overcoming or circumventing technical impediments. Such individuals, Law argued, are singularly important in the development and propagation of technological systems.[97]

Just as there is evidence of technological determinism in military history, the pattern of social influences on technological development is also evident. When researching American technological innovation in the military following World War II, Thomas Mahnken concluded, "the [US military] services molded technology to suit their purposes more often than technology shaped them."[98] Similarly, the historian Williamson Murray emphasized the social influences on military technology, observing that it is the combination of "technology and potent management skills" that produces change.[99]

Sheehan's account of US ICBM development offers a prime example of a social constructivist account of military technological innovation. Within his narrative, Sheehan cast Schriever as a master strategist deftly outmaneuvering a manned bomber bureaucracy allied against him, while simultaneously surmounting an array of scientific and technological hurdles and operating within the constraints of a budget-conscious political administration wary of burdensome military expenditures.[100] Sheehan concluded that without Schriever's "intellectual bent and the foresight to see the implications for the future," the development of a US ballistic missile force would have failed.[101] Indeed, for Sheehan, the history of US ballistic missile development is a history of Schriever—a heterogeneous engineer triumphing over technical and social adversities.[102]

Like most "great man" narratives, the ICBM development story is both interesting and appealing, involving colorful individual personalities drawn together by unique and trying circumstances.[103] For example, Sheehan cited the importance of the appointment of the hard-drinking and paper-chewing Trevor Gardner as the special assistant to the secretary of the Air Force for research and development and his subsequent selection of Schriever to lead the Air Force's ICBM efforts.[104] He also cites the nontraditional yet successful efforts of the ICBM proponents to secure a National Security Council briefing in front of President Eisenhower,[105] the decision by an Air Force engineer to subvert a cruise missile program to support ballistic missile rocket engine development,[106] and even Schriever's prowess as a golfer as all being critical to the ICBM effort.[107]

According to Sheehan and the social constructivist argument, the fabric of history would have undoubtedly unfurled differently absent any one of these meetings, decisions, or personal attributes. However, the development of a Soviet ICBM force discounts Sheehan's position that without Schriever, the US Air Force's foray into ballistic missiles was destined to fail. While there is no denying that Schriever's skills certainly influenced the quick realization of ICBM technology, it is possible that another individual could have taken up the torch, and technology would have continued marching along.

... And Hughes's Link between the Two

Therefore, history supports both the technological determinist and the social constructivist arguments. Hughes's theory of technological momentum steps between the two and offers an alternative to the Manichean perspectives that have unnecessarily polarized past historical analyses. For Hughes, "a technological system can be both a cause and an effect; it can shape or be shaped by society."[108] Thus, the theory of technological momentum "does not contradict the doctrine of social construction of technology, and it does not support the erroneous belief in technological determinism."[109] Hughes suggested that as technological systems acquire momentum by amassing "technical and organizational components," they exhibit a pattern of behavior that appears to be "autonomous," yielding an image of technological determinism.[110] This description, however, rests on a razor's edge. Despite Hughes's unwillingness to declare his acceptance of the tenets of technological determinism, his description of momentum still acknowledged the significant influence technology could exert on society.

Within his theory of technological momentum, Hughes credited an important role to time, suggesting that technology's influence on society, and its reciprocal, is "time dependent."[111] Granted, time itself is not sufficient for technologies to develop momentum, but it is necessary to allow technological systems to "grow larger and more complex" and to become "more shaping of society and less shaped by it."[112] Based on this observed relationship, Hughes claimed that "the social constructivists have a key to understanding the behavior of young systems; technical determinists come into their own with the mature ones."[113]

Applying Hughes's theory of technological momentum to the earlier description of the military's relationship with technological systems yields the model in figure 1. New, revolutionary technological systems like the Wright brothers' aircraft, the machine gun, and the ICBM are initially dominated by socially constructed influences and are typically frustrated by technological skepticism and bureaucratic resistance. If the skepticism is surmounted and the technological system allowed to mature over time, the technology acquires momentum and begins to exert an influence over the bureaucracy corresponding to the technological determinist position. Furthermore, mature technological systems are often reinforced by evolutionary innovation and improvements, further adding to the momentum and the institutionalization of the technological system. While technological exuberance can exist at any stage of the development process, it typically dominates once the technology has acquired momentum.

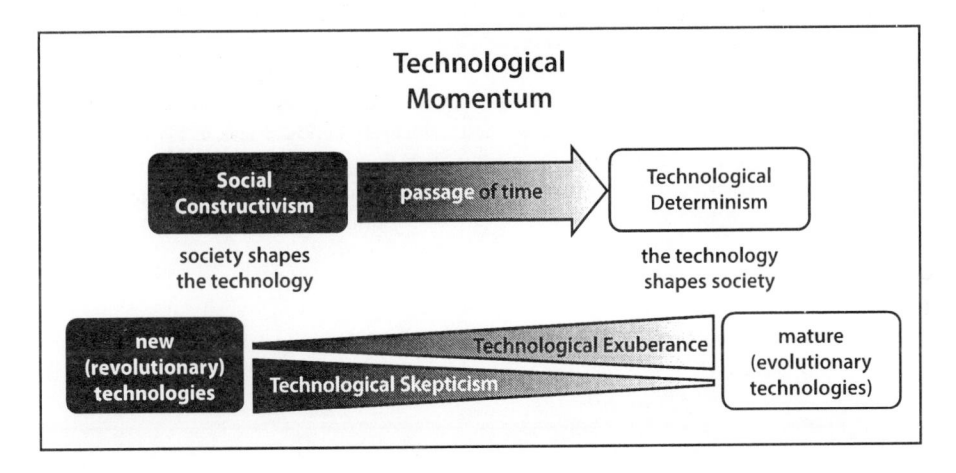

Figure 1. Technological momentum. (*Created by* the author)

While Hughes's theory of technological momentum offers hope for reconciling the discrepant deterministic and constructivist analyses of technological history, upon closer inspection it reveals itself to be also imperfect and too reductionist. Although Hughes acknowledged that the "phases in the history of a technological system are not simply sequential," his theory presumes that the transition from social constructivism to technological determinism is unidirectional.[114] His theory therefore tends to focus historical analysis on characterizing the transition from technological adolescence to maturity—from when society dominates the technology to when the technology begins to dominate society. The model of technological dislocations explored in the next chapter addresses the consequences of this limitation.

Notes

1. Quoted in Medhurst, *Dwight D. Eisenhower*, 191–92.
2. Ambrose, *Eisenhower*, 536.
3. McDougall, *Heavens and the Earth*, 230. McDougall cites Claud Witze's "How Outer Space Policy Evolved."
4. Ibid., 229.

5. Ibid., 5, 436, and 443. McDougall defined *technocracy* as "the institutionalization of technological change for state purposes, that is, the state-funded and -managed R & D [research and development] explosion of our time." Further describing the US transition to a technocratic ideology, McDougall continued, "Technocratic *ideology* captured the country only after Sputnik, when a new willingness to view state management as a social good and not a necessary evil turned a quantitative change into a qualitative one. . . . 'Scientific' management only seduced its practitioners into thinking themselves objective" (emphasis in original).

6. York, *The Advisors*, ix (emphasis added).

7. Other scholars have noted the attempt to apply technological solutions to ill-defined strategic problems. Singer cited retired Marine officer T. X. Hammes: "We continue to focus on technological solutions at the tactical and operational levels without a serious discussion of the strategic imperatives or the nature of the war we are fighting. I strongly disagree with the idea that technology provides an inherent advantage to the United States." Singer deemed Hammes's comments noteworthy because of their uniqueness within the US military establishment. Singer, *Wired for War*, 213.

8. Smith, "Introduction," 4.

9. Singer, *Wired for War*, 238.

10. Arnold's autobiography *Global Mission* is cited by Spires, *Beyond Horizons*, 8; and Sheehan, *A Fiery Peace*, xvi.

11. Sheehan, *A Fiery Peace*, xvi.

12. Ibid.

13. Ibid., 121 (emphasis added).

14. Barlow, *Revolt of the Admirals*, 46. Air Force general Tooey Spaatz announced in October 1945, "The aeronautical advance of the past few years has ushered in the 'Air Age.' Its primary force is Air Power. As sea-power was the dominant factor in the destiny of nations in the nineteenth century, so today the dictate is Air Power." Barlow summarized the mood of the nation's defense establishment: "air power had become the nation's dominant force" and "the first line of defense for the United States."

15. Spires, *Beyond Horizons*, 79. According to Spires the Air Force desperately sought a manned space presence, especially after President Eisenhower's 1959 decision to transfer the manned space mission and the responsibility for developing "superbooster" rockets like the Saturn V to NASA.

16. McDougall, *Heavens and the Earth*, 340.

17. Ibid.

18. Ibid., 341.

19. Ibid.; and Spires, *Beyond Horizons*, 124–26.

20. The principles of cybernetic warfare are discussed in Bousquet, *Scientific Way of Warfare*, 123 and 137. Bousquet characterized cybernetic warfare as "the shift from traditional notions of command to that of 'command and control,' the reduction of war to a set of mathematical functions and cost-benefit calculations susceptible to optimization through the techniques of operations research and systems analysis, and the increasing modeling and simulation of conflict." Reflective of Eisenhower's "scientific-technological elite," Bousquet noted that the cyberneticists sought to reduce "war to a complex equation to be resolved by a technoscientific priesthood." See also Lonsdale, *The Nature of War*.

21. Quoted in Bousquet, *Scientific Way of Warfare*, 136. Bousquet also cited Paul Edwards's critique of Westmoreland's speech: it "epitomizes the 'vision of a closed world, a chaotic and dangerous space rendered orderly and controllable by the powers of rationality and technology.'"

22. Randolph, *Powerful and Brutal Weapons*, 47; and Bousquet, *Scientific Way of Warfare*, 156–57.

23. Randolph, *Powerful and Brutal Weapons*, 47; and Bousquet, *Scientific Way of Warfare*, 126. Bousquet noted that it is "not surprising that the military embraced computers as the panacea to the eternal problem of uncertainty and unpredictability on war."

24. Randolph, *Powerful and Brutal Weapons*, 47.

25. Ibid., 47–48; Bousquet, *Scientific Way of Warfare*, 157–58; and Lonsdale, *Nature of War*, 83. Critiquing the US efforts, Bousquet noted, "The North Vietnamese were being treated [by US officials] as a cybernetic system which could be steered toward the desired behavior by a selective input of information in the form of targeted aerial bombardment." By treating "the war as a purely technical problem to be solved through overwhelming application of materiel according to a scientific methodology, these [US] officials failed to grasp the sheer determination of their opponents and the extent of the success of their political strategy." Lonsdale cautioned that "unbridled confidence in the robustness of RMA [cybernetic] capabilities to countermeasures should not go unchallenged. . . . Every weapon system is countered eventually to some degree."

26. Quoted in Mahnken, *Technology and the American Way*, 150.

27. "I [Gates] have noticed too much of a tendency toward what might be called next-war-itis—the propensity of much of the defense establishment to be in favor of what might be needed in a future conflict." Shanker, "Gates Says New Arms Must Play," A18; and Walker, "Air Force Firings," 61.

28. Holmes, "Why USAF's Top Two Were Forced Out," 8.

29. Dreazen, "Gates Ousts Top Leaders," A1. Citing "a pattern of poor performance," Gates ousted the Air Force's secretary, Michael Wynne, and its chief of staff, Gen Michael Moseley, on 5 June 2008. Although the "immediate trigger for the resignations" was the accidental shipment of ballistic missile fuses to Taiwan, several Washington pundits believed the firings to be "the culmination of a broader dispute between Mr. Gates and the Air Force's leadership over the service's strategic direction. The biggest source of tension has been the Air Force's insistence on buying hundreds of expensive, state-of-the-art F-22 fighter jets . . . despite opposition from Mr. Gates who has argued that the planes aren't needed for prosecuting America's current wars."

30. Singer, *Wired for War*, 54. Robert Finkelstein's firm developed software for an unmanned F-4 Phantom target drone. Describing the evolution of the technology and the oppressive bureaucratic skepticism, Finkelstein commented, "The new software began to beat pilots consistently, and the idea grew to use it as an advanced teaching tool for fighter pilots. But it never came to be. The program was too much, too soon, and most important too good for its own sake."

31. Ibid., 252. Singer cited Andrew Krepinevich, "a former Defense Department analyst who is now executive director of the Center for Strategic and Budgetary Assessments."

32. Ibid., 253.

33. Mahnken, *Technology and the American Way*, 114; and Randolph, *Powerful and Brutal Weapons*, 194–95. Randolph described the drones' success: "During 1972 the drones flew a total of 498 missions, losing 52 aircraft. The missions targeted a total of 6,335 high-priority points for photos, succeeding with 2,543 of these."

34. Singer, *Wired for War*, 61.

35. Mahnken, *Technology and the American Way*, 202.

36. Singer, *Wired for War*, 35.

FOUNDATIONS OF TECHNOLOGY

37. Morrow, *Great War*, 5. Morrow noted that the 1903 "abject failure" of Samuel Langley's $50,000 airplane project sponsored by the War Department's Board of Ordnance and Fortification "made the War Department wary of future winged projects."

38. Hughes, *American Genesis*, 101.

39. Morrow, *Great War*, 6.

40. Hughes, *American Genesis*, 100–104.

41. Von Neumann and Teller "predicted that by 1960 the United States would be able to build a hydrogen bomb that would weigh less than a ton but would explode with the force of a megaton, i.e., eighty times the power of the simple atomic or fission bomb that had blown away Hiroshima." Sheehan noted that "these two attributes were the *sine qua non* for the building of a practical intercontinental ballistic missile." Sheehan, *Fiery Peace*, 178.

42. Ibid., 223. LeMay also "predicted that the Atlas [ICBM] would turn out to be an extravagant boondoggle. It would never perform as anticipated."

43. Ibid., 412–13.

44. Ibid., 415. In 1955 LeMay stated that he would "consider the ICBM 'the ultimate weapon' worthy of inclusion in SAC's [Strategic Air Command's] inventory when one could be created with a capability of instantaneous launch and with acceptable reliability, accuracy, and yield." Three years later, after receiving the briefing on the Minuteman missile, LeMay "swung around to the three-star deputy chiefs of staff sitting in the rows behind him and asked: 'Do you agree it's a go?'"

45. Ellis, *Social History of the Machine Gun*, 16. In 1862 Richard Jordan Gatling produced a crank-operated gun that fired an impressive steady stream of 200 rounds per minute. Twenty-two years later, Hiram Maxim developed an automatic firing mechanism. By 1892 William Browning had produced a gun that used its own muzzle gasses to operate an automatic firing mechanism.

46. Ibid.

47. Ibid., 17; and Hughes, *American Genesis*, 105. Hughes noted that one irrational argument for denigrating the machine gun was that it "could not be supplied rapidly enough with ammunition in the field."

48. McNeil, *Pursuit of Power*, 142.

49. Ibid.

50. Ibid., 226.

51. Despite the revolutionary technologies associated with manned spaceflight, the X-20 Dyna-Soar could be considered an evolutionary technology, as it was envisioned as an extension of the Air Force's vision of manned aircraft delivering atomic weapons as part of a strategic bombing campaign. This contrasts with the revolutionary ICBM technological system, which dramatically changed the concept of warfare. Similarly, it is difficult to suggest that Reagan's SDI program followed an evolutionary trend—it was also revolutionary. The difference between evolutionary versus revolutionary technologies has been treated extensively in the literature: see Hughes, "Evolution of Large Technological Systems," 57 and 59; Dosi, "Technological Paradigms and Technological Trajectories," 158; Constant, *Origins of the Turbojet Revolution*, 4; and Luttwak, *Strategy*, 234–36. Hughes offered his interpretation: "Inventions can be conservative or radical. Those occurring during the invention phase are radical because they inaugurate a new system; conservative inventions predominate during the phase of competition and systems growth, for they improve or expand existing systems." Describing the rationale for technological skepticism toward revolutionary technologies, Hughes continued: "Large organizations vested in existing technology rarely nurtured inventions that by their nature contrib-

uted nothing to the momentum of the organization and even challenged the status quo in the technological world of which the organization was a leading member. Radical inventions often deskill workers, engineers, and managers, wipe out financial investments, and generally stimulate anxiety in large organizations." Dosi described the difference between " 'incremental' innovation versus 'radical' innovation" in his article. Similarly, Constant noted the importance of identifying the nature of the technological change—"the relative importance of incremental versus discontinuous or revolutionary changes"—when assessing society's reaction to the new technology. Luttwak discussed the "bureaucratic aversion to new [military] equipment that does not fit the established order of things."

52. Hughes, *Human-Built World*, 1.

53. Ibid., 2; and Smith, "Technological Determinism," 7. Smith noted that "the belief that in some fundamental sense technological developments determine the course of human events had become dogma by the end of the [nineteenth] century." However, as evident in James P. Boyd's 1899 *Triumphs and Wonders of the 19th Century* (cited by Smith), *progress* was a diffuse entity and not specifically linked to the term *technology*. "It may be said that along many of the lines of invention and progress which have most intimately affected the life and civilization of the world, the nineteenth century has achieved triumphs and accomplished wonders equal, if not superior, to all other centuries combined."

54. Hughes, *Human-Built World*, 4.

55. Examples of processes representative of technological innovation include Henry Ford's assembly line and Frederick Taylor's principle of scientific management. Similarly, the US interstate freeway system could be considered a technological innovation, despite its lack of any high-tech gadgetry.

56. Hughes, *American Genesis*, 460.

57. Hughes, "Hydrogenation," 106–32.

58. Ibid., 106. The Nuremberg charges were later dismissed.

59. Ibid., 116–17.

60. Ibid., 112.

61. Ibid., 122.

62. Ibid., 131.

63. Ibid., 131–32.

64. Hughes, *American Genesis*, 460.

65. Ibid.

66. Ibid.

67. "Sunk cost—in accounting, a cost that grows out of a past, irrevocable decision. A typical example is a fixed asset, such as a machine, that has become obsolete and whose book value therefore cannot be recovered." Ammer and Ammer, *Dictionary of Business and Economics*, 449; and Brauer and Tuyll, *Castles, Battles, and Bombs*, 78.

68. Thoreau, *Walden*, 29.

69. Quoted in Smith, "Technological Determinism," 29.

70. Smith and Marx, "Introduction," x. The Civil War death total is from McPherson: "More than 620,000 soldiers lost their lives in four years of conflict—360,000 Yankees and at least 260,000 rebels. The number of southern civilians who died as a direct or indirect result of the war cannot be known; what *can* be said is that the Civil War's cost in American lives was as great as in all of the nation's other wars combined through Vietnam." McPherson, *Battle Cry of Freedom*, 854 (emphasis in original).

71. Smith and Marx, "Introduction," x.

72. "Eurasian crowd diseases evolved out of diseases of Eurasian herd animals that had become domesticated . . . [and] played a key role in decimating native peoples." Diamond, *Guns, Germs, and Steel*, 212–13.

73. Smith and Marx, "Introduction," x.

74. Taleb, *Black Swan*, 63.

75. Ibid. (emphasis in original).

76. Ibid., 64 (emphasis in original).

77. Hughes, "Technological Momentum," 102. Hughes defined *technological determinism* as "the belief that technical forces determine social and cultural changes" and social constructivism as "presum[ing] that social and cultural forces determine technical change."

78. Smith and Marx, "Introduction," xii. Edgerton directly challenged the deterministic assertion by offering a "use-based" theory of technological development: "A central feature of use-based history, and a new history of invention, is that alternatives exist for nearly all technologies: there are multiple military technologies, means of generating electricity, powering a motor car, storing and manipulating information, cutting metal or roofing a building. Too often histories are written as if no alternative could or did exist." Edgerton, *Shock of the Old*, 7.

79. Kurzweil described the effects of Moore's Law: "The result is that every two years, you can pack twice as many transistors on an integrated circuit. This doubles both the number of components on a chip as well as its speed. Since the cost of an integrated circuit is fairly constant, the implication is that every two years you can get twice as much circuitry running at twice the speed for the same price. For many applications, that's an effective quadrupling of the value. The observation holds true for every type of circuit, from memory chips to computer processors." In a nod to the determinist camp, Kurzweil noted that this "remarkable phenomenon has been driving the acceleration of computing for the past forty years." Kurzweil, *Age of Spiritual Machines*, 21.

80. For a description of the US/USSR H-bomb competition, see Rhodes, *Dark Sun*, and Rhodes, *Arsenals of Folly*.

81. Sheehan, *Fiery Peace*, 405.

82. On von Braun and German V-2 rockets, see Neufeld, *von Braun*.

83. MacKenzie, *Inventing Accuracy*, 162. The remark is ironic because MacKenzie declared early in his book that he is a social constructivist. Indeed, he purports that his book is a counter to the assumption that technological determinism drove ICBM guidance system development.

84. Dosi, "Technological Paradigms," 148.

85. Kuhn challenged the notion that "scientific development" is "the piecemeal process" by which a "constellation of facts, theories, and methods" are "added, singly and in combination, to the ever growing stockpile that constitutes scientific technique and knowledge"—the interpretation that science is a naturally evolving process. In its place, Kuhn postulated that science can be divided into two phases: "normal science" operating within an established "scientific paradigm," and "revolutionary science" that evolves from investigating anomalies during "normal" scientific experiments and which yields new scientific paradigms. By addressing the social components of paradigm development and acceptance within the scientific community, Kuhn established the foundation for a socially based analysis of scientific history. Kuhn, *Structure of Scientific Revolutions*, 1–2, 10, and 84.

86. Dosi, "Technological Paradigms," 152. Besides drawing on Kuhn's notion of "paradigms," Dosi also drew a corollary to Kuhnian "normal science" by describing a "technological trajectory as the pattern of 'normal' problem solving activity (i.e., of 'progress') on the grounds

of a technological paradigm." On "normal science," see Kuhn, *Structure of Scientific Revolutions*, 24–34.

87. Dosi, "Technological Paradigms," 153 (emphasis in original).
88. MacKenzie, *Inventing Accuracy*, 167.
89. Dosi, "Technological Paradigms," 153 (emphasis added).
90. MacKenzie, *Inventing Accuracy*, 168. MacKenzie noted that economist Brian Arthur found a similar pattern of increasing returns; Arthur's discovery is explained within the context of complexity science in Waldrop, *Complexity*.
91. MacKenzie, *Inventing Accuracy*, 168. However, MacKenzie's argument does not address the possibility that people's increased investment in a particular technology is a manifestation of a particular technology exerting influence over society through technological determinism.
92. Hughes, "Evolution," 73.
93. MacKenzie, *Inventing Accuracy*, 79–80.
94. Ibid.
95. Ibid., 9.
96. Bijker, Hughes, and Pinch, "Common Themes," 13.
97. Law defined *heterogeneous engineering* as "the association of unhelpful elements into self-sustaining networks that are, accordingly, able to resist dissociation. . . . 'Heterogeneous engineers' seek to associate entities that range from people, through skills, to artifacts and natural phenomena." Law, "Technology and Heterogeneous Engineering," 114 and 129.
98. Mahnken, *Technology and the American Way*, 219.
99. Murray and Millet, *Military Innovation*, 1.
100. Sheehan, *Fiery Peace*, 139–40, 223, and 412. Gen Curtis LeMay typified the Air Force's manned bomber bureaucracy opposed to the ballistic missile. Sheehan quoted LeMay, "These things [ICBMs] will never be operational, so you can depend on them, in my lifetime." Sheehan also described LeMay as being "vociferously opposed" to the ICBM because it "would divert funds from aircraft production." While parrying LeMay's attacks, Schriever had to develop strategies to overcome problems with rocket propulsion, guidance, and delivery vehicles and operate under a tightening military budget as part of Eisenhower's "new look" strategy. Although the new look strategy's emphasis on "security with solvency" favored the purported cost benefits associated with nuclear weapons and the "intercontinental reach of LeMay's nuclear bombers" over the expense of conventional military forces, Schriever's nuclear missile capabilities were viewed as an unknown quantity and therefore a risky venture.
101. Sheehan, *Fiery Peace*, xviii. Sheehan suggested that Schriever was also bolstered by a prodigious personal charge delivered by Gen Hap Arnold to facilitate the mobilization of "science and technology into air power's service." Sheehan continued, "Whatever his reasons, [General] Arnold had summoned the right man."
102. See note 97 for a discussion of John Law's heterogeneous engineer.
103. For another example of a "great man" historical narrative, see Cherny, *Candy Bombers*, 404. Describing the individual actions of Pres. Harry Truman and Gen Lucius Clay during the Berlin Airlift, Cherny wrote: "There was never a clearer refutation of the canard that it is simply the current and not the captain that guides humanity past the shoals."
104. Sheehan noted, "Gardner had a serious drinking problem. He kept it under control during the day, although a couple of double-shot Old Forester bourbons with ginger ale, his standard portion at lunch, made him more aggressive back at the office in the afternoon." Gardner also had a penchant to tear off a corner of yellow legal pad paper, "roll it into a wad with his thumb and forefinger . . . toss it into his mouth," and "begin chewing it, all the while continuing

FOUNDATIONS OF TECHNOLOGY

to listen" to the topic of discussion. Describing Schriever's selection for the job, Sheehan recounted Schriever's tentative acceptance: "I'll [Schriever speaking] take the job . . . provided I can run it—completely run it—without any interference from those nitpicking sons of bitches in the Pentagon." Sheehan, *Fiery Peace*, 197–98 and 228.

105. Ibid., 268–78, and 299. Schriever had to lobby aggressively to secure a National Security Council briefing in front of the president. Less than two months after the briefing, President Eisenhower signed NSC Action no. 1433, designating the ballistic missile "a research and development program of the highest priority above all others" to be built with "maximum urgency."

106. Ibid., 246–47. Sheehan described how Air Force lieutenant colonel Ed Hall, working at the Air Development Center, devised a strategy to "use the requirements for adequate engines for the Navaho booster [an intercontinental cruise missile then in development but destined to be cancelled] as a cover to acquire a rocket engine for an intercontinental ballistic missile." Coincidentally, Sheehan noted that this was not Hall's first time skirting Air Force regulations—Hall and a friend previously planted a fraudulent intelligence report of a massive Russian rocket engine to help ensure sufficient funding for Hall's rocket engine development office.

107. Ibid., 253 and 260. Sheehan described how part of Schriever's campaign to win over a dissatisfied superior, the "win over Tommy Power" campaign, included "arranging the schedule so that their get-togethers were also an opportunity for the general to play [golf] with a partner in top form." The campaign and the golf worked; Sheehan noted that Schriever's first fitness report signed by General Power characterized Schriever as possessing "excellent staying qualities when the going gets rough."

108. Hughes, "Technological Momentum," 112.

109. Hughes, "Evolution," 80.

110. Ibid., 76.

111. Hughes, "Technological Momentum," 102.

112. Ibid., 112.

113. Ibid.

114. Hughes identified the phases of technological system evolution as "invention, development, innovation, transfer, and growth, competition, and consolidation." Indicative of the sequential process, Hughes described "mature systems" as acquiring a "high level of momentum" that "often causes observers to assume that a technological system has become autonomous." Hughes, "Evolution," 56 and 76.

Chapter 3

Technological Dislocations

If technological determinism implies that "technical forces determine social and cultural changes" and social constructivism suggests that "social and cultural forces determine technical change," then Hughes's theory of technological momentum provides a conceptual bridge between the two opposing perspectives.[1] It also helps explain how a technology can go from being shunned to being exuberantly embraced by a bureaucratic institution. Yet, Hughes's theory requires further refinement. The alternate theory of technological dislocations addresses the limitations of Hughes's theory and provides a more useful lens with which to study the role and process of innovation within the Air Force and the military in general.

A Technological Tipping Point?

While not specifically subscribing to the technological determinist position, Hughes conceded that mature systems possessing technological momentum invite perceptions of determinism.[2] The more momentum a technological system acquires, the more it can influence society in a deterministic fashion.[3] While acknowledging "that technological momentum, like physical momentum, is not irresistible," Hughes noted that effecting change in a technological system that possessed significant momentum would require a Herculean effort directed across a "variety of its components."[4] In short, "shaping is easiest before the [technological] system has acquired political, economic, and value components."[5] According to Hughes, these "value components" tighten a technology's grip on its surrounding environment. As bureaucratic institutions devoted to the technology begin to flourish, they provide the necessary funding and procedural regimens that reinforce the technology's growing influence on society. After sufficient time, the technological system may cement itself within the society's collective psyche. One popular example of this phenomenon is the story of the gasoline-powered automobile, which, after an initially cool reception, now exerts a dominant influence on American society. Thus, within Hughes's construct, time plays a significant role in technological development. Although rarely sufficient, time is necessary for momentum to build and for technology to evolve from society-shaped to society-shaping—from social constructivism to technological determinism.[6]

Hughes's theory is conceptually convenient. However, the unidirectional evolutionary process suggested by his theory is not without complications. Specifically, if a successful technology transforms from being socially constructed to being deterministic, then that transformation should be marked by a transition point—a tipping point—that divides the two influences (fig. 2). While Hughes did not explicitly treat the notion of a discrete technological tipping point in his writings, other scholars have investigated the phenomenon.

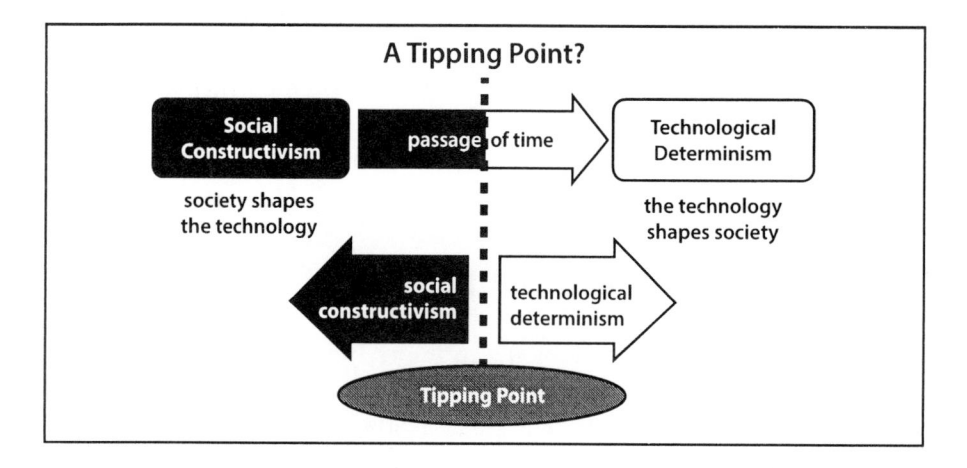

Figure 2. The putative tipping point between social constructivism and technological determinism. (*Created by* the author)

One such author, Malcolm Gladwell, used the notion of "tipping points" to describe how products and ideas spread through society. Defining a tipping point as the "dramatic moment in an epidemic when everything can change all at once," Gladwell examined, among others, the 1995 popular resurgence of Hush Puppies shoes, the almost overnight decline in New York City's crime rate in 1992, and the 1987 proliferation of low-priced fax machines.[7] Regarding the fax machine, Gladwell reported that after Sharp introduced the first low-priced fax machine in 1984, sales remained relatively flat and unimpressive for the next three years. In 1987, however, business suddenly and unexpectedly boomed. At that point, "enough people had faxes that it made sense for everyone to get a fax"; the low-priced fax machine crossed a tipping point.[8] There is a link between Gladwell's tipping point and Hughes's technological momentum. Using Hughes's parlance, in 1987 fax machines assumed sufficient technological momentum to influence a substantial segment of society to forgo any lingering skepticism and purchase the machines; the technology began to shape society in a deterministic fashion.

Identifying when the tipping point for low-priced fax machines was crossed is relatively easy—1987. Describing the causal factors that led to the tipping point is more difficult. In fact, Gladwell provided none, other than the raw sales numbers.[9] While a more practical fax machine model, a lower unit cost, or a favorable review in a business journal may have contributed to the sudden explosion in the fax's popularity, according to Gladwell's theory of tipping points, it was not necessarily a combination of these factors, nor should it simply be attributed to a steadily growing level of acceptance. All of a sudden, something relatively minor happened, and society was profoundly affected.

The transition from society-shaped to society-shaping is rarely as black and white as Gladwell asserts. For example, during the development of the ICBM, President Eisenhower's 1955 decision to declare the ballistic missile "a research and development program of the highest priority" could be regarded as the tipping point that catalyzed future US reliance on ICBMs.[10] Similarly, LeMay's 1958 acceptance of the Minuteman ICBM and the implicit organizational legitimacy that it granted may be regarded as a more appropriate tipping point.[11] However, one could also argue that the development of US ICBMs and their consequent role in national defense strategy was assured when Schriever was selected to head Air Force ICBM development in 1954, or when Pres. Harry Truman decided in 1950 to pursue the H-bomb, or when von Braun launched his first successful V-2 rocket from Peenemünde in October 1942.[12] These examples illustrate the difficulty with trying to identify an individual technological tipping point, even retrospectively.

Thus, while Hughes's theory of technological momentum and Gladwell's theory of tipping points are plausible at a macro level, when finely applied to a specific, complex technological system like the ICBM, they quickly lose their appeal. Neither author provides adequately descriptive terminology—Hughes for the transition between social constructivism and technological determinism, Gladwell for the causal factor that manifests as the technological tipping point. Both theories are too reductionist and fail to adequately address the complex nature of technological development.

Unlike the idealized model (fig. 2), there is often no clear, time-dependent technological metamorphosis that separates a society-shaped technology from a society-shaping technology; the two forms coexist throughout the technology's lifetime.[13] This observation marks a distinct departure from Hughes's theory. Hughes stated that "a technological system can be both a cause and an effect; it can shape or be shaped by society." His interpretation was based largely on the unidirectional transition from one form to another.[14]

While Hughes acknowledged that changes can still be made even after the technology had acquired momentum, his theory fails to provide a descriptive mechanism to address those later-in-technological-life changes. Similarly, the theory of technological momentum fails to address the society-shaping influences that even a nascent technology may exert.

A more holistic appraisal of the nature of technological change suggests that technologies often begin to exert deterministic tendencies early in their development process. It also suggests that social pressures can influence technological development even after a deterministic trajectory has been realized. MacKenzie, recognizing that Hughes's artificial restriction of society's impact on mature technologies discounted the later influence of individual events and the power of historical contingency or chance, championed the latter point.[15] MacKenzie argued that it is a fallacy to suggest that a technological system is only "social up to the point of invention and self-sustaining thereafter. Its conditions of possibility are always social."[16] For example, the Cuban missile crisis reflected a social influence that reinforced the need to develop a sufficient strategic deterrent force, consequently accelerating the missile race and profoundly influencing future strategies of international brinksmanship. However, many would agree that ICBM technology had already begun to shape strategic policy in a deterministic fashion prior to October 1962.

In his zeal to emphasize the social element of technological development, MacKenzie's constructivism-based critique goes to the opposite extreme and fails to recognize the sometimes-deterministic influences of technology. As cited previously, even MacKenzie had to acknowledge that "the United States built its missile arsenal without any agreed understanding—even within elite circles, much less among the general population—of why it was doing so."[17] Collectively, these inconsistencies suggest that Hughes's theory of technological momentum—with its reliance on a seemingly discrete transition from technological adolescence and social constructivism to system maturity and technological determinism—requires refinement.

Theory of Technological Dislocations

The alternative conceptual perspective provided by the theory of technological dislocations facilitates a better understanding of the mechanisms that contribute to technological development and military innovation.[18] Rather than trying to identify and characterize a technology's transition from socially constructed to technologically deterministic, it is more useful to recognize that the two characterizations may be inextricably intertwined within a technological system. Unlike social constructivism, the theory of technological dislocations acknowledges the potential existence of an orderly, technologically deterministic framework operating beneath the surface of popular history. And, unlike technological determinism, the theory of technological dislocations provides for the introduction of perturbations caused by changing social contexts that alter both nascent and mature technologies' otherwise logical evolutionary patterns.

The theory of technological dislocations builds upon the above discussion of Hughes's theory of technological momentum and a metaphor drawn from solid-state physics. Invoking the scientific metaphor, at the atomic level solid materials are made up of an ordered array of interlocking atoms. Frequently, though, that order is interrupted: an atom may go missing; the wrong type of atom may be inserted in the wrong place; or in some instances, a whole sheet of atoms may interpose and alter the structure (fig. 3). When the last occurs, it is referred to as a dislocation. Dislocations form whenever the developing crystalline structure is subjected to some form of stress, either nonmechanical stress caused by nonuniform heating or the presence of chemical impurities, or mechanical stress caused by physical damage.[19] Despite the disruption to the atoms immediately surrounding the dislocation, the lattice structure usually does not collapse in disarray. Rather, the structure quickly adapts and reassumes an ordered pattern, although the new structure differs slightly from the crystalline structure that existed before. This scientific metaphor helps one to better visualize the process of technological innovation and development.

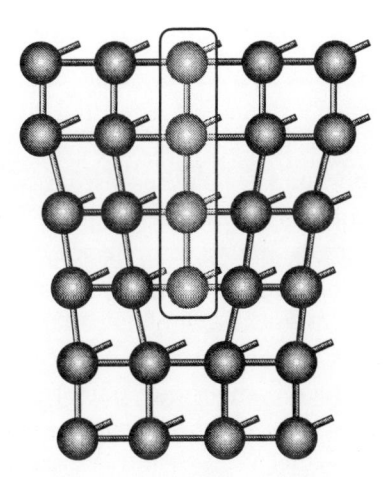

Figure 3. Edge dislocation. (*Adapted from* J. S. Blakemore, *Solid State Physics*, 2nd ed. [Cambridge: Press Syndicate of the University of Cambridge, 1960], 78.)

All but the most stringent technological determinists acknowledge the significant role socially constructed influences play in the birth of a technological system. As Hughes pointed out, fruitful technologies are rarely the product of a single "Eureka!" moment but more often result from the determined labors of a small cadre of inventors, financiers, and marketers.[20] Law similarly emphasized the distinctly social nature of an emerging technology with his concept of heterogeneous engineers and their knack for associating disparate entities to spur technological progress.[21] These collective social influences can either nurture or stymie the embryonic technology. If the social influences suppress the technology through skepticism or bureaucratic resistance, further development halts, and the technology typically withers away. But if cultivated by its surrounding social context, the budding technology may blossom.

Almost immediately, a technological trajectory develops within the emerging technological paradigm. According to Dosi, this "technological paradigm" channels the efforts of the organization in a precise direction to propagate a "technological trajectory," often to the exclusion of possible alternatives.[22] Thus, the technology quickly begins to exert a shaping influence on society. Invoking the solid-state physics metaphor, the crystalline solid begins to take shape, and additional growth aligns itself to the underlying pattern. In the military realm, the new technology begins to shape the bureaucratic institutions, either through the addition of a new directorate tasked with monitoring or promoting the new technology or the assignment of responsibility for the new technology to an existing directorate. Referencing Hughes, the technological system begins to gain momentum. However, this early trajectory and its metaphorical structural influence on society do not imply that the technology cannot thereafter succumb to bureaucratic neglect or mounting skepticism. Rather, it illustrates that nearly from its inception a technology begins to shape its surroundings in a somewhat deterministic fashion according to a logical technological trajectory.

As the technology continues to mature—as the solid crystal grows—socially induced stressors may interpose and introduce a technological dislocation, disrupting the logical technological trajectory. The dislocation jars the bureaucracy from the technological rut that previously constrained revolutionary innovation. Such stressors might include a competing alternative technology, a changed political agenda or economic environment, or a looming scientific stumbling block.[23] The magnitude of these stressors may vary. Consequently, the disruptiveness of the dislocation and the significance of the departure from the previous technological trajectory may also vary. For example, McNamara's decision to cancel the Air Force's Dyna-Soar program in 1964 effectively crushed the technological trajectory that was leading toward an independent military-manned presence in space. Other dislocations need not be so calamitous.

Consider the effect of Sputnik on the Soviet and American space programs. Prior to the 4 October 1957 launch, both countries' space and missile programs endeavored toward a common vision made apparent by von Braun. However, immediately following Sputnik, increased political pressures in the Soviet Union and the United States resulted in an altered trajectory for both nations. In the Soviet Union, Nikita Khrushchev's insatiable demand for propaganda victories led to highly publicized launches of dubious scientific value; in the United States, public outcry invigorated American space efforts and placed a high priority on manned missions.[24] Sputnik therefore represents a technological dislocation that disrupted the US and USSR space and missile technological trajectories; it forced both nations to reconsider their preconceived notions of space-related progress and reorient their efforts.

Another advantage of the theory of technological dislocations is that it provides a conceptual basis for understanding how different technological systems can develop interdependently. Much like the three-dimensional crystalline lattice structure in the physical realm, technologies can become linked to one another in the social realm. For example, if the American and Soviet space and missile technologies are recognized to be competitive and therefore mutually reinforcing, then they can be aggregated into a broader space and missile technological system. The model of technological dislocations allows for a single dislocation like Sputnik to influence the linked technologies (fig. 4). Similarly, as the following case study will demonstrate, the Air Force's guided air-to-air missile technology can be aggregated into a broader air-to-air armament technological system comprised of the missiles and the aircraft built to carry them, and stressors associated with the American air combat experience during Vietnam can be interpreted as introducing a technological dislocation into the system.

Most significantly, the theory of technological dislocations provides a conceptual model and a practical, descriptive vocabulary that aids analysis by describing how societal influences can affect a technological system at any time during its life. There is no putative, binary tipping point that illogically separates social constructivism from technological determinism. Immature technologies may be greeted with skepticism; their supporting bureaucratic institutions may exuberantly embrace mature technologies. Throughout, socially constructed contexts, or even historical contingencies, always threaten to perturb the otherwise established technological trajectories that guide technological development.

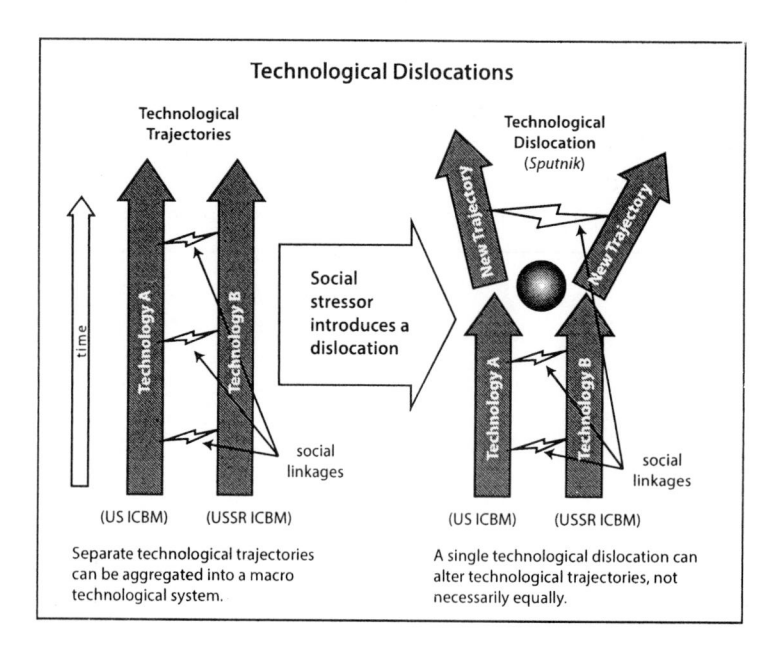

Figure 4. Technological dislocations. (*Created by* the author)

Furthermore, there can be numerous dislocations during the life of a technological system. In the ICBM example, Schriever's efforts helped garner technical and organizational legitimacy for the new technology, thereby dislocating the dominant technological trajectory that had earlier denigrated ICBMs in favor of massive fleets of manned nuclear bombers. Later, the trajectory toward more lethal ICBM targeting that spurred the development of the multiple independently targetable reentry vehicles (MIRV) encountered a dislocation in 1993 when the United States agreed to dismantle its MIRV warheads as part of the Strategic Arms Reduction Treaty II.[25] Attempting to discern the frequency and character of the numerous dislocations in the life of a technological system becomes a philosophical question of agency.

A Question of Agency

How much influence does any one individual and his or her actions have on society? Does it matter if one individual decides to ride a bike to work instead of driving a car? Can a single e-mail sent from one individual to another have important societal ramifications? Do the identities of the individuals in question matter? Certainly, one individual electing to ride a bike to work will not cut down on pollution, but a thousand individuals making independent decisions to ride their bikes to work may. Similarly, if the US president sent the e-mail to the prime minister of Great Britain, then the e-mail would likely be considered important.

It is difficult to determine agency in real time and absent context. What is expected to have significance often does not, and what is occasionally seen as innocuous can quickly become momentous. For example, within the realm of technology, nuclear power was initially seen as a potential solution to the world's burgeoning energy demands. General Electric (GE), Westinghouse, Babcock and Wilcox, and Combustion Engineering all established nuclear reactor development facilities in the 1950s, supported and subsidized by the federal government. Reflective of the national enthusiasm, Hughes reported that "a GE executive promised a young man entering the company that within ten or twenty years the company's nuclear-power business would be larger than the entire company in the 1950s."[26] Thomas E. Murray, atomic energy commissioner in 1953, proclaimed, "The splitting atom . . . is to become a God-given instrument to do the constructive work of mankind."[27] Despite this fanfare, nuclear energy fizzled.

Conversely, when Henry Ford introduced his Model T automobile on 1 October 1908, a virulent "anti-auto mood" already pervaded the nation.[28] One author noted that "the horseless carriage's arrival [nearly a decade earlier] had left more people behind than it carried along, offering the less fortunate no choice but to watch and yearn."[29] Using slightly stronger language, a *Breeder's Gazette* from 1904 described automobile owners as "a reckless, blood thirsty, villainous lot of purse-proud crazy trespassers."[30] Nevertheless, despite the initially hostile public attitude, by 1923 Ford was producing two million cars and trucks annually.[31]

These failed predictions about nuclear energy and the automobile suggest that analysis of technological development is best conducted after the fact. Study aided by the concept of technological dislocations is no exception; it is also limited to descriptive analysis used to inform decision makers, not to accurately predict the utility and practicality of a particular technology.

Even then, determining where to draw the line between the significant and the insignificant is difficult. The clash between technological determinism and social constructivism has roots in this question of agency, as it affects the historian's interpretation of technological transformations (fig. 5). Social constructivists impart high agency to individual actions; strict technological determinists grant no agency. There clearly should be bargaining room between the two. Hughes's theory of technological momentum offered one compromise by suggesting that high agency dominated immature technologies and low agency ruled mature technologies. The theory of technological dislocations takes Hughes's theory one step further and eliminates the purported distinction between immature and mature technologies.

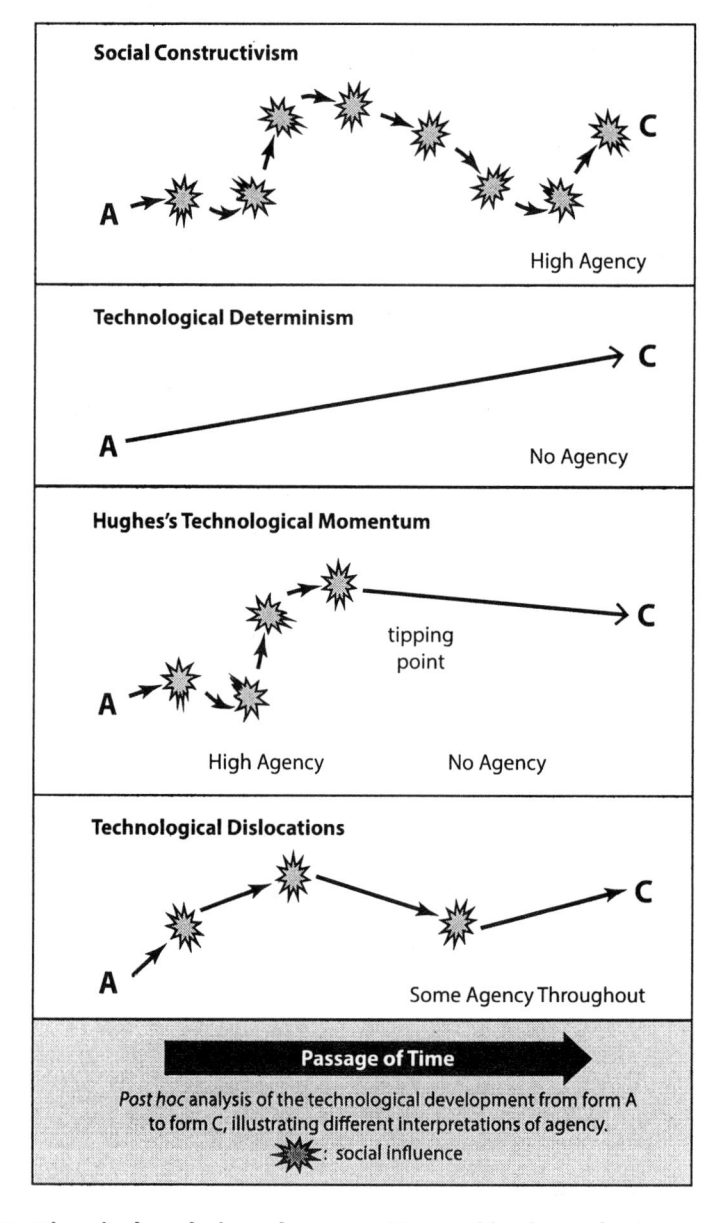

Figure 5. Historical analysis and agency. (*Created by* the author)

Issues of scale also confound the assessment of agency. "In a large technological system there are countless opportunities for isolating subsystems and calling them systems for purposes of comprehensibility and analysis," Hughes noted.[32] If historical research is narrowly focused on an individual technological system, then the level of agency imparted to particular individuals and events typically rises. For example, if studying American ICBM development, Sheehan's story of Air Force lieutenant colonel Ed Hall and the unauthorized diversion of funds from a languishing Air Force cruise missile project to help with ICBM rocket engine development are noteworthy.[33] However, if the scope of investigation addresses the role of rocketry in strategic posturing between the United States and the Soviet Union, as in McDougall's text, then Hall's actions are robbed of much of their significance—it no longer makes sense to extend agency that far down the ladder. Thus, scale and agency may be in inverse proportion—as the scale widens, agency narrows, and vice versa. Unfortunately, if neither is adequately defined, the resulting historical analysis quickly devolves into a teleological mess.

Applying the Theory of Technological Dislocations

Yet there must be some limiting principle that precludes the possibility of making a mountain out of every historical molehill. Alas, there is none, except the historian's own judgment. It is up to the historian to present a convincing analysis that portrays the past in relevant, useful terms.

In light of this objective, this author asserts that through the 1950s and 1960s, the allure of guided air-to-air missile technology entranced the Air Force. Blinded by technological exuberance, the Air Force failed to recognize that the assumptions guiding the development of its air-to-air armament were faulty. Even after those faults were laid bare by combat experiences in Korea, the Air Force continued to pursue missile and aircraft development in accordance with the dominant technological trajectory. That path demanded more complex missiles capable of targeting higher and faster-flying bomber aircraft at the expense of pursuing alternative forms of air-to-air armament optimized for different target sets.

If not for the efforts of a handful of determined individuals, the Air Force might never have introduced an air-to-air gun on the F-4 Phantom prior to the conclusion of Operation Rolling Thunder in November 1968. Furthermore, because the introduction of the old technology in an innovative fashion challenged the dominant culture within the Air Force and the prevailing technological trajectory, the new technology was initially greeted with intense

TECHNOLOGICAL DISLOCATIONS

skepticism. Fortunately, the individual agents overcame this bureaucratic resistance. The resulting technological dislocation had wide-ranging implications that extend to today.

The following historical case study and the articulation of a theory of technological dislocations are not simple pedantry. By understanding how a specific technological dislocation was generated, decision makers gain insight into the nature of technological development. They also gain a contextual appreciation for the methods that historically have helped organizations dislocate the powerful technological trajectories that favor incremental evolution over truly creative and revolutionary innovation.

Notes

1. Hughes, "Technological Momentum," 102.
2. Hughes stated: "Technological systems, even after prolonged growth and consolidation, do not become autonomous; they acquire momentum. They have a mass of technical and organizational components; they possess direction, or goals; and they display a rate of growth suggesting velocity. A high level of momentum often causes observers to assume a technological system has become autonomous. Mature systems have a quality that is analogous, therefore, to inertia of motion. The large mass of a technological system arises especially from the organizations and people committed by various interests to the system." Hughes, "Evolution of Large Technological Systems," 76.
3. As technological systems "grow larger and more complex, systems tend to be more shaping of society and less shaped by it." Hughes, "Technological Momentum," 102, 112.
4. "A system with great technological momentum can be made to change direction if a variety of its components are subjected to the forces of change." Ibid., 112–13.
5. Ibid., 112.
6. Ibid., 102.
7. Gladwell identified three characteristics of social change—"one, contagiousness; two, the fact that little causes can have really big effects; and three, that change happens not gradually but at one dramatic moment.... Of the three, the third trait—the idea that epidemics can rise or fall in one dramatic moment—is the most important, because it is the principle that makes sense of the first two and that permits the greatest insight into why modern change happens the way it does." Gladwell, *Tipping Point*, 7 and 9.
8. Gladwell offered the cell phone revolution as another example of a tipping point: "Through the 1990s, they got smaller and cheaper, and service got better until 1998, when the technology hit a Tipping Point and suddenly everyone had a cell phone." Ibid., 12.
9. The fax machine was not a prime case study within Gladwell's book, which helps explain the omission of potential causal factors. In other sections, Gladwell offered two lessons for fomenting a tipping point: "Starting epidemics requires concentrating resources on a few key areas.... The Band-Aid solution is actually the best kind of solution because it involves solving a problem with the minimum amount of effort and time and cost." Second, because "the world—much as we want it to—does not accord with our intuition," those "who are successful at creating social epidemics do not just do what they think is right. They deliberately test their intuitions.... What must underlie successful epidemics, in the end, is a bedrock belief that

change is possible, that people can radically transform their behavior or beliefs in the face of the right kind of impetus." Ibid., 255–59.

10. Quoted in Sheehan, *A Fiery Peace*, 299. President Eisenhower's decision was articulated in NSC Action no. 1433 following a National Security Council briefing given by von Neumann and Schriever.

11. Ibid., 415. One of the officers that briefed General LeMay in 1958 on the Minuteman Missile Program recollected that LeMay was captivated by the "massiveness of the scheme. The thought of hundreds and hundreds of rockets roaring out of silos was LeMay's vision of how to frighten the Russians and then to reduce the Soviet Union to cinders if it did come to nuclear war."

12. Ibid., 195; Rhodes, *Dark Sun*, 401–2; and Neufeld, *Von Braun*, 137.

13. Law referenced Edward Constant's notion of coevolution as "an attempt to grapple with the interrelatedness of heterogeneous elements and to handle the finding that the social as well as the technical is being constructed." Law, "Technology and Heterogeneous Engineering," note 5.

14. Hughes, "Technological Momentum," 112.

15. MacKenzie, *Inventing Accuracy*, 79–80; and Hughes, "Evolution," 73. While Hughes's concept of "reverse salients" offers some redress to the criticism, the notion of correcting "laggard components" implicit in the description of a reverse salient discounts the other opportunities for social influences to alter technological systems that have otherwise established momentum.

16. MacKenzie, *Inventing Accuracy*, 4.

17. Ibid., 162.

18. The term *theory* is used in the social science construct. Martel's exploration of the different interpretations of theory in the political science and international relations realm provides a basis for the present discussion of the theory of technological dislocations: "For [David] Easton, theory should also provide 'guidance to empirical research' by serving as an 'incentive for the creation of new knowledge.' ... For Brecht, theory is 'one of the most important weapons in the struggle for the advance of humanity,' because correct theories permit people to 'choose their goals and means wisely so as to avoid the roads that end in terrific disappointment.' ... The real test of a theory, for international relations theorist Hans Morgenthau, is for it to be 'judged not by some preconceived abstract principle or concept unrelated to reality, but by its purpose: to bring order and meaning to a mass of phenomena which without it would remain disconnected and unintelligible.'" Martel, *Victory in War*, 90–92. Jervis offered a similar interpretation: "A theory is necessary if any pattern is to be seen in the bewildering and contradictory mass of evidence." Jervis, *Perception and Misperception*, 175.

19. Blakemore, *Solid State Physics*, 78–80; Campbell, *Science and Engineering of Microelectronic Fabrication*, 17–18; and West, *Solid State Chemistry*, 320 and 340–55. This type of dislocation is known as an *edge dislocation*—"one of the planes of atoms terminates, resembling a knife blade stuck part way into a block of cheese." For information on dislocation formation in semiconductor crystals grown for electronics applications, see Swaminathan and Macrander, *Materials Aspects of GaAs*, 57–61 and 450–56.

20. Hughes, "Evolution," 57.

21. "'Heterogeneous engineers' seek to associate entities that range from people, through skills, to artifacts and natural phenomena. This is successful if the consequent heterogeneous networks are able to maintain some degree of stability in the face of the attempts of other entities or systems to dissociate them into their component parts." Law, "Heterogeneous Engineering," 129.

22. Dosi, "Technological Paradigms," 147–62.

23. An example of an alternative technology is the introduction of steam power into the sail-powered British Navy in the early nineteenth century; see McNeil, *Pursuit of Power*, 226. Eisenhower's "New Look" defense policy and the resultant shift from a large Army and Navy towards a leaner defense establishment reliant upon Air Force nuclear bombers illustrate the effects of a changing political and economic agenda on military technology; see Barlow, *Revolt of the Admirals*. On the role of scientific obstacles or "presumptive anomalies," see Constant, *Origins of the Turbojet Revolution*.

24. McDougall, *Heavens and the Earth*, 205, 249, and 295. McDougall described the implications of NSC-5918, "U.S. Policy on Outer Space," signed by the president on 12 January 1960: "It was therefore the American objective, among others, 'to achieve and demonstrate an overall U.S. superiority in outer space without necessarily requiring U.S. superiority in every phase of space activities.' To minimize Soviet psychological advantages, the United States should select and stress projects that offer the promise of obtaining a demonstrably effective advantage, and proceed with manned spaceflight 'at the earliest practicable time.'" Concerning the Soviet approach to space, McDougall first cites Khrushchev's memoirs: "Of course, we tried to derive the maximum political advantage from the fact that we were the first to launch our rockets into space." He then notes, "But [Khrushchev's] quest for 'maximum political advantage' led to the espousal of a 'party line' that not only hindered rapid and rational development of [Soviet] space technology but encouraged a dangerous deception in military policy as well.... [Khrushchev's] radical error was hyperbole, for he plotted the Soviet curve in the Space Age as hyperbolic, when it in fact was parabolic. After straining upward on a dizzy slope, Space Age communism slowed, then arced downward ... to a premature end."

25. MacKenzie, *Inventing Accuracy*, 214–16; and Schmemann, "Summit in Moscow," A1.

26. Hughes, *American Genesis*, 439.

27. Ibid., 441.

28. Brinkley, *Wheels for the World*, 114.

29. Ibid., 113.

30. Ibid., 114–15. Woodrow Wilson, then president of Princeton University and within seven years president of the United States, characterized the automobile as "a picture of the arrogance of wealth, with all its independence and carelessness."

31. Hughes, *American Genesis*, 208. Similar false starts have been observed in the scientific community. When Martin Fleischmann and Stanley Pons announced that they had achieved cold fusion in 1989, news reporters heralded the discovery as an astounding scientific accomplishment. However, efforts to reproduce the experiment flopped, and the story is now but an inglorious footnote in scientific history. Browne, "Fusion in a Jar," C1; and Browne, "Physicists Debunk Claim," A1. Conversely, the discovery of C60 went largely unnoticed but has since spawned the nanotechnology craze that dominates university laboratories today. "Soccer Ball Molecules," C9.

32. Hughes, "Evolution," 55. However, Hughes warned that in "isolating subsystems, ... one rends the fabric of reality and may offer only a partial, or even distorted, analysis of system behavior."

33. Sheehan described how Air Force lieutenant colonel Ed Hall, working at the Air Development Center, devised a strategy to "use the requirements for adequate engines for the Navaho booster [an intercontinental cruise missile then in development but destined to be cancelled] as a cover to acquire a rocket engine for an intercontinental ballistic missile." Sheehan, *Fiery Peace*, 247.

Chapter 4

Rise of the Missile Mafia

There will be a gun in the F-4 over my dead body.
—Gen William Momyer, USAF

Like Gen Hap Arnold before him, Gen William Momyer was a technology zealot. Serving as director of operational requirements for the Air Force from 1961 to 1964, Momyer was in a unique and powerful position to define the role of technology in the Air Force, especially after the Kennedy administration decided to revitalize the nation's nonnuclear force structure. Momyer's purview extended to the development of Air Force air-to-air armament, both the guided missiles and the aircraft designed to carry and employ them. In this position, one Air Force officer noted that Momyer had "just one feeling . . . and that was to exploit technology to its fullest; . . . if it didn't fly faster or higher, [it was] a step backwards."[1] In a 1977 interview, Maj Gen Frederick "Boots" Blesse described Momyer's particular affinity for missile technology.

> General Momyer, bless his heart, was one of the fuzzy thinkers in that [air-to-air missiles] area. He was in Requirements in the Pentagon. He was determined that the missile was the name of the game, guns just did not have any part in anything from then on. . . . In fact, I went to see General Momyer when he was a full colonel, I was a major at the time, in early 1953 or 1954. His statement to me was, "You goddamn fighter pilots are all alike. You get a couple of kills with a gun and you think that the gun is going to be here forever. Why can't you look into the future and see that the missile is here and the guns are out? There is no need for a gun on an airplane anymore."
>
> I said, "But Colonel Momyer, it is like a guy who has a pistol or it is like a guy who has a rifle fighting against another guy who has a knife. Now if you had a knife and a rifle and you threw the knife away, and you were fighting this guy near a phone booth, obviously the best weapon would be the rifle. However, if he somehow got you inside the phone booth, you would be in deep serious trouble. And that is what the gun is, the gun is the knife in the phone booth. It is for close-in protection. The missile goes off and does not even arm itself for about 1,500 feet. Now I am talking about a range within 2,000 feet; when you get to turning, you are inside that range and you cannot get away. The first guy who turns away is going to get knocked down. You just need to have a gun for those close-in times."
>
> The response to that was, "There will not be any close-in times because you will die long before you get to the missile [sic]." I said, "That is if the missile works, sir." He said, "All the missiles work."[2]

Momyer's faith in missiles proved to be without basis during Vietnam, as aptly illustrated in the dismal performance of Speedo and Elgin flights' missiles on 14 May 1967. However, Momyer was not alone in his faith in missiles, nor was he the first to promote the promise of long-range air-to-air missiles in future air combat. His attitude was reflective of a common one-dimensional understanding of future air combat that would be fought primarily against Soviet bomber aircraft and the trend toward technological exuberance that underpinned Air Force weapons decisions in the 1950s and 1960s. During that period, the Air Force's embrace of air-to-air missiles established a technological trajectory that subsequently exerted a deterministic influence on Air Force weapons development, blinding Air Force leaders to potential alternatives in the character of future conflicts and the technologies required for success therein.

Air-to-Air Missile Development

The Air Force's fascination with high-speed, air-to-air guided missiles blossomed during the closing stages of World War II. The Airmen of the Army Air Forces, intrigued by the performance of German V-1 and V-2 missiles, sought to apply the developments in modern rocketry to the emerging "air-to-air combat problem" presented by faster, higher-flying aircraft.[3] Beginning in 1948, students at AU's Air Tactical School (ATS) at Tyndall AFB, Florida, received a one-hour lecture on the armament problem. The lesson's stated purpose was to "acquaint the student with the need for air-to-air guided missiles and with some of the problems associated with their development and operational use."[4] The lesson plan focused on two issues.

The first was "the effect of the high speed on the pilot."[5] While newer, faster aircraft subjected the pilot to the increased physiological stresses of higher altitude flight and greater G-forces, the lesson focused instead on the cognitive limitations the pilot would encounter in the faster-paced environment. In this new age, the Air Force determined most of its pilots would be unable to autonomously process information quickly and accurately enough to complete an air-to-air intercept to a position from which they could employ existing weapons.[6]

The second issue of jet-age air combat was characterized by the limited effectiveness of air-to-air cannon technology at high airspeeds. "New 50 caliber machine guns can fire 1,000 to 1,200 rounds per minute with a muzzle velocity of 2,700 feet per minute, but the range at which the average pilot can expect to obtain telling hits is very short. In fact, even using the A-1 [gun] sight, he will still have to get within 800 yards of the target to obtain hits.... The way aircraft are being built these days," the lecture continued, "it would be a very lucky round indeed that might destroy another ship."[7] Consequently, only air-to-air guided missiles offered the prospect of "enabl[ing] a pilot to stand off at least 10,000 feet away and fire at a target with fatal results to that target."[8]

Summarizing the promise of the new missile technology, the lesson concluded, "As presently visualized, the missile has the following advantages over armament now mounted in our aircraft:

1. Much longer effective range
2. Controllable all the way to the target
3. Powerful enough to insure a kill."[9]

By the time the ATS lesson was introduced in April 1948, the Air Force already had gathered valuable air-to-air missile experience. The first Air Force air-to-air missile, the JB-3, boasted a massive 100-pound warhead, a top speed of 600 miles per hour, a range of five to nine miles, and an ability to attack aircraft at altitudes of up to 50,000 feet.[10] Designed by Hughes Aircraft according to a January 1945 Army Air Forces contract, the missile, nicknamed "Tiamet" after the "goddess in Assyrian-Babylonian mythology," was guided toward the target by an internal FM radar homing device.[11] Ironically, the first Tiamet launch occurred on 6 August 1945—the same day the United States ushered the world into the atomic age, which would consequently place a greater premium on an aircraft's ability to defend the nation from future higher and faster Soviet bombers threatening atomic attack. However, according to Air Defense Command's (ADC) *History of Air Defense Weapons, 1946–1962*, "none of the first ten [Tiamet] missiles tested showed much promise," and the "very cumbersome" 625-pound missile—"essentially a 100-pound bomb with wings"—was terminated in September 1946.[12] The Air Force instead rededicated and accelerated its efforts toward acquiring a more "'practical' air-to-air missile that could be developed within two years."[13]

One ADC historian described the ensuing effort: "Missile development contracts sprouted like spring flowers immediately after the war."[14] Several contracts were issued, including two separate contracts each for a fighter-launched missile (to attack bombers) and a bomber-launched missile (to attack fighters). However, when President Truman drastically curtailed the national defense budget, the windfall in missile spending quickly evaporated and the newly independent Air Force allowed several contracts to wither and die in 1947–48.[15] By the end of 1948, only two Air Force air-to-air missile contracts remained: Ryan's Firebird missile, designed for use by fighter aircraft; and Hughes Aircraft's Falcon missile, designed for use by bomber aircraft.[16] Further budgetary pressure led to the realization that the "distinction between bomber-launched missiles and fighter-launched missiles had blurred to the point where the two were interchangeable," and the Air Force adapted its contracts to reflect the need for only a single air-to-air missile that would enable "use as an offensive weapon for interceptor aircraft and for defensive use by bombers."[17] Finally in April 1949, the Air Force terminated Ryan's Firebird program and devoted all of its air-to-air missile funds and energy to Hughes's Falcon missile program.[18]

The first version of the Falcon missile was radar-guided.[19] It relied on the interceptor aircraft to use its fire-control radar to illuminate the target aircraft. Once the missile was launched, the seeker within the GAR-1 Falcon sensed the radar energy reflected off the target, measured the relative change in line-of-sight between the missile body and the radar reflections, and steered itself using hydraulic servos that actuated its control fins to zero-out the relative changes in line-of-sight to create a collision intercept. These principles of radar guidance allowed the interceptor to launch the missile in any weather condition—even if the interceptor pilot could not see the target—and from any direction (aspect) relative to the target.[20] However, it also required the interceptor aircraft's radar to remain locked to the target while the missile was in the air—easy against a large nonmaneuvering target but exceedingly difficult against a small maneuvering one. Therefore, successful GAR-1 employment demanded flawless performance from both the interceptor radar and the missiles. It proved to be a high and often unachievable standard.

The ambitious project was also hampered by continued bureaucratic skepticism and technical difficulties. Despite being the sole Air Force air-to-air missile project, funding for the Falcon continued to deteriorate, the victim of tightening defense budgets and bureaucratic coffer scavenging to fund the Air Force's focus on strategic bombing. In 1949 the Air Force set aside a puny $200,000 emergency fund for the program, lest all development work be halted if the program's funds completely disappeared.[21] Funding was eventually restored, but the influx of money did little to address the performance failures plaguing the missiles.

The weapons system was extremely complex. The missile relied on 72 notoriously unreliable radio vacuum tubes; the interceptor aircraft's radar relied on countless more.[22] Persistent technical problems resulted in numerous production delays, forcing Hughes to slip the promised delivery date for the missile from June 1954 to October 1954 and finally to August 1955.[23] The first GAR-1 Falcon-equipped squadron of F-89H Scorpion aircraft was not declared operational until March 1956, almost two years after the first scheduled delivery date.[24]

Hughes addressed some of the performance limitations of the GAR-1 missile with its follow-on version, the GAR-1D. Notably, the GAR-1D increased the missile's performance against high-altitude targets from a 50,000-foot maximum target altitude to 60,000 feet.[25] The GAR-1D, however, did not remarkably improve the reliability of the GAR-1. ADC's *History of Air Defense Weapons* recorded, "Although the F-89H and F-102A and the GAR-1D missiles, which were their primary armament, were available to ADC in appreciable quantities by the end of 1956, the missiles were not usable at that time. While the fire control systems (R-9 and MG-10) designed for use in connection with the Falcon missile were far from reliable, the missiles themselves also failed to live up to expectations."[26] For example, the Air Force Weapons Center in Yuma, Arizona, determined that "37.5 percent of the Falcons in storage failed to meet operational standards upon initial inspection. A later check showed another 16.5 percent to be unfit for use. Firing tests resulted in a large proportion of near misses even when the fire control system was operating normally."[27]

Based in part on these failures, the Air Force removed the GAR-1D missiles from its operational inventory in January 1957. The missiles returned to service six months later after Hughes corrected some of the deficiencies.[28] Reminiscing about the difficulties associated with early guided missile development, Fred Darwin, then executive secretary of the Department of Defense's Guided Missiles Committee, lamented, "Day-by-day, then with increasing acceleration, I became convinced of something I considered important: THESE THINGS WILL NEVER BE OPERATIONALLY USEFUL. Even Should We Make Them Perfect."[29]

Hughes's infrared-guided (heat-seeking) variant of the Falcon, the GAR-2, suffered from an equally tumultuous development process. The GAR-2 missile was initiated in November 1951 and Air Force officials hoped the GAR-2 missile would complement the radar-guided GAR-1.[30] Indeed, the GAR-2 offered multiple advantages over the GAR-1. According to a 1956 Air Force evaluation report, the "GAR-1B [GAR-2] can be used at lower levels (no ground clutter); against multiple targets (it will select a target); and it has greater accuracy since the missile homes on a point source of heat rather than seeing the entire target. Additional advantages are that it is a passive seeker, it is immune to electronic countermeasures, and it can be launched with less specialized fire control equipment."[31]

Unfortunately, the GAR-2 and its improved variant, the GAR-2A, performed miserably during low-altitude tests conducted in 1959.[32] Nevertheless, a "single success after universal failure" during the testing buoyed the Air Force's and Hughes's "hopes that something might, after all, be done to make the GAR-2A useful at low altitudes."[33] In this instance, the optimism was deserved; Hughes successfully designed an improved infrared guidance unit and solved many of the low-altitude guidance problems.[34] By 1961 the GAR-2A provided the primary punch for the F-102A and served as secondary armament on the F-101B.[35]

As Hughes struggled to work the kinks out of its guided missile systems, the Air Force hedged and looked toward unguided rockets as an interim air-to-air armament solution. Ironically, the Air Force turned to the Army's Ordnance Department for a viable system. The Army obliged and began transforming the German World War II two-inch R4M unguided rocket into a "2.75-inch spin-stabilized rocket expected to have a range of about 2,000 yards."[36] Although very different from the 10,000-yard range the Air Force desired, the Army's 2.75-inch folding-fin aerial rocket (FFAR) promised to help "increase interceptor firepower until the guided missiles were ready."[37]

However, the effectiveness of the unguided rockets was questionable. "In a case famous at the time [in 1956], two F-89s equipped with a total of 208 rockets fired all of them, but failed to shoot down an F6F Hellcat drone that had drifted off course and was threatening to crash on Los Angeles. [The wayward drone] eventually ran out of fuel and crashed harmlessly. The rockets did more damage. Several started brushfires, and one errant missile hit a pickup truck in the radiator but failed to detonate."[38] Unguided rockets were still in use as air combat armament in 1961, but confidence in their utility remained low. One Marine Corps pilot remarked, "The plan was to fire a salvo of four 19-shot pods on a 110-degree lead-collision course, with a firing range of 1,500 feet. Whether or not we would have hit anything on a regular basis is a matter for conjecture, but I think not."[39]

Hughes continued to improve the Air Force's Falcon guided missiles, eventually developing an upgraded GAR-1D radar-guided missile, designated the GAR-3, and an enhanced GAR-2 infrared-guided missile, designated the GAR-4. Announcing the development of the GAR-3 in 1958, the *New York Times* described the new missile as having "a longer, higher, and deadlier reach than that of any other air-to-air missile." In the same article, Roy Wendahl, vice president of Hughes's airborne systems group, claimed that the GAR-3 could "climb far beyond the altitude capabilities of the interceptor and destroy an enemy H-bomber in any kind of weather."[40]

In 1961 the Air Force reclassified its missile programs, and the GAR-1 through 4 Falcon missile designations were subsumed under the AIM-4 label.[41] Besides now sharing a common designation, the family of Falcon missiles also shared a notorious deficiency. Because the missiles were specifically designed to be paired with the F-102A Delta Dagger under the new aircraft-missile weapon system construct, the missile's dimensions were restricted by the size of the F-102A's internal weapons bay. After allotting space within the missile body for the complex and bulky array of vacuum radio tubes needed for missile guidance, there was disappointingly little room left for the missile warhead and fusing assembly, rendering the Air Force's desire for "a kill even from a one-hundred-foot miss" laughable.[42] Instead of a 300-pound missile warhead, the Air Force eventually settled on Hughes's puny 2.8-pound warhead, later increased to a whopping five pounds.[43] To detonate the Falcon's miniscule warhead, the missile relied on a contact fuse mounted on the leading edge of the missile fins, which meant that the missile had to hit the target to explode.[44]

Like the Air Force, the Navy also pursued development of both radar- and infrared-guided air-to-air missiles for its fighter aircraft. And like the Air Force, the Navy's guided missiles were initially greeted with technological skepticism. William McLean, overseeing the Navy's Sidewinder guided air-to-air missile program while working at the Naval Ordnance Test Station at China Lake, California, described the constraints they encountered:

> Every time we mentioned the desirability of shifting from unguided rockets to a guided missile, we ran into some variant of the following list of missile deficiencies:
>
> Missiles are prohibitively expensive. It will never be possible to procure them in sufficient quantities for combat use.
>
> Missiles are impossible to maintain in the field because of their complexity and the tremendous requirements for trained personnel.
>
> Prefiring preparations, such as warm-up and gain settings required for missiles, are not compatible with the targets of surprise and opportunity which are normally encountered in air-to-air and air-to-ground combat.
>
> Fire control systems required for the launching of missiles are complex, or more complex, than those required for unguided rockets. No problems are solved by adding a fire control computer in the missile itself.
>
> Guided missiles are too large and cannot be used on existing aircraft. The requirement for special missile aircraft will always result in most of the aircraft firing unguided rockets.[45]

The Navy's radar-guided missile, the Sparrow, evolved from a 1947 contract with Sperry Gyroscopic Laboratory. Sperry's Sparrow I saw limited fleet use beginning in September 1952; widespread deployment throughout the fleet began in May 1954.[46] However, because of design limitations in the Sperry missile, the Navy pursued two alternate versions of the Sparrow:

Douglas Aircraft's Sparrow II and Raytheon's Sparrow III. A series of missile fly-offs between the three versions led to a 1957 Navy decision to award its future contracts exclusively to Raytheon and its Sparrow III design.[47] Unlike Sperry's beam-rider missile, which steered its control fins to keep the missile in the center of a radar beam pointed at the target aircraft, Raytheon's Sparrow III relied on a semiactive seeker that guided the missile body toward radar energy reflected off the target, similar to the guidance system used by the Air Force's radar-guided Falcon.[48] The Sparrow, never designed to be carried internally in a particular aircraft, was significantly larger than the Falcon, measuring 12 feet in length compared to the Falcon's six feet, and packed a considerably larger wallop with a 65-pound warhead.[49] The Navy set sail with the Sparrow III in July 1958.[50]

The Navy's infrared missile, the Sidewinder, was developed in-house by engineers at China Lake. Despite being denied the level of resources devoted to radar-guided missiles, the Sidewinder beat the Sparrow to the fleet by almost two years, becoming operational in 1956.[51] The genius of the Sidewinder lay in its relative simplicity. Whereas the Air Force's infrared Falcon missile variant required 19 technicians just to maintain the missile's test equipment, which in turn occupied 40 feet of wall space, the Navy designed the Sidewinder for the harsh and cramped conditions on an aircraft carrier.[52] Moreover, the Sidewinder generally performed better than the Falcon. The disparities were too great to ignore, and in 1957 the Air Force reluctantly decided to co-opt the Navy's Sidewinder project.[53]

In contrast to the Air Force's Falcon missiles that relied solely on a contact fuse to detonate the warhead, the Navy's Sparrow and Sidewinder missile designs incorporated both a contact and a proximity fuse. Thus, even if the Navy missile did not hit the target, if the missile flew close enough to it, the warhead would still detonate, hopefully causing enough damage to disable the enemy aircraft. However, the addition of a proximity fuse necessitated a greater minimum firing range—approximately 3,000 feet of separation between the interceptor and the target—to preclude the possibility of the missile inadvertently fusing off the launching aircraft. At the time, few pilots recognized that the minimum ranges of the missiles roughly corresponded to the maximum effective range of existing aircraft cannons.[54]

The poor reliability of the Air Force's Falcon missiles and the greater minimum ranges of the Navy's Sparrow and Sidewinder missiles were not the only limitations of the new air-to-air missiles. Launching a radar-guided missile entailed a time-consuming and complex procedure involving multiple switch actuations and dial manipulations to configure the aircraft radar, acquire the target with the radar, and select and launch the appropriate missile.[55] After launch, the pilot had to ensure that the radar remained locked on the target to provide the constant radar illumination that the missile required for guidance. Loss of the radar lock resulted in the missile veering wildly off course. Furthermore, early aircraft fire-control radars had difficulty acquiring and tracking targets that operated below the interceptor and close to the ground due to a problem known as *ground clutter*—the radar could not distinguish the low-altitude target aircraft from the terrain features on the ground.[56]

Infrared missiles had their own set of limitations. Whereas infrared missiles did not require a radar lock, they did require the pilot to maneuver the interceptor aircraft into a small 30-degree cone directly aft of the target.[57] This was the only region where the infrared seeker on the missile could observe and track the target's hot jet exhaust; outside of the cone, the missile was incapable of detecting the target's heat source. To defeat a heat-seeking missile prior to launch, the enemy only had to aggressively turn the aircraft to keep the interceptor aircraft outside of the cone. Under the same premise, a similarly aggressive turn could also defeat a Sidewinder missile already in flight.[58]

Although Air Force and Navy officials recognized many of these limitations, they were not deemed significant in the next anticipated conflict. Air Force and Navy officials expected pilots to have ample time to acquire the targets, actuate switches, and maneuver their aircraft into position to employ a radar-guided missile or, if necessary, an infrared-guided missile. Few challenged these assumptions during missile testing. Rather than conducting missile tests against small, maneuverable, fighter-like aircraft, both services concentrated the majority of their air-to-air missile testing on intercepting high-flying, nonmaneuvering targets, reflective of their anticipated combat against massive formations of large Soviet bombers en route to attack western Europe and the United States. There was no need to worry about targeting the Soviet fighters that might accompany the bombers to the target because there would not be any fighters; they did not have sufficient fuel for the bomber-escort mission. Similarly, the majority of US fighters faced the same fuel limitations and would be unable to escort American bombers to their targets within the Soviet Union. Logic therefore suggested that American interceptor aircraft need only be concerned with attacking high-flying, nonmaneuvering Soviet bomber aircraft.

This general assessment of the threat was clearly reflected in the Air Force's decision to acquire nuclear-armed air-to-air unguided rockets and guided missiles for its interceptor aircraft. Having determined that "existing and programmed armament [was] deficient" and cognizant of the need for weapons that would "assure a high degree of kill probability," on 31 January 1952 ADC issued a requirement for a nuclear interceptor missile capable of "cut[ting] a wide swath of destruction through a formation of enemy bombers."[59] However, at that time, no nuclear warhead existed that was small enough for use in a fighter-interceptor missile. ADC reissued its requirement on 23 March 1953 and stressed the urgent need for a "lightweight atomic warhead of lowest possible cost with yields within the range of 1–20 KT [kilotons]."[60] The Joint Chiefs of Staff (JCS) approved development of a nuclear-armed air-to-air rocket a year later, and the MB-1 Genie, an unguided rocket complete with nuclear warhead, was test fired by an F-89J Scorpion in July 1957 over the desert north of Las Vegas, Nevada.[61] Partly because the unguided MB-1 did not fit within the F-102A internal weapons bay, but also reflective of the Air Force's fascination with guided missiles, the Air Force ordered Hughes to develop a nuclear variant of the Falcon, the GAR-11, which was test fired without a warhead on 13 May 1958.[62]

From its inauspicious beginnings as the JB-3 Tiamet in 1946, the air-to-air guided missile underwent a major technological transformation in the ensuing 15 years, overcoming much of the early bureaucratic skepticism and its "rhetoric of denial."[63] Although still suffering from significant employment limitations and questionable reliability, by the time of the Korean War armistice in 1954, guided missiles were considered up to the task of inflicting considerable damage on the ominous hordes of Soviet bombers should the opportunity present itself. Reinforcing that assessment, the Air Force elected to remove the guns from its interceptor versions of the F-86 (F-86D),[64] the F-89 (F-89D),[65] the F-94 (F-94C),[66] and its newly designed F-102A interceptor.[67]

Gun development continued within the service until 1957, but only in an air-to-ground context and only for aircraft designed for fighter-bomber applications such as the F-100 Super Sabre, the F-101 Voodoo, and the F-105 Thunderchief. The GE 20 mm M61 Vulcan Gatling gun, capable of firing 6,000 rounds per minute, armed the Thunderchief.[68]

For its air-to-air armament, the Air Force focused exclusively on developing its guided missiles—optimized for attacking large, nonmaneuvering aircraft—despite its experiences in the Korean War struggling to wrest air superiority from a determined foe armed with small, maneuverable MiG fighters.[69] For example, the Air Force's 1957 post-Korea requirements for the F-106, a follow-on to the F-102A, addressed the need for "carry[ing] one MB-1 air-to-air atomic rocket and four GAR-3/GAR-4 Falcons, launchable in salvo[s] or in pairs."[70] Reflecting the opinion of the day, Secretary of Defense McNamara reportedly quipped, "In the context of modern air warfare, the idea of a fighter being equipped with a gun is as archaic as warfare with bow and arrow."[71]

The Phantom II

In light of this fixation on guided missiles, it is not surprising that the Navy's F-4 Phantom II (then designated the F4H-1F), once deemed the "classic modern fighter of the free world" by aviation historian and former Smithsonian Air and Space Museum director Walter Boyne, entered the fleet in December 1960 bristling with missiles but missing an internal cannon.[72] Originally proposed to the Navy as a follow-on to the F3H Demon in September 1953, McDonnell Aircraft's design morphed several times during the next two years as the Navy waffled between requesting a fighter-interceptor and an aircraft optimized for ground attack.[73] During the attack-aircraft phase, McDonnell reengineered the F4 design into the AH-1, a twin-engine, single-seat aircraft armed with four 20-mm Colt Mark-12 guns or 56 two-inch unguided rockets.[74] However, in April 1955 the Navy finally announced that it would pursue acquisition of a two-seat, all-weather, fighter-interceptor. McDonnell responded and began manufacturing several F4H-1 test aircraft, which eventually evolved into the F4H-1F version destined for fleet use.[75]

After settling on a fighter-interceptor design, the Navy had to address the aircraft's armament requirements. A series of Sparrow missile tests conducted in August 1955 convinced Navy engineers "that missiles provided a better interception system than a combination of cannon and aircraft."[76] In short, the F-4 engineers believed that "guns were . . . a thing of the past, . . . [and] guided missiles were the wave of the future," and they quickly moved to incorporate the missiles and the necessary accompanying fire-control radar equipment into the aircraft design.[77]

Still, the transition to an all-missile configuration took several design iterations. Initially in 1955, the Sparrow missiles were added only as a supplement to the already planned cannon and rockets. Less than a year later, Navy engineers designated Raytheon's Sparrow III missile the aircraft's primary weapon. By 1957 the internal cannon was completely removed from the F-4 design.[78] According to Marshall Michell III, the "lack of a cannon did not appear to unduly disturb the F-4 aircrews; in fact, many supported it."[79] Glenn Bugos described the rationale behind the armament decision:

> There were four main reasons for dedicating the F4H-1 to guided missiles. First, the missiles were lighter than the cannons they replaced. Second, they were much cheaper than aircraft, which, if carrying cannons or rockets, would need to get more dangerously close to the enemy. Third, self-guided missiles reduced the workload of the aviators, who simply pushed a button in response to symbols on a computer screen rather than engaging in the extensive dogfighting maneuvers needed with cannons or rockets, though the aviators saw this as being de-skilled by the missiles. Finally, the use of guided missiles

allowed a more flexible reconstruction of the F4H-1's interception system. . . . Unlike rockets or cannons, there was an electromagnetic umbilical cord between the Sparrow III in flight and the F4H-1. This meant McDonnell engineers could decide which tasks—how much guidance or speed—should be built into the missile and which built into the aircraft, and how these tasks should be shifted between the aircraft and the missile as the technologies changed.[80]

Contrary to popular lore, McNamara did not mandate that the Air Force adopt the Navy's F-4. The Air Force by October 1961 had already expressed interest in acquiring an Air Force version of the Navy F4H-1, which they would label the F-110 Spectre, the next designation in the Air Force's century series of fighters.[81] But the secretary of defense did pressure the Air Force to cancel its next version of the F-105, the F-105E, in favor of procuring additional Navy Phantoms for Air Force use. Emphasizing commonality and cost effectiveness, McNamara also urged the Air Force to accept the new Navy fighter with little modification.[82] Finally, the secretary, "preoccupied with standardization of things both technical and nomenclatural," demanded that the services accept a common designation for the aircraft. Thus, the Navy's F4H-1 test aircraft became F-4As, the F4H-1F production aircraft became Navy F-4Bs, and the Air Force's F-110 aircraft became Air Force F-4Cs.[83]

Modifications of the Navy's F-4B for Air Force use as the F-4C were limited to enhance "the notion of commonality and . . . [maintain] the program schedule."[84] The Air Force requested only seven changes: (1) an improved radar display; (2) an autonomous inertial navigation system (INS) similar to the type installed in Strategic Air Command (SAC) bombers; (3) a larger oxygen supply to support transoceanic flights; (4) a refueling receptacle compatible with Air Force boom-equipped aerial refueling aircraft; (5) a cartridge-based engine-starting system for use at remote locations without adequate ground support; (6) larger, softer main landing gear tires to better distribute the aircraft's weight on concrete runways (vice the Navy's steel carrier decks); and (7) a full set of flight controls for the rear cockpit.[85] The lack of an internal cannon and the aircraft's total reliance on air-to-air missiles was not an item of concern for most Air Force procurement officials despite recognizing the variety of missions—ranging from ground attack to counterair—the Air Force expected its newest multirole fighter aircraft to perform.[86]

As the Air Force F-4C began to materialize, a handful of determined officers tried to alert the Air Force leadership that the decision to forego a gun that could complement the guided missile armament hinged on faulty assumptions. However, they met stiff resistance. According to Maj Gen John Burns, the prevalent attitude within the Pentagon at the time was that aircraft guns were "anachronisms, throwbacks to earlier, bygone days, . . . that the day of the gun was gone, and that the day of the maneuvering fighter was gone, and that air combat would consist entirely of a radar detection and acquisition and lock-on, followed by a missile exchange."[87]

Working at the Pentagon in Air Force Operations as a colonel in the early 1960s, Maj Gen Richard Catledge recounted his Pentagon experience with the antigun sentiment and Momyer:

> I realized this two-star, General "Spike" Momyer[,] ran the Air Staff—very strong-minded individual, very knowledgeable individual, who did his homework on everything.... It was his belief and his concept that future airplanes would not have guns in them. There was no need for guns. I couldn't believe this when I came across it in the Pentagon.
>
> So I built a flip chart briefing, with my convictions, why we needed guns, more for air-to-air than for air-to-ground.... Anyway, I found it was an uphill fight. That every colonel, every major, in requirements, whose business I was getting into, believed as their boss did. So I really went uphill.
>
> I built my chart, got my ducks all lined up, and went to my boss, [Major General] Jamie Gough, and gave him that briefing. He said, "Well, it's a good story, ... [but] you are going to have to run this by Spike Momyer, and I'm not going with you."...
>
> So I went up, got the appointment, put my stand in front of his [Momyer's] desk, and started in telling him why we needed guns in airplanes. Well at one point in this—he stopped me several times and gave me a few words on why we did not, and [that] essentially missiles had taken over. Missiles had taken over for air-to-air ... and other kinds of munitions [had taken over] for air-to-ground, so there really was no need [for obsolete guns].
>
> Well, I thought I had a pretty good argument, but [I] didn't convince him. I remember he'd beat on his table and say, "There will be a gun in the F-4 over my dead body." That was his attitude.[88]

The Air Force's first YF-4C prototype was delivered on 27 May 1963, 65 days ahead of schedule. On 1 August 1964, the 558th Tactical Fighter Squadron of the 12th TFW at MacDill AFB, Florida, conducted a "limited evaluation ... to determine the practical capabilities, deficiencies, and limitations of the F-4C aircraft."[89] Unfortunately, air-to-air testing was a "relatively low test project priority." Of the 46 scheduled Sparrow shots, only 17 sorties were flown and, of those, only four successfully launched the test missile. All four test launches were later "termed non-productive" due to failure of the telemetry scoring system. No Sidewinder missiles were launched during the test. Despite the inconclusive findings, the evaluation report was optimistic, declaring, "The F-4C [air-to-air] delivery capability is somewhat apparent."[90] The Air Force F-4C entered operational service at MacDill AFB on 20 November 1963, armed with Navy Sparrow III radar-guided and Sidewinder infrared-guided missiles but no gun.[91]

The effects of the Air Force's fascination with guided missiles began to manifest in another area—aircrew training. One aviation historian accused the Air Force of placing "more emphasis on its capital equipment throughout the late 1950s and 1960s than it did on preparing its pilots for aerial combat."[92] Indeed, Blesse characterized the Air Force between 1956 and 1963 as being dominated by an overriding and unhealthy concern for aircraft safety: "Safety became more important than the tactics, more important than gunnery, more important than anything. Safety was king."[93] For example, following two Phantom training accidents, Tactical Air Command (TAC) imposed strict limits on aircraft maneuvering, relegating the F-4 crews "to train for aerial combat using a flight regimen confined to unrealistically high airspeeds and low angles of attack."[94]

Many senior Air Force leaders justified the tight restrictions on air-to-air training because they believed there was no need to practice aggressive aircraft maneuvering for an intercept mission that would only entail taking off, climbing to the altitude of the Soviet bomber targets, selecting the appropriate missile, and pulling the trigger.[95] This idealistic vision of air combat extended to the Navy. One Navy pilot reminisced, "F-4 squadrons, being state-of-the-art in equipment and doctrine, seldom bothered with 'outmoded' pastimes such as dogfighting. Besides, they had no guns and consequently felt little need to indulge in ACM [air combat maneuvering]."[96]

Thus, at the beginning of the 1960s, technological exuberance for air-to-air missiles exerted a profound influence over the Air Force. Fascination with the promise of air-to-air guided missile technology, optimized to defend the nation from Soviet nuclear bombers, blinded Air Force leaders to the shift in Soviet strategy from manned bombers to ICBMs. Even after intelligence assessments confirmed the Soviet strategic swing, Air Force leaders failed to adapt their vision of future air combat to the new strategic context. They deemed the missile technology "too promising to discard" and continued to focus missile development against the preexisting target set.[97] The assumption that the missiles would attack large, high-flying, nonmaneuvering targets went unchallenged.

American missile technology and American pilots were "expected to dominate air combat" upon entering the Vietnam War.[98] In the words of Momyer, "All the missiles work."[99] Unfortunately, the reality in the skies over Vietnam did not match the rhetoric.

Notes

1. Hildreth, oral history interview, 29.

2. Blesse, oral history interview, 59–60. There is a discrepancy in Blesse's narrative. Blesse states that he confronted then-colonel Momyer in 1953–54, but this would have been prior to Momyer's assignment at the Pentagon. It is possible that Blesse encountered Momyer in the specified period while Momyer was serving on the Air War College faculty at Maxwell AFB, Alabama, and Blesse was assigned to the Fighter Gunnery School at Nellis AFB, Nevada. Blesse provided no further clarification on the meeting's timing elsewhere during the interview. In a touch of irony, Blesse and Momyer would meet again to discuss the practicality of installing guns on fighter aircraft; as the 366 TFW deputy commander for operations, Blesse needed the Seventh Air Force commander, Momyer, to approve his proposed aircraft modification.

3. McMullen cited the influence of the German V-1 and V-2 weapons on American air-to-air missile development. The 1948 ATS lesson plan described the "air-to-air combat problem": "At present, it would appear that our faster aircraft . . . may be fine to carry a pilot from one point to another in a great hurry but may be of little or no use in air-to-air combat." McMullen, "History of Air Defense Weapons"; and ATS lecture manuscript, "Air-to-Air Guided Missiles," 1.

4. ATS, "Air-to-Air Guided Missiles," 1.

5. Ibid.

6. Ibid. The lesson plan used the following example of an air-to-air intercept to illustrate the geometric problem: "The B-45 flying at 500 miles per hour is travelling south. The P-51 [propeller-driven aircraft] and P-88 [jet-powered aircraft] flying 450 miles per hour and 677 miles per hour respectively are flying north about one mile west of the B-45 flight path. When these two fighters sight the bomber, it is two miles away. Now let both fighters attack using a curve of pursuit. Both will fly so that the acceleration on the pilot never exceeds 4 Gs. The P-51 flies around and may be able to get in a short burst at fairly long range. If the pilot miscalculated slightly, he will never come within firing distance of the B-45. The P-88 will find itself several miles from the B-45 when it has arrived at the same heading as the bomber. It will be unable to fire a single round and will be practically out of identifying sight of the bomber. Had the P-88 been an aircraft flying at 1,200 miles per hour, the problem would be even more acute. At this speed, the radius of curvature becomes 4.63 miles. When the 1,200 miles per hour aircraft comes to the same heading as the bomber, it will be 8.26 miles to the east of where the B-45 was originally and about four miles astern. Therefore, as aircraft speeds rise, it will become more and more difficult for fighters to attack other aircraft."

7. Ibid., 2. There is an interesting parallel between Edward Constant's notion of a *presumptive anomaly* that led to the turbojet revolution and the presumed necessary shift from cannon to missile armament predicated by the same turbojet revolution. Constant proposed, "Presumptive anomaly occurs in technology, not when the conventional system fails in any absolute or objective sense, but when assumptions derived from science indicate either that under some future conditions the conventional system will fail (or function badly) or that a radically different system will do a much better job. No functional failure exists; an anomaly is presumed to exist; hence presumptive anomaly." Constant, *Origins of the Turbojet Revolution*, 15. In this instance, the scientific limitations associated with gunpowder and bullets were presumed to limit their effectiveness in the jet-age future. The limitations of guns in air combat can also be interpreted as an example of Hughes's notion of a "reverse salient"—a laggard system component that "holds up technical progress." Hughes, "Evolution of Large Technological Systems," 73; and MacKenzie, *Inventing Accuracy*, 79–80.

8. ATS, "Air-to-Air Guided Missiles," 2.

9. Ibid.

10. Wildenberg, "A Visionary Ahead of His Time," 6.

11. Ibid.; and ATS, "Air-to-Air Guided Missiles," 4.

12. McMullen, "History of Air Defense Weapons," 12. The ATS lecture stated, "The project [JB-3] was cancelled when it was decided that the missile no longer met the requirements of an air-to-air missile because it was too large in size and lacked sufficient maneuverability." ATS, "Air-to-Air Guided Missiles," 4.

13. Wildenberg, "A Visionary Ahead of His Time," 6.

14. McMullen, "History of Air Defense Weapons," 44. Illustrative of the relative importance granted to advanced armament following the war, "item three on the revised [1947] AAF [Army Air Forces] priority list specified the need for 'greatly improved defense armament for bombers,' and [the recommendation] that the bomber launched air-to-air missile should proceed on a high priority." Wildenberg, "A Visionary Ahead of His Time," 8.

15. Wildenberg, "A Visionary Ahead of His Time," 8; and Cherny, *Candy Bombers*, 231. For example, Truman demanded that the military limit its spending to an inflexible $15 billion for fiscal year (FY) 49.

16. McMullen, "History of Air Defense Weapons," 47.

17. Ibid.; and Wildenberg, "A Visionary Ahead of His Time," 8.

18. McMullen, "History of Air Defense Weapons," 88.

19. Hughes's family of Falcon missiles underwent several changes in designation during its almost 40-year life. Initially, the Air Force assigned aircraft type designations to its guided missiles; as an interceptor missile, the Falcon missile became the F-98. However in 1955, the Air Force changed its missile designations to use a GAR (guided air rocket) prefix, and the radar-guided Falcon became known as the GAR-1 (alternate versions of the Falcon became the GAR-2, GAR-3, and GAR-4). In 1963 under the secretary of defense's direction, the services standardized the nomenclature, adopting the AIM (air intercept missile) prefix for guided missiles, and the Falcon family of missiles assumed the AIM-4 designation. Windelberg, "A Visionary Ahead of His Time," 8.

20. Air Research and Development Command (ARDC), *Evaluation Report*, 2.

21. McMullen, "History of Air Defense Weapons," 88–89.

22. Westrum, Sidewinder, 28.

23. McMullen, "History of Air Defense Weapons," 157 and 277.

24. Ibid., 277.

25. Ibid., 284; and ARDC, *Evaluation Report*, 1–2. The follow-on GAR-3 also raised the acceptable aircraft to launch a missile from Mach 1.3, based on the F-102, to Mach 2.0, based on the Air Force's faster F-106.

26. McMullen, "History of Air Defense Weapons," 278; and ARDC, *Evaluation Report*, 8, 12. The ARDC report prophetically warned that "a chain is not stronger than its weakest link and even though the missile itself may be highly reliable, the fire control system, because of its complexity, may cause trouble." The sentence's reference to a "highly reliable" missile is suspect. The report later noted, "The probability of hit for each missile is 0.25 giving a 0.578 probability of hit for a salvo of three missiles."

27. McMullen, "History of Air Defense Weapons," 278; and Holloman Air Development Center, *Test Report on GAR-1*, 5. The 29 October 1956 test performance report alerted that "out of 48 missiles launched from 1 January 1956 until 1 September 1956, only seven intercepted their target."

28. McMullen, "History of Air Defense Weapons," 278–80.

29. Westrum, *Sidewinder*, 34 (emphasis in original).

30. McMullen, "History of Air Defense Weapons," 280.

31. ARDC, *Evaluation Report*, 2. The 1956 report referred to the GAR-2 infrared missile as the GAR-1B. The missile changed designations on 1 March 1956, shortly before the report's release. The evaluation report also reflected the Air Force's dominant vision of air combat against fleets of invading Soviet bombers. Curiously absent from the assumed advantage that the infrared missile "will select a target" was the criteria that the missile actually guide toward the target that it was fired against; apparently, any target would do, and there would supposedly be plenty of them in the sky. *Ground clutter* occurs anytime the radar is pointed below the horizon; radar energy reflected off the ground often masks the target return. Stimson, *Introduction to Airborne Radar*, chapter 22.

32. McMullen, "History of Air Defense Weapons," 283.
33. Ibid., 284.
34. Ibid.
35. Ibid.
36. Ibid., 47.
37. Ibid., 88.
38. Westrum, *Sidewinder*, 30.
39. Ibid.
40. "New Missile Ready," 11.
41. Wildenberg, "A Visionary Ahead of His Time," 8.
42. Westrum, *Sidewinder*, 28.
43. Ibid.; and ARDC, *Evaluation Report*, 1.
44. ARDC, *Evaluation Report*, 1–2. Additionally, the small size of the warhead meant that its effects on the target would be negligible unless the missile hit the target.
45. Westrum, *Sidewinder*, 32.
46. Ibid., 44; and Bugos, *Engineering the F-4 Phantom II*, 78–79.
47. Westrum, *Sidewinder*, 44–45.
48. Ibid.; and Bugos, *Engineering the F-4 Phantom II*, 79. Bugos elaborated on the design limitations of Sperry's Sparrow I: "The missile homed and maneuvered best when the fins were in the x position, but it carried and launched best with the fins in the + position. However, just that one-eighth of a turn disrupted the gyros. Also, the pilot had to power up the homing head as long as he suspected enemy aircraft nearby, causing reliability problems. Furthermore, beam riding presented a problem whenever the launch aircraft and the target aircraft maneuvered relative to each other. The missile intercepted the changing beam as a curve rather than as a new direct route between itself and the target, and it spent its thrust following that curve."
49. Michell, *Clashes*, 15; ARDC, *Evaluation Report*, 1; and Westrum, *Sidewinder*, 44–46.
50. Westrum, *Sidewinder*, 132.
51. Ibid., 130.
52. Ibid., 138–39; and ARDC, *Evaluation Report*, 12. The 1956 evaluation report was prophetic, noting, "This complex weapon system [the Falcon] will require large numbers of trained airmen." According to Westrum, the Sidewinder's ruggedness was illustrated in dramatic fashion during a Navy demonstration of the Sidewinder for Air Force representatives at Holloman AFB, New Mexico, 12–16 June 1955. Asked by Hughes's engineers if they wanted to store their Sidewinder test missiles in a temperature- and humidity-controlled room with the Falcon test missiles, the Navy engineers shrugged and instead elected to store their missiles "on a mattress in the bed of a pickup truck."
53. Westrum, *Sidewinder*, 161, 176–78, and 186–87. The Air Force initially adopted the Navy's AIM-9B version but then elected to develop subsequent versions of the Sidewinder

independent of the Navy. During Vietnam, F-4 pilots like Maj James Hargrove, pilot of Speedo 1 on 14 May 1967, lobbied hard to discard the Air Force's AIM-9E missiles in favor of the Navy's AIM-9D. During an interview on 19 September 1967, Hargrove commented, "The A[IM-]9D is something that we could probably have like within a couple of weeks, if we made the decision to get it, and it would give us a lot better missile capability for close-in fighting like we're doing up there with MiGs." Hargrove, oral history interview, 15. Bugos compared the Sidewinder with the Sparrow, noting, "The Sparrow III was a high-cost solution to relieving dogfighting duties. . . . But its complex radar made the Sparrow expensive and unreliable. At twice the cost, its success rate in test flights was half that of Sidewinder." Bugos, *Engineering the F-4 Phantom II*, 89.

54. "It was not noticed that the minimum range of the missile was the beginning of the effective envelope of aircraft cannon, which were more effective the closer the range." Michell, *Clashes*, 16.

55. Poor cockpit design and complex armament switchology plagued the F-4 design. Lt Col Steve Ritchie, an Air Force Vietnam war ace, derided the F-4's "cockpit arrangement—particularly the positioning of the master arm and several other vital switches." Ritchie, "Foreword," 6. Anderegg described F-4 pilots' attempts to improve functionality in the cockpit: "Some pilots went to their crew chief and asked for a piece of the stiff plastic tubing the maintenance troops used to take oil samples from the engines. The pilot would then cut a two-inch length of the tubing and slip it over the missile select switch. . . . [That way], if [the pilot] quickly needed a [different type of missile], he could swat the plastic tubing down with his left hand." Anderegg, *Sierra Hotel*, 12.

56. Michell, *Clashes*, 15; see note 31 for information on ground clutter.

57. Ibid., 13. Early AIM-9s did not use a cooled infrared seeker; later versions did, enabling better target discrimination and tracking. The AIM-9 also suffered from severe employment restrictions—the interceptor had to be flying at less than 2-Gs when the missile was launched or the missile would fail to guide. Westrum, *Sidewinder*, 177; and Davies, *USAF F-4 Phantom II*, 18.

58. Capt John R. Boyd noted that the attacker's task became more difficult "when employing [an] AIM-9B [Sidewinder] against a maneuvering target, [because] the cone not only diminishes in size, it also changes in shape." Boyd, "Aerial Attack Study," 42

59. McMullen, "History of Air Defense Weapons," 158.

60. Ibid., 290. For comparison purposes, the atomic bomb dropped on Hiroshima detonated with the force of 12.5 kilotons. Rhodes, *Making of the Atomic Bomb*, 711.

61. Schaffel, *Emerging Shield*, 234; and McMullen, "History of Air Defense Weapons," 294. The Joint Chiefs of Staff set 1 January 1957 "as the target date for air defense forces to become operational with nuclear weapons." According to Schaffel, "To prove the weapon safe for air defense over populated areas, several volunteers stood directly below the detonation in the Nevada desert."

62. McMullen, "History of Air Defense Weapons," 294–96.

63. "The rhetoric of denial seemed to provide powerful arguments against wasting time on the expensive and complicated guided missiles for use in air-to-air combat." Westrum, *Sidewinder*, 32.

64. Knaack, *Post–World War II Fighters*, 69. The F-86D, which became operational in April 1953 more than two years behind schedule, shared only a common wing design with its predecessor of Korean War fame. The F-86D relied on "interception radar and associated fire-control systems" that "could compute an air target's position, guide the fighter on to a beam-

attack converting to a collision course, lower a retractable tray of 24 rockets (2.75-inch [Navy-designed] Mighty Mouse, each with the power of a 75-mm shell) and within 500 yards of the targets fire these automatically in salvos."

65. Ibid., 83–97. After several production fits, the Air Force elected to replace the 20 mm nose-mounted cannon armament of the F-89C with "104 2.75-inch folding-fin aerial rockets (FFAR), carried in permanently mounted wing-tip pods" in the follow-on F-89D, which became operational on 7 January 1954. In March 1954 during the F-89D production run, the Air Force elected to modify the F-89 wingtip pods to incorporate 42 standard FFARs and six Falcon missiles. The modified aircraft became the F-89H. The final model of the F-89 earned a new designation, the F-89J, based solely on the significance of its armament—"two Douglas-built, unguided, air-to-air MB-1 Genie [nuclear-armed] rockets."

66. "The success of the F-94C's all-rocket armament hinged on rocket accuracy and interceptor performance reliability. The F-94C and its rockets had neither." Ibid., 108.

67. Ibid., 159. The F-102, originally dubbed the "'1954 Interceptor' for the year it was expected to become operational," was specially designed to combat the expected speed and altitude capabilities of new Soviet intercontinental jet bombers. The aircraft did not enter operational service until April 1956. The F-102 was the first aircraft developed under the weapons system concept, which married the development of the aircraft and its accompanying Falcon armament into a weapon system, thereby theoretically ensuring that each component retained compatibility with the other components. As noted earlier, this imposed significant size constraints on the F-102's Falcon missiles.

68. Mets, "Evolution of Aircraft Guns," 225–26; and Michel, *Clashes*, 11 and 158. Michel noted that the Air Force "stopped the development of guns for fighter aircraft in 1957 (fortunately not until after the M-61 Vulcan was developed)."

69. According to Thomas Hone, the faster-paced, jet-powered air combat over Korea confirmed the Air Force's armament worries of the late 1940s: "Air-to-air combat in Korea was different than in World War II. Jet fighters approached, engaged, and disengaged at much higher speeds. Firing opportunities were brief and fleeting. Neither the MiG nor the Sabre (but especially the MiG) had armament or gunsight suited to this cascading, turbulent form of combat. As a result, losses on both sides were lower, given the number of aircraft sortied, than during comparable battles in World War II." Hone, "Korea," 496. Unfortunately, the Air Force failed to adapt based on its experiences. Maj Gen Emmett O'Donnell's statement to Congress in 1951 aptly summarized the prevalent attitude within the Air Force during the Korean War: "I think this is a rather bizarre war out there, and I think we can learn an awful lot of bad habits in it." Crane noted, "Perceived success provides little incentive for improvement, and because of this confidence [following the Korean War] and SAC's focus on general war, most of the lessons about airpower in limited wars were lost or deemed irrelevant. They would have to be relearned again, at high cost, in the skies over Vietnam." Crane, *American Airpower Strategy*, 60, 170.

70. Knaack, *Post–World War II Fighters*, 210.
71. Michel, *Clashes*, 16.
72. Boyne, *Phantom in Combat*, 10.
73. Ibid., 32; and Bugos, *Engineering the F-4 Phantom II*, 1.
74. Bugos, *Engineering the F-4 Phantom II*, 24.
75. Ibid., 25.
76. Ibid., 27–28.
77. Boyne, *Phantom*, 32; and Bugos, *Engineering the F-4 Phantom II*.

78. Bugos, *Engineering the F-4 Phantom II*, 27–28.
79. Michell, *Clashes*, 13.
80. Bugos, *Engineering the F-4 Phantom II*, 3 and 28.
81. Knaack, *Post–World War II Fighters*, 333.
82. Enthoven and Smith, *How Much Is Enough?*, 263; Thornborough, *USAF Phantoms*, 11; Bugos, *Engineering the F-4 Phantom II*, 120; Crane, *American Airpower Strategy*, 172; and Hannah, *Striving for Air Superiority*, 30. Multiple interpretations of the Air Force's F-4 procurement decision exist. Enthoven and Smith noted that, "the [Defense] Secretary's decision in 1962 to stop the F-105 and to procure the Navy's F-4 for the Air Force—over the strong official objections of the Air Force—was based on a cost-effective analysis." Thornborough suggested that McNamara "brought pressure on the Air Force" to select the F-4 over the F-105; such a decision would "maintain US service modernization rates while capitalizing on longer production runs and lower joint-service lifecycle costs to reduce unit prices and keep the budget watertight," which were key McNamara priorities. However, Bugos characterized the acquisition decision as being informed more by Air Force analysis than by secretary of defense meddling. Bugos described the F-4/F-105 fly-off in November 1961 at Nellis AFB, Nevada, and the subsequent decision process: "The two aircraft performed equally well, and the choice once again became a matter of policy. Several considerations added up in the Phantom's favor. First, the Air Force also needed a fast, low-flying aircraft for tactical reconnaissance. The F-4 flew low and fast, and, once McDonnell removed the APQ-72 radar, the Air Force could add lots of cameras, radars, and other sensors. . . . Second, [President] Kennedy was increasing the number of nuclear warheads in the Air Force inventory, and General William Momyer . . . thought TAC could only compete, politically, with the Strategic Air Command if TAC flew a fighter like the F-4 that could also drop nukes. Most importantly, Lt General Gabriel Dissoway, the deputy chief of staff for Programs and Requirements at Air Force Headquarters, praised the Phantom's flexibility for the cost." The identity crisis and insecurity gripping TAC at the turn of the decade stemmed from a SAC-dominated Air Force bureaucratic structure. Crane noted that TAC's focus "primarily on nuclear strikes in support of NATO was a sure way to garner budget support and force structure in the national security environment of the mid-1950s, but it skewed the focus of USAF tactical airpower away from limited and conventional wars. [General] Weyland and his TAC successors struck a Faustian bargain with the atomic Mephistopheles, transforming the organization into a 'junior SAC.'" Hannah quoted from Caroline Ziemke's PhD dissertation, "In the Shadow of the Giant: USAF Tactical Air Command in the Era of Strategic Bombing, 1945–1955": "By the late 1950s, the command [TAC] perceived itself primarily as an extension of nuclear deterrence—a sort of massive retaliatory capability on the regional rather than global level. Other missions, especially air-ground and air-air operations, fell into neglect as TAC became an increasingly specialized strike command. Like Dorian Grey, TAC had sold its soul in exchange for vitality, and in Vietnam, the world got a look at its aged and decrepit conventional structure." Hannah described the repercussions: "By becoming a miniature version of SAC, TAC entered the air war in Vietnam with aircraft that were ill suited for aerial combat with the small, highly maneuverable MiG fighters."
83. Bugos, *Engineering the F-4 Phantom II*, 2.
84. Ibid., 121.
85. Ibid., 122–23; and Thornborough, *USAF Phantoms*, 11.
86. Knaack, *Post–World War II Fighters*, 265. The multirole mission of the Air Force F-4, "covering the entire tactical mission—close air support, interdiction, and counter air"—was detailed in Specific Operational Requirements 200, dated 29 August 1962.

87. Burns, oral history interview, 3.
88. Catledge, oral history interview, 31–32.
89. History, 12th Tactical Fighter Wing, vol. 1.
90. TAC, *F-4C Limited Evaluation*, 55–59. The evaluation report noted that the disparity between the number of sorties flown and the number of missiles launched was due to "aircraft or target problems."
91. Knaack, *Post–World War II Fighters*, 266. Early requirements for the F-4C to support the Air Force's AIM-4 Falcon missile were dismissed to avoid delaying production.
92. Hannah, *Striving for Air Superiority*, 94.
93. "The fuzzy thinkers thought that was great. [In their minds,] it was a hell of a lot better to fly three hours with drop tanks than it was to fly an hour and 20 minutes in a very productive mission that involved doing a lot of different things with the airplane." Blesse, oral history interview, 61–65.
94. Hannah, *Striving for Air Superiority*, 95.
95. Michel, *Clashes*, 160. Additionally, because the Air Force used the F-4 for both air-to-air and air-to-ground missions, its crews had to be qualified and trained for both. As a result, air-to-air training, especially training emphasizing dogfighting skills, was virtually nonexistent.
96. Hannah, *Striving for Air Superiority*, 97.
97. "By the time intelligence assessments revealed the Soviet ICBM emphasis in the early 1950s, most air-to-air missile programs were well on their way. The technology was too promising to discard." Westrum, *Sidewinder*, 29.
98. Michell, *Clashes*, 13.
99. Blesse, oral history interview, 59–60.

Chapter 5

The Gun Resurrected

We were voices in the wilderness in those days.

—Maj Gen John Burns, USAF

In 1963, as the specter of air combat over Vietnam grew, the Air Force hurriedly organized an internal assessment of its aircraft capabilities for a non-nuclear, limited war. Completed in January 1964, the resulting secret report, *Project Forecast*, concluded that the majority of the Air Force's tactical fighter fleet was unprepared and ill-equipped for the pending conflict. The one ray of hope lay in the Air Force's newest fighter, the F-4C, which, according to the report, "has an equal or better capability than present interceptors against the same air targets. . . . In addition, the F-4C [is] useful against fighter and recce [reconnaissance] aircraft."[1] The first engagements between the USAF F-4Cs and the North Vietnamese MiG-17s in 1965 seemed to confirm the enthusiastic assessments trumpeted in *Project Forecast*. Unfortunately, the report proved exceedingly optimistic. Over the next three years, the gross inadequacies of the Air Force's air-to-air missile armaments in modern, fighter combat would become all too apparent, as would the Air Force's penchant for technological exuberance.

Early Air Combat

After a grueling transpacific flight, 18 F-4C aircraft from the 555th Tactical Fighter Squadron (TFS), 12th TFW, MacDill AFB, Florida, touched down on the southwestern edge of Okinawa on 10 December 1964.[2] As the first F-4Cs to deploy to the Pacific region, the members of the "Triple Nickel" squadron were tasked with "establish[ing] transoceanic deployment procedures and test[ing] aircraft maintainability" for the Air Force's barely one-year-old weapons system "away from the luxuries of home."[3] The deployment paved the way for the bevy of F-4s that would eventually provide almost 30 percent of the tactical aircraft fleet in Southeast Asia (SEA) in 1968.[4] That influx began in earnest in April 1965 when the 15th TFW's 45th TFS, also from MacDill, sent 18 of its F-4C aircraft to Ubon Royal Thai AFB, Thailand.[5] Over the next year, the number of F-4Cs in theater would increase more than tenfold, from 18 in 1965 to 190 by the end of 1966. The Air Force concentrated its F-4s at three bases: the 8th TFW at Ubon; the 12th TFW at Cam Ranh AB, South Vietnam; and the 366th TFW at Da Nang AB, South Vietnam.[6]

The USAF F-4C Phantom II first drew MiG blood on 10 July 1965.[7] On that day, a flight of four 45th TFS F-4Cs engaged and destroyed two MiG-17s who were harassing a flight of F-105 Thunderchiefs attempting to a tack the Yen Bai ordnance and ammunition depot 30 miles outside Hanoi.[8] In what the Phantom flight lead, Maj Richard Hall, later described as "a schoolbook exercise," the F-4Cs, armed with the standard complement of four Sparrow and four Sidewinder missiles each, fired eight Sidewinder missiles at the two MiGs during the four-minute engagement.[9] The next day back in Thailand, each victorious two-person F-4 crew was awarded a Silver Star; the aircrews from the accompanying F-4s received Distinguished Flying Crosses.[10]

Although Hall's confident assessment of the engagement did not address it, American missile and aircraft performances that afternoon were far from perfect. In one aircraft piloted by Capts Kenneth Holcombe and Arthur Clark, the violent maneuvering during the engagement caused their radar to fail, instantly rendering their Sparrow missiles worthless for the remainder of the flight. Additionally, two of their four Sidewinder missiles failed to launch when fired. Fortunately, the remaining two Sidewinders did function properly and brought down a MiG: one missile "produced a large fireball at or slightly to the right of the MiG"; the other "detonated slightly to the right of the MiG."[11]

Capts Thomas Roberts and Ronald Anderson, flying in an accompanying F-4, had a similarly frustrating experience. Their first Sidewinder "streaked past the [enemy's] tail and detonated four to six feet from the left wing tip." However, the MiG kept flying, "rolling slowly to the left in a bank." Flustered, Roberts "hastily" launched a second Sidewinder missile without a valid missile tone (a growl in the aircrews' headsets indicating that the missile had acquired the target); it also "proved ineffective." Roberts's third Sidewinder "tracked well and exploded just short of the MiG's tail," but because he "saw no debris emitting from the aircraft," he launched his last Sidewinder missile. Roberts and Anderson could not observe their last missile's flight path because they came under AAA fire that forced them to initiate aggressive defensive maneuvers.[12]

This first F-4C versus MiG-17 engagement foretold many of the problems the F-4C fleet would face in the coming years: unreliable electronic equipment, faulty missiles and imprecise weapons employment (e.g., firing a Sidewinder without acquiring a valid tone), and the difficulty of engaging a MiG while also defending against ground-based air defenses like AAA and SAMs.[13] Yet the engagement also validated, in some Air Force leaders' minds, earlier appraisals that the 1950s-era Soviet-built MiGs were no match for the Americans' modern F-4C fighter.

One problem that drew attention that day was the significant impact of the United States' restrictive rules of engagement (ROE) governing the F-4C weapons system and its aircrews. To reduce the possibility of airborne fratricide, aircrews were required to positively identify their target before firing a missile. Unfortunately, Air Force fighters such as the F-4C lacked reliable means to do so electronically, thereby often necessitating a visual identification of the suspected enemy aircraft.[14] Writing after Vietnam, General Momyer, who served as the Seventh Air Force commander responsible for all tactical air operations in Southeast Asia during the war, described the ROE's impact: "The necessity for a visual identification of the enemy hindered successful shoot-downs by reducing the frequency of opportunities for employing, for example, the Sparrow. . . . We forfeited our initial advantage of being able to detect a MiG at thirty to thirty-five mile range and launch a missile 'in the blind' with a radar lock-on from three to five miles. Many kills were lost because of this restriction."[15] A *New York Times* article detailing the 10 July 1965 engagement reported that most F-4 pilots "were not too happy with the requirement for visual identification . . . [but] that they preferred this to shooting down one of their own aircraft by mistake."[16]

Pilot reports and interviews after the July engagement also alluded to the F-4's need for better short-range armament. Whereas the North Vietnamese MiG adversaries, often armed solely with air-to-air cannons, had earlier proven the continued viability of the gun in jet combat, several members of the victorious 10 July 1965 F-4 flight dismissed the combat potential of a gun on the F-4. For example, Holcombe warned that adding a gun to the F-4 "will just get people into trouble" by tempting aircrews to get dangerously "low and slow" with the MiGs.[17] Holcombe's concerns echoed the conclusions of the Air Force's 1965 Feather Duster program, which warned that trying to outmaneuver the smaller MiG aircraft was an F-4 air combat "no-no."[18] Thus, instead of entertaining the potential of an antiquated-but-proven-effective system, many aircrews longed for better, more advanced missiles that would allow them to exploit the F-4's overwhelming thrust advantage and high-speed capability when attacking the more maneuverable MiGs at close range.

The next nine months following the July shoot-down witnessed only sporadic MiG activity as the North Vietnamese Air Force retooled the country's air defense system. Central to the upgrade were new ground-controlled intercept (GCI) procedures to vector their MiG-17 and recently acquired and more sophisticated MiG-21 fighters into favorable positions against US aircraft and the deployment of large numbers of SAMs such as the SA-2 across the theater.[19] The new arrangement proved formidable. The United States did not claim another MiG until mid-April 1966. By then, the MiGs had claimed four more US fighters and had harassed numerous F-105 fighter-bombers, forcing them to jettison their ordnance while defensively reacting to the attacking MiGs. Additionally, the North Vietnamese SAMs levied a heavy toll on the American fighters.[20]

The next F-4C MiG kill occurred on 23 April 1966; four F-4Cs engaged four MiG-17s and destroyed two of them after firing seven missiles—five Sparrows and two Sidewinders. Reminiscent of the missile problems that frazzled the F-4C aircrews on 10 July 1965, of the five Sparrows launched one was fired inside its minimum range, two missiles' motors never ignited after launch, one guided but missed the target, and one hit and downed a MiG. Of the two Sidewinders launched, one was fired without a valid tone, and the other hit and destroyed the second MiG.[21]

In the F-4C's first two successful engagements, four MiGs were downed at a cost of 15 missiles. Of the 15 missiles fired, four failed to launch properly (27 percent), and three were launched outside of parameters (20 percent). But those numbers only accounted for missile shots during engagements that resulted in a kill. For example, that same day—23 April 1966—two F-4Cs were dispatched to intercept a pair of MiG-21s en route to attack an orbiting Douglas EB-66 electronic jamming aircraft. Unfortunately, the two F-4Cs came up empty-handed, but not for lack of effort; the two Phantoms fired a total of six Sparrow and Sidewinder missiles against the MiGs to no avail.[22]

Despite the missiles' lackluster performance in these and other engagements, the earlier antigun sentiment expressed by Holcombe persisted. One of the pilots from the successful 23 April engagement commented, "The need for [an] F-4 gun is overstated, although it would be of value if it could be obtained without hurting current radar and other systems performance. If you are in a position to fire [the] gun, you have made some mistake. Why, after a mistake, would a gun solve all [your] problems? Also, having a gun would require proficiency at firing, extra training, etc. [We] have enough problems staying proficient in [the] current systems. If the F-4 had guns, we would have lost a lot more [F-4s], since once a gun duel starts, the F-4 is at a disadvantage against the MiG."[23]

Missile performance was markedly better three days later when Maj Paul Gilmore and 1st Lt William Smith scored the Air Force's first MiG-21 kill. Gilmore fired three Sidewinders at the MiG. His first Sidewinder severely crippled the MiG, and the pilot ejected from the aircraft. However, Gilmore thought that the first missile had missed the target and, not seeing the pilot eject, repositioned and fired another missile; that second missile clearly missed the target. "After missing twice," Gilmore explained, "I was quite disgusted. I started talking to myself. Then I got my gunsights on him and fired a third time. I observed my [Sidewinder] missile go directly in his tailpipe and explode."[24] As a *New York Times* article describing the combat noted, "It was only then that Major Gilmore's wingman, who had temporary radio failure, was able to radio him that the first missile had hit and that the pilot had ejected and parachuted."[25] Following the kill, the two F-4Cs attempted to engage a second MiG-21, but Gilmore's last Sidewinder missile missed the target, and now low on fuel, Gilmore's flight of F-4s decided to return home.[26]

Air Force leaders greeted Gilmore's MiG-21 victory with enthusiasm. Early analyses concluded that the F-4 was at a significant disadvantage relative to the modern Soviet MiG-21. The Southeast Asia Counter-Air Alternative (SEACAAL) study, forwarded to the secretary of the Air Force a few weeks later on 4 May 1966, predicted that the Air Force "should expect to lose three F-4s for each MiG-21 . . . shot down."[27] The results from Gilmore's 26 April engagement seemed to refute that analysis. It also proved that, while side-by-side comparisons of aircraft energy-maneuverability diagrams could help inform American pilots of where their aircraft were expected to perform best against the MiG fighters, actual air combat was too fluid to draw definitive categorizations.[28] Aircrew experience, area radar coverage, environmental factors, and chance all played a significant role in dictating who would return home to paint a star on the side of his or her aircraft.

As MiG activity increased during the remainder of April and May 1966, several American pilots continued to follow the Feather Duster advice and tried to avoid entering a turning engagement with the MiGs. However, sometimes during the course of an engagement, attacking MiGs could force the F-4 pilots to defend themselves with a series of aggressive, defensive turns. In these situations, the Phantom crews had no choice but to discard the approved combat solution.

Despite this emerging combat reality, many pilots let their faith in missile technology and published tactics color their opinions of air-to-air armament. Most continued to categorically dismiss the potential value of a gun on the F-4. Following a successful engagement on 29 April 1966 in which an F-4C downed a MiG-17 with a Sidewinder missile, one Air Force pilot commented, "It would be undesirable and possibly fatal for an F-4 to use a gun in fighting with a MiG because the MiG is built to fight with guns and the F-4 is not."[29]

However, attitudes began to change a month later. According to Michel, "By the end of May, Air Force F-4 aircrews reported losing much of their confidence in the Sparrows."[30] Additionally, several F-4 aircrews reported that

THE GUN RESURRECTED

many times in combat they could have dispatched an enemy MiG with a gun, if only they had had one.[31]

Because the F-4C did not have a gun, nor were there any plans to add a gun to the platform, the Air Force focused its efforts on improving the "poor" performance of the F-4's missile armament.[32] The uninspiring combat results were difficult to ignore. From April 1965 through April 1966, the primary armament of the F-4, the AIM-7 Sparrow—the weapon that had guided the aircraft's design and development—had accounted for only one kill, downing a MiG-17 on 23 April 1966.[33] To address the problem, the Air Force appointed a special team of Air Force and F-4/Sparrow specialists to travel to SEA to personally review the weapon system's combat performance and "recommend the required actions necessary to enhance success of future Sparrow/Sidewinder firings." Unfortunately, the team concluded that even "assuming proper maintenance of both aircraft and missiles, the probability of kill with the Sparrow can be expected to be low."[34] The team found that during the period from 23 April to 11 May 1966, Air Force F-4Cs fired 13 AIM-7s (and tried to fire an additional three which never left the aircraft) to down a single MiG—a 6 percent hit rate.[35] Whereas some failures could be attributed to faulty missile maintenance and aircraft loading or improper pilot performance, the team noted that "four of the Sparrows launched during the period 23–24 April were fired under ideal conditions and missed" for inexplicable reasons.[36]

In spite of these compelling anomalies, the Air Force remained committed to its dominant paradigm and deployed the newest version of the AIM-7, the AIM-7E, to the theater in mid-1966. Unfortunately, the new version did not appreciably improve the combat statistics, adding only one more victory to the F-4's tally by the end of 1966.[37]

The Sidewinder's performance was markedly better—a 28 percent hit rate over 21 shots in April and May 1966—but still less than what aircrews had expected based on earlier, euphoric test reports that had predicted a 71 and 68 percent hit rate for the Sparrow and the Sidewinder, respectively.[38] Additionally, aircrews complained about the Sidewinder's restrictive launch envelope, both relative to the target's position, range, and angle-off, and the 2-G limit when launching the heat-seeking missile. One frustrated Air Force pilot, Maj Robert Dilger, quipped in a July 1967 interview, "The Sidewinder—this is the AIM-9B—totally hopeless in the air-combat environment. It's a reliable missile and it will work most of the time. It has a good Pk, probability of kill, if launched within its parameters. Well, the trouble is you can't launch it in the ACT [air combat tactics] environment within its parameters. It's always going to be out-G'd, just about; so the only thing that we can do with a Sidewinder is use it as a scare tactic or if the MiGs don't know we're there."[39] Not all pilots shared Dilger's opinion. While acknowledging the missile's restrictive launch envelope, MiG-killer Maj William Kirk of the 433rd TFS concluded, "It's a damn fine little missile if you can get the thing launched under the right parameters."[40]

The problem was that the Sparrow and Sidewinder missiles were neither designed nor tested for fighter-versus-fighter combat. They were designed to shoot down high-altitude, nonmaneuvering, bomber-type targets.[41] Sidewinder engineers never envisioned a requirement to attack small, low-altitude, maneuverable fighters. Sparrow engineers counted on their missile being launched, in Momyer's words, "in the blind," with the target still three to five miles away.[42] The 8th TFW's *Tactical Doctrine* manual, dated 1 March 1967, called pilots' attention to the disparity between the anticipated F-4 combat environment and 1967 reality in Vietnam:

> The F-4C/APQ-100/APA-157 weapons control system and associated armaments, the AIM-9B and the AIM-7E, are designed to be employed in a non-maneuvering environment using close control. This close control coupled with the long ranges of the armament provide an element of surprise and thus a high probability that the target will be in a non-maneuvering state. Further, the system was designed more as a defensive rather than an offensive system. *The chances of employing the system in this manner in SEA are very remote.*
>
> The system as employed in SEA is in an offensive role in the enemies [sic] environment. Therefore, the enemy has the advantages since he can employ radar and fighters in defense against the F-4C system. The enemy knows more about us than we know of him in this type of environment. The F-4C now becomes the hunted as well as the hunter. Further, due to saturation in the battle areas, visual identification is necessary prior to armament launch. In order to positively identify the target, the F-4C must move into visual acquisition range and the chances are very good that the enemy will see the F-4C at the same time, since the enemy has knowledge of approaching aircraft through ground radar control. Once the attackers' presence is known to the enemy, it becomes a battle of aircraft maneuvering for advantageous firing position.[43]

The Air Force's decision to limit aircrews to a single 100-mission tour unless they volunteered for a second also began to take its toll on the F-4C's combat performance. As the Vietnam War dragged on, the personnel policy created an insatiable appetite for fighter aircrews. Responding to the demand, the Air Force "simply lowered standards, brought in more students, and graduated more pilots from pilot training."[44] The Air Force allowed, and then eventually required, pilots with little or no tactical fighter experience to transition to fighter aircraft like the F-4 and fly a combat tour.

Regardless of prior tactical experience or lack thereof, new Phantom pilots completed a six-month training program at a replacement training unit

THE GUN RESURRECTED

(RTU). However, air-to-air combat training at the RTU was limited; aircrews had to be trained for every potential F-4 mission, including basic skills such as how to take off and land the aircraft, in only six months. The Air Force's "corporate belief that air combat maneuvering among inexperienced pilots would lead to accidents," combined with the dominant culture that prioritized safety over training, also thwarted efforts to prepare the new aircrews for actual, ongoing air-to-air combat.[45] Navy pilot and Vietnam-ace Randy "Duke" Cunningham characterized the Air Force's aircrew training program as "an out-and-out crime."[46]

The F-4 units in SEA felt the effects. One Air Force pilot commented in July 1967, "Some of our pilots are terrific. I mean they're really top drawer, aggressive, well-trained, well-motivated people. Some of our pilots fall short of these standards, and part of the problem [is] that—through no fault of their own, in a lot of cases—they just don't have the background. [An] 80-hour training course like they get in the RTU program, if they have no previous fighter time, fighte background, fighter tactics, is just not quite enough to bring them up to par."[47]

Despite declining aircrew proficiency and the shortcomings in armament, the F-4C was performing remarkably well in air combat against the MiGs. The first 18 months of combat saw only four F-4Cs lost due to MiG action out of 69 total F-4C losses. In return, the F-4Cs downed nine MiG-17s and five MiG-21s.[48] One Air Force pilot summed up the F-4C's early performance: "With no gun and two types of missiles whose reliability was about ten percent, you'd have to rate the F-4C's abilities as a fighter as low. Still, I'd take that F-4 ride into Hanoi over the F-105 any day!"[49]

More deadly than the MiGs, though, was the heavy concentration of ground defenses the North Vietnamese hid around their lucrative target areas. With mounting losses to SAMs and AAA threatening the Air Force's ability to attack targets in NVN, in October 1966 the Air Force responded by deploying the QRC-160 electronic countermeasures (ECM) jamming pod, which was designed to confuse the enemy SAM and AAA fire-control radars.[50]

Initially, the ECM pods were loaded on the F-105 fighter-bombers so that they could attack heavily defended targets. "But after the F-105s started carrying the [ECM] pods," a 31 December 1966 SEACAAL report stated, "the [accompanying] F-4s', having neither jamming nor warning equipment, began to suffer unusually heavy losses to SA-2s. As a consequence, the F-4s were restrained from flying into SA-2 areas—which were also the MiG areas—until protective equipment was available." The report noted that the North Vietnamese quickly took advantage of the F-4s' absence—"MiG activity has surged this past month and they have enjoyed appreciable success in harassing our aircraft."[51]

Still, the SEACAAL report was optimistic. "Adaptor pylons [to mount the ECM pods] have been airlifted to SEA so that by 1 January 1967, some F-4s can also be pod equipped." But, reflective of the true Catch-22 situation, the report's next sentence read, "The pods are in short supply at present so they can be used on F-4s only by taking them off F-105s."[52] The aircraft shared the valuable pod resources, relying on special formations that maximized ECM protection for all flight members, until production could catch up with demand, which occurred in mid-1967.[53] As the Air Force scrambled in 1966 to deal with the emerging SAM and AAA threat, it also renewed its efforts to address the poor performance of the F-4's air-to-air armament.

A Focus on Technical Solutions

Michel described the air-to-air results of Rolling Thunder as a "Rorschach test for the US Air Force and Navy." True to the test, "the two services drew almost exactly the opposite conclusions from their battles with the MiGs." Whereas the Navy "decided that lack of training was the problem," which led to the establishment of their famed Top Gun Fighter Weapons School in 1969; the Air Force, gripped by the promise of technology, "looked at its losses to MiG-21s . . . and decided that the problem was a technical one."[54] The Air Force consequently went to great lengths to address the technical deficiencies of its missiles and its aircraft.[55]

The Air Force, in partnership with the Navy, first sought to improve Sparrow performance. Their initial answer was the AIM-7E Sparrow, which entered the fray in mid-1966. Sporting only minor improvements over the earlier AIM-7D, the AIM-7E failed to address many of the Sparrow's shortfalls. The next AIM-7 version, the AIM-7E-2, was introduced in August 1968. Hailed as the "dogfight Sparrow," Air Force and Navy officials believed the new AIM-7E-2 missile would provide the necessary edge for F-4 aircrews in the tight-turning, high-G, close-range air-to-air engagements that typified combat in the skies over Vietnam. Boasting a "minimum-range plug" that "(in theory) gave the AIM-7E-2 a minimum range of 1,500 feet instead of 3,000 feet, better fusing, and better capability against a maneuvering target," the missile saw only limited use and contributed no additional MiG kills before Rolling Thunder ended three months later.[56] Renewed MiG action in 1971 provided the missile with another opportunity to prove itself, but ultimately the missile failed to live up to the hype. During the course of the Vietnam War, 281 AIM-7E-2 missiles were fired, yet the missiles scored only 34 kills—a dismal 12 percent success rate.[57]

THE GUN RESURRECTED

Whereas the Air Force and Navy elected to address the Sparrow's faults collectively, albeit without notable success, the Air Force abandoned the Navy's efforts to improve the Sidewinder in favor of readying its own AIM-4D Falcon, offspring of the 1960s' Hughes GAR-4 air-to-air missile.[58] Accompanying the Air Force's new D-model variant of the F-4 Phantom to the 8th TFW at Ubon in late May 1967, the AIM-4D, although promising better combat performance against fighter aircraft, was not well received by the aircrews. First, the missile retained its 1950s' contact-only fusing system and small warhead. Second, in a horrible misunderstanding of the nature of fighter-versus-fighter air combat, engineers designed the Falcon with only enough cooling supply for two minutes of operation. Compounding matters, "the sequence of switches to start the coolant flow was complicated," and once started, "the coolant flow to the seeker head . . . could not be stopped."[59] Hence, if the missile was not launched two minutes after it was first armed and cooled, then it became a "blind, dead bullet—derisively called the 'Hughes Arrow'—which had to be carried home and serviced before it could be used again."[60] Thus, "the F-4D pilot had a choice: either arm the AIM-4D early in the engagement and hope he would get a chance to use it within the next two minutes, or wait and try to remember to arm it after the fight began and when there was a target available. In a turning dogfight where shot opportunities were fleeting, such restraints on a missile clearly were unacceptable."[61]

In a postwar interview, Brig Gen Robin Olds, World War II ace and former 8th TFW commander credited with 16 air-to-air victories, derided the Air Force's AIM-4D Falcon missile:

> They gave us another weapon called the AIM-4 Falcon built by Hughes for air defense and my only comment on that weapon was that it was no good. It was just no good. In assuming that everything worked just as advertised, which it seldom did, the missile had only 2 ¾ pounds of unsophisticated explosive in it, and it had a contact fuse so the missile had to hit what you're aiming at for this little firecracker to go off. . . . Too many times, time and time again, the missile would pass right through the hottest part of the exhaust plume of the MiG-17 which is about a 12-foot miss and that and, you know, five cents will get you a bad cup of coffee.
>
> Secondly, its launch parameters were much too tight, not as advertised, but as changed once they got the things to the theater. Then they sent in the wire and said what your minimum firing range was under altitude, overtake, G conditions. And it turned out that if you were at 10,000 feet in a 4 G turn, the minimum altitude at which that weapon was any good was 10,500 feet. The maximum range of the little son-of-a-b_ _ _ _ was 12,000 feet or something on that order.
>
> So it's just no good. I mean, maybe, if one of the MiGs would be very accommodating and sort of hold still for you out here, you know, that would be fine. . . . There may have been some occasions, when yes, you could use it, but I never ran into one. In summary,

I didn't like the AIM-4. I don't think it's worth a d_ _ _. Nor do I think it has any growth potential.[62]

Less than three months after the Falcon's introduction to the theater, officers at Pacific Air Forces (PACAF) informed Headquarters USAF in Washington, DC, that it intended to replace the AIM-4D Falcons on its F-4Ds with AIM-9B Sidewinders. The process was more complicated than simply slapping the old Sidewinder missiles back on the aircraft; the F-4D had to be rewired to accept the new, old missiles.[63] The F-4D units would have preferred to upgrade to the Navy's new Sidewinder missile; but instead of modifying its missile rails to accept the Navy's AIM-9D, the Air Force—smacking of technological hubris—elected to design its own Sidewinder, which became the AIM-9E. Development delays ensured the AIM-9E would not reach the theater until after Rolling Thunder concluded, and even then its performance was significantly lacking relative to the Navy's AIM-9D.[64]

In addition to addressing the limitations of its air-to-air missiles, the Air Force addressed some of the problems inherent in the F-4C airframe. Unable to make many design changes to the Navy's F-4 early in the program, the Air Force quickly began drafting requirements for an updated, Air Force–tailored F-4 Phantom. In 1964 the Air Force, working through the Navy, issued a contract to McDonnell Aircraft for a new F-4D.[65] Stemming from the Navy's original F-4 fighter-interceptor configuration, the majority of the Air Force's proposed changes were intended to bolster the F-4's multirole capability. For example, by installing a new "GE AN/ASG-22 servoed Lead Computing Optical Sight Set (LCOSS), which replaced the old, fixed, manually depressed gunsight, and the AN/ASQ-91 automatic Weapons Release Computer System (WRCS)," the F-4D was able to perform "a brand new radar-assisted visual bombing mode known as 'dive-toss,' which increased bombing accuracy and crew survivability in one fell swoop."[66] Engineers also addressed some of the F-4's air-to-air deficiencies, although not all of the changes were successful—aptly illustrated by the AIM-4D debacle. Additionally, engineers designed the LCOSS gunsight with an available air-to-air mode, but since the F-4D lacked an internal gun, the capability went unappreciated and unused when the new Phantom model reached combat in May 1967.

Rhetoric and Reality Converge

By mid-1966, the Air Force finally began to acknowledge North Vietnam's inconvenient refusal to adhere to the American idealistic vision of air combat upon which the Air Force's entire fleet of air-to-air missiles had been built. A

THE GUN RESURRECTED

PACAF *Tactics and Techniques Bulletin* discussing "F-4C Fighter Screen and Escort," dated 14 July 1966, noted that since the ideal F-4 engagement— "obtain[ing] long range radar contacts and establish[ing] an optimum attack position within the launch envelope for AIM-7 firing"—was often unachievable, "close-in fighting may become necessary."[67] The report issued by the summer 1966 Heat Treat Team—the Air Force and F-4/Sparrow contractor team tasked with improving missile reliability—echoed the apparent inevitability of close-in maneuvering during MiG engagements and the lack of a viable short-range weapon: "The MiG/F-4C encounters thus far have resulted in close-in maneuvering engagements. Missiles were intentionally fired out of designed parameters in hopes of achieving a 'maybe' hit since guns were not available for the close-in maneuvering."[68] The 31 December 1966 SEACAAL report noted, "The lack of guns on the F-4 is considered one of the factors for the low kill rate in the MiG encounters."[69] Most tellingly, by mid-1966, Air Force mission debriefings implied that North Vietnamese pilots were starting to exploit the disparity in short-range weapons, especially the " 'safe zone'— the approximately one-half mile in front of a Phantom created by the lack of a cannon."[70] And, there was no longer any denying that, when push came to shove, the cannon on the F-105 Thunderchief was proving effective.[71] The pressure to equip the Air Force's newest fighter with a 1950s-era gun was reaching a crescendo.

According to Bugos, "as early as October 1963, the Air Force's TAC had suggested an F-4E version, with a built-in gun, to fly as a tactical strike fighter." He also noted, "Air Force pilots anticipated more situations where a gun would be useful."[72] One of those officers was Catledge, the then-colonel who had set up his flip charts in front of Momyer and pleaded for a gun in the F-4. Undeterred by Momyer's brush-off, Catledge persisted and eventually secured funding for a podded gun system.[73]

Another gun proponent was Col John Burns. Tasked in 1964 with helping develop requirements for the Air Force's next-generation F-X fighter, Burns and the other members of the group recommended a new fighter design. The group also suggested "the installation of an internal gun in the F-4, because we became concerned that we [the Air Force] were putting too much reliance on missiles alone."[74] In a 1973 interview, Burns described the advantages of the gun:

> There is only one countermeasure to a gun and that is better performance in the gun platform. . . . . If you've got superior air combat maneuvering performance and you've got a gun—you stick the gun in the guy's ear. There is no countermeasure for that.
>
> So our view, then, was that relying on missiles alone was a serious mistake, which means that you don't need the synergism of a very fine and superior air combat vehicle that

gave you the performance bedrock, and the avionics system to exercise that performance through a *complete and proper complementary set of armaments: radar missiles, IR missiles, and a gun.* . . . We were voices in the wilderness in those days.

. . . We had OST—Office of Science and Technology—and the President's Scientific Advisory Committee all over our backs, and in 1965, arguing about why we don't just put a better radar and better missiles in the F-4. . . . [But by April 1966,] there were many, many [MiG] engagements, and the capabilities and serious limitations of missiles were very amply demonstrated. . . . From then on, the things that we argued about—sanctuaries, maneuvering performance, *the need for guns as well as missiles*—seemed very well demonstrated over North Vietnam.[75] (emphasis added)

On 18 October 1966 the Pentagon announced its intention to purchase "99 improved Phantom jets equipped for the first time with a built-in gun and designed to give the United States clear superiority over Russian-made MiG-21s in Vietnam."[76] Based on a more detailed press release issued the following month, the *New York Times* proclaimed that the "new model of the McDonnell Phantom fighter plane recently ordered by the Air Force will incorporate some new features as a result of lessons learned in the air over North Vietnam and Laos." Leading the discussion of the aircraft improvements was the description of "a 20-mm internally mounted gun with a rate of fire of 6,000 rounds a minute [which] will complement the plane's missile capability and should give it superiority in both long-range action and close combat." Later, the article outlined the combat-demonstrated requirement for the gun: "While the Phantom has the performance and weapons to stay out of range of the MiG[-21] and shoot it down, it is often difficult in a few seconds at high speeds to maneuver into firing position. The lack of internally mounted guns has sometimes meant the escape of a MiG. Although the United States missiles outrange the Soviet missiles, the Sidewinder and Sparrow cannot be fired from close in; they will not 'arm' in time to detonate."[77]

The new F-4E was to be armed with the GE M61 20-mm Vulcan Gatling gun, the same gun that had equipped the F-105D in the 1950s and that had been produced in podded form due in part to Catledge's advocacy within the Pentagon.[78] Bugos noted that "integrating this gun into the Phantom airframe, however, caused considerable problems."[79] Lacking space within the airframe, McDonnell officials decided to lengthen the nose of the F-4 and mount the gun on its underside. Because the nose also housed the aircraft's sensitive electronics, including the already finicky radar, McDonnell and GE had to design a special system of shock mounts to isolate the equipment from the 100-G instantaneous vibrations that rattled the jet when the gun began firing its six rounds per second.[80] Aviation historians Anthony Thornborough and Peter Davies described an additional complication: "The original gun muzzle

THE GUN RESURRECTED

shroud configuration . . . did not dissipate gun gasses adequately, frequently resulting in heart-palpitating engine flame-outs. And, without engine power, the sleek F-4 shared the same flying characteristics as a brick."[81]

The other major planned air-to-air improvement for the F-4E was a radical new radar system that boasted of an unparalleled ability "to filter out ground clutter at low level so that moving targets, such as a fleeting, low-level MiG, would be picked out and presented as a clear, synthesized target symbol."[82] Unfortunately, Hughes's Coherent-on-Receive Doppler system (CORDS) outpaced the capability of premicrochip electronics, and the radar system failed to sufficiently mature in time. The Air Force cancelled the CORDS program on 3 January 1968. The CORDS decision put the whole F-4E program in jeopardy; when the F-4E was originally conceived, the Air Force determined that if CORDS failed to materialize in a timely fashion, the F-4E program would be scrapped and the procurement of the F-4D model extended.[83] Fortunately, the Air Force elected to continue F-4E development using an alternative, but less advanced, AN/APQ-120 radar set.

The first F-4E entered operational testing on 3 October 1967 while the CORDS program was still in turmoil. Further production delays and requirements revisions delayed the F-4E's deployment to SEA until November 1968. Additional aircraft problems slowed the influx of the newer Phantoms, such that by mid-1971 only 72 F-4Es were in-theater.[84]

Air Force pilots yearned for the F-4E's arrival.[85] Kirk commented, "Eventually we're going to have the E-model airplane with the internal gun. That's the answer. That's obviously the answer. I think the Air Force has learned its lesson. We'll never build another fighter without an internal gun. I'm convinced of that, or at least I hope to God we don't."[86] General Olds had a slightly different perspective, "Putting the gun in the F-4E doesn't automatically make out of that aircraft an air superiority fighter. You haven't changed that airplane one damn bit except now you've made a fighter out of it from what the F-4 was before; sort of a fish or fowl thing."[87]

Ironically, for all the Air Force's development efforts, the F-4E's gun would eventually account for only 12 percent of the total number of MiGs downed by 20-mm fire by the end of the Vietnam War. The jury-rigged gun system developed at Da Nang in May 1967 was responsible for more than double that figure.[88]

Notes

1. *Project Forecast*, IV-9. In drawing their conclusion, the *Project Forecast* team emphasized the long-range detection and engagement capability of the F-4 and its Sparrow and Side-

winder missiles: "The F-4C equipped with the APQ-100 radar and Sparrow missiles is now entering the operational inventory. This radar has an 85 percent probability of detection of a five square meter target at 34 miles and the same probability against a 10 square meter target at 45 to 50 miles. The attack course provided by the fire control system is a lead pursuit course; in practice, however, a constant bearing course is flown, with a conversion to a pursuit course shortly before reaching the firing range. This aircraft also carries the GAR-8 [Sidewinder] infrared missile. . . . This system has a good area defense capability (the purpose for which it was designed), and it also has a capability for fighter-to-fighter combat."

2. History, 12th Tactical Fighter Wing, vol. 1, 19. As part of the Lima-Mike tasking, the 555th TFS launched 24 F-4Cs (20 primary and 4 spare aircraft) from MacDill AFB, Florida, on 8 December 1964. After a brief stopover at Hickam AFB, Hawaii, 20 aircraft (18 primary and two spare aircraft) launched toward Okinawa.

3. Thornborough and Davies, *Phantom Story*, 92.

4. Hone, "Southeast Asia," 521.

5. History, 15th Tactical Fighter Wing, vol. 1; Thornborough and Davies, *Phantom Story*, 92; and Hone, "Southeast Asia," 521. The JCS-directed deployment, Two Buck, began on 4 April 1965.

6. Hone, "Southeast Asia," 521 and 524. The 12th TFW moved to Cam Ranh in November 1965, and the 8th TFW set up shop at Ubon in December 1965. The 366th TFW eventually established itself at Da Nang in March 1966.

7. Futrell, *Aces and Aerial Victories*, 4–5. NVN MiG-17s scored the first aerial victories of the war. On 4 April 1965 MiG-17 fighters, armed only with air-to-air cannons, successfully shot down two "heavily loaded F-105's orbiting over the target waiting for their turn to attack." The MiG-17s sped away before the F-100 escort fighters could even the score. The first F-4 victories of the war went to the Navy's F-4Bs when they downed two MiG-17s with Sparrow missiles on 17 June 1965.

8. Ibid., 22–25.

9. "Maj. Richard Hall . . . said the downing of the MiG-17's was almost a schoolbook exercise because the two United States aircraft were able to turn inside their slower but highly maneuverable enemies for the kill with heat-seeking Sidewinder missiles." "Pilots Describe Downings of MiG's," 3.

10. Futrell, *Aces and Aerial Victories*, 25–26.

11. Ibid., 24–25.

12. Ibid., 25.

13. Peck, e-mail; Bugos, *Engineering the F-4 Phantom II*, 134; and AFHRA, "1965–10 July; Holcombe and Anderson." (Each AFHRA Aerial Victory Credit folder contains a narrative summary and aircrew personal statements and/or memoranda to the "Enemy Aircraft Claims Board" that described the MiG engagement. Hereafter, unless otherwise indicated, the cited information came from the narrative summary within the AFHRA folder.) Electrical problems plagued the early F-4Cs, especially in SEA. One former combat F-4 pilot, Peck, described the electrical problems: "Sometimes weird and unexpected things happened for either no reason at all or one thing happened when another thing was directed. By this, I mean things falling off the jet unexpectedly or when a different station was commanded to release." According to Bugos, Air Force and McDonnell engineers later determined that the heat and humidity in Southeast Asia were causing the potting compound used "to seal the backs of [electrical] wire-bundled connectors from water and motion . . . to revert to a viscous, tarry gum. . . . Each F-4 had six hundred potted connectors," all of which eventually had to be replaced, requiring over "$40,000

and two thousand man-hours per aircraft." Further illustrative of the equipment problems encountered on 10 July 1965 (in addition to Holcombe and Clark's radar failing during the engagement), one aircrew reported that their radar operation was degraded, and another aircrew reported that their radar failed to operate in search before finally shutting down altogether during the flight. The aircraft whose radar failed to search also encountered radio problems that prohibited the two pilots in the aircraft from communicating with each other. On a more humorous note, following the engagement one of the pilots suggested, "An ash tray would be desirable in [the] F-4."

14. For example, although the MiGs on 10 July were detected at a range of 33 miles, by the time the F-4s could positively visually identify the aircraft as hostile MiG-17s, they were too close to employ their Sparrow missiles. AFHRA, "1965–10 July; Holcombe and Anderson."

15. Momyer, *Airpower in Three Wars*, 156.

16. "Pilots Describe Downings," 3.

17. AFHRA, "1965–10 July; Holcombe and Anderson." Not every member of the 10 July flight agreed with Holcombe's assessment; Captain Anderson (Captain Roberts's GIB) commented that he "would like [an] internal gun."

18. Davies, *USAF F-4 Phantom II*, 9; and Michel, *Clashes*, 19. To derive air combat lessons, the Air Force's Feather Duster program pitted F-4Cs against Air National Guard F-86H Sabres simulating the smaller MiG-17 fighters. The test program "showed the folly of getting 'low and slow' in a turning fight" with the MiG. Michel summarized the Feather Duster conclusions: "Overall, the Feather Duster tests suggested some rather pessimistic projections about US fighter performance against the MiG-17 and another, more advanced Soviet fighter, the MiG-21, which the North Vietnamese was expected to receive soon. The final report said that both MiGs would out-turn and generally outperform all US fighters at 0.9 Mach and below, and, the slower the speed, the greater their turn advantage against the F-4 and F-105."

19. Futrell, *Aces and Aerial Victories*, 26; and *Project Red Baron II*, 8–9.

20. "During this period, American crews shot down five MiGs, while four fighters were lost to the enemy's aircraft." Futrell, *Aces and Aerial Victories*, 26.

21. AFHRA, "1966–23 April; Cameron and Evans." The narrative noted, "The flight had prebriefed to fire missiles on the identification pass even though there was little probability of aircraft making the identification getting a hit. Past history had been that the MiGs were always on the offensive, and any action that could be taken to put them on the defensive would be beneficial to the F-4C flight." Ultimately, three missiles—two Sparrows and one Sidewinder—were launched with known low probability of hit on the head-on identification pass. The cumbersome nature of the F-4C cockpit was aptly illustrated in the ensuing dogfight. One of the pilots noted, "When the MiG aircraft selected afterburner after my first missile firing, I attempted to select HEAT on my missile panel to fire an AIM-9B Sidewinder. My inertial reel [seatbelt] was locked and I had difficulty releasing the inertial lock so I could reach the panel and change the switch. Since the MiG was starting to evade, I elected to remain in the radar position and fire another AIM-7D Sparrow."

22. Davies, *USAF F-4 Phantom II*, 11.

23. AFHRA, "1966–23 April; Cameron and Evans." The Navy experienced similar issues with its missiles. However, the use of F-8 Crusader aircraft in the air-to-air role to escort the F-4B fighters, which were primarily tasked with performing strike missions, mitigated some of the negative effects. Although it lacked a radar and therefore the capability to employ the Sparrow missile, the F-8 was designed to be an air-to-air fighter and was armed with Sidewinder missiles and four 20-mm Colt Mark 12 cannons. More importantly, the F-8 crews were able to focus

their attention on air-to-air combat and routinely practiced the "type of dogfighting that became the norm over North Vietnam." The results were telling; by the end of the Vietnam War, the F-8 boasted the highest MiG kills per engagement, leading Michel to conclude, "The F-8 pilots were the best air-to-air pilots in the theater during Rolling Thunder." Michel, *Clashes*, 11 and 161.

24. Futrell, *Aces and Aerial Victories*, 28–29.

25. "US Flier Says 2," 3.

26. Futrell, *Aces and Aerial Victories*, 29.

27. PACAF, SEACAAL, iii, I–5; and Hiller, SEACAAL, 2. The assumption that air combat would take place above 30,000 feet altitude gave an advantage to the MiG-21. A follow-up SEACAAL report published on 31 December 1966 noted that this assumption was incorrect; in fact, the majority of air combat occurred below 20,000 feet, a regime where the F-4 enjoyed a slight advantage over the MiG-21. A PACAF briefing at HQ USAF in Washington, DC, concluded that the May 1966 SEACAAL report had an ulterior motive: The "study was devoted to providing a rationale for striking the [North Vietnamese] airfields. The [first SEACAAL] study emphasized—quite correctly—the rapid growth of the NVN Air Force, the gun defenses, the SA-2s, and the GCI system. It painted the MiGs not only as a threat to our strike aircraft over NVN but also to our bases in SVN."

28. Michel, *Clashes*, 20. The initial comparisons were based on US fighters whose performance was thought to match the MiGs'; more accurate relative comparisons "would have to wait until the US had real MiG-17s—and especially real MiG-21s—to test."

29. AFHRA, "1966–29 April; Dowell and Gossard." Two Sidewinders were fired during the engagement. The first was fired outside of parameters to distract the MiG from prosecuting an attack on an F-4 in the flight. The next Sidewinder fired "went up the tail of [the] MiG, exploded, [and the] pilot ejected with [the] aircraft on fire and corkscrewing." As the flight egressed the area, they encountered another MiG. The F-4s launched two more Sidewinders, but the missiles were fired at too great a range and failed to down the second MiG.

30. Michel, *Clashes*, 43. One pilot's comments recorded after shooting down a MiG-17 on 30 April 1966 confirm Michel's assessment: "Confidence in Sparrow was low at this point; there had been 13 firings with no hits in the previous week." AFHRA, "1966–30 April; Golberg and Hardgrave."

31. One pilot commented that he "didn't think [the] Sparrow could ever have been used in this encounter because all [the] attacks were diving at the ground and were never in the proper range band." Another noted, "After the initial attack . . . [we] were never able to achieve the necessary conditions for an ideal missile attack. The nearness to the ground negated much of the missile effectiveness." Within the flight, two pilots noted that an F-4 gun would have been valuable, commenting that "a gun would have been useful—could have gotten into gun range," and "an internal gun could have been used very effectively in this environment." AFHRA, "1966–29 April; Dowell and Gossard." Following an engagement on 30 April, another pilot commented, "A gun would be nice in the F-4C as long as it was clearly understood that it was only a weapon of last resort. Soviet fighters are more capable than US aircraft inside gun range." AFHRA, "1966–30 April; Golberg and Hardgrave." Still another F-4 pilot remarked after an engagement on 16 September 1966, "[An] air-to-air weapon with close range [is] required, down to 1,500 to 1,000 feet. Could have used a gun in several instances." AFHRA, "1966–16 September; Jameson and Rose."

32. "The overall performance of the guided missiles system has proven to be poor. Many missiles either would not fire or, once fired, failed to guide and function correctly." PACAF, SEACAAL, H-12.

33. AFHRA, "1966–23 April; Cameron and Evans." Even though the Sparrow proved successful in this instance, the pilot stated that we wanted to launch a Sidewinder but were unable to reach the missile selector switch during the intense engagement. See n21 for the pilots' description of the engagement.

34. PACAF, "F-4C Fighter Screen," 10. Extracts from the Heat Treat team's SEA trip report were distributed as an attachment to the article.

35. Ibid., 6; and Davies, *USAF F-4 Phantom II*, 17.

36. PACAF, "F-4C Fighter Screen," 6. In an Air Force interview conducted after the war, Brig Gen Robin Olds alluded to the difficulty of maintaining the F-4 radar, critical to Sparrow success. "We had to continually keep the radars peaked and when you're flying a bunch of airplanes—those that are available to you—an average of 85 to 90 and sometimes more airframe hours—hours of utilization, per bird, per month—this turnaround rate is pretty high and you just don't have time to peak up all the little systems with all the exactness that it takes to make this system [the Sparrow] work well." General Olds, oral history interview, 68–69.

37. Michel, *Clashes*, 150.

38. Ibid., 43 and 156.

39. Dilger, oral history interview, 25–26.

40. Kirk, oral history interview, 3. Olds, former 8th TFW commander, echoed Kirk's assessment in a later interview, "Sidewinder. A wonderful little weapon. Limited tactically, yes. Its fire cone was somewhat limited. 2½ Gs, certain range, a minimum range. . . . However, it was reliable, it was simple to maintain. . . . And . . . it was lethal. . . . I was personally quite happy with the Sidewinder." General Olds, oral history interview, 69–70.

41. Davies noted that "pre-war tests [were] held in ideal conditions at high altitude against non-maneuvering targets." Davies, *USAF F-4 Phantom II*, 19.

42. Michel, *Clashes*, 156–58; and Momyer, *Airpower in Three Wars*, 156.

43. Eighth TFW, "Tactical Doctrine," March 1967, 78 (emphasis added). Close control occurs when an individual fighter is directed against an individual target via precise vectors provided by either a ground- or air-based radar operator.

44. Michel, *Clashes*, 163.

45. Ibid., 165.

46. Cunningham remarked, "When I went into combat I had over 200 simulated dogfights behind me. By way of comparison, in Da Nang, I met an Air Force C-130 pilot who had just transitioned to F-4s. He went through a total of 12 air combat training flights, then he was going up North to fights MiGs! I considered this situation an out-and-out crime." Hannah, *Striving for Air Superiority*, 89.

47. Dilger, oral history interview, 6–7.

48. Davies, *USAF F-4 Phantom II*, 16. One MiG kill occurred when the MiG pilot flew into the ground while trying to evade an F-4 attack.

49. Ibid. The pilot also described the F-4C's problems in combat: "We were having a tussle fighting 1950s-era MiGs. The only real advantage we had was to accelerate out of the fight. I'd trade that for turn performance any day. Turning with a MiG-17 was suicidal. You could do pretty well turning with a MiG-21, but he was so small that it was tough keeping him in sight. We were twice the size of the MiGs and had that big smoke trail [from our engines]."

50. Michel, *Clashes*, 62; and *Project Red Baron II*, B-3. The increase in SAM activity earned notice in the *Red Baron II* report's chronology of Rolling Thunder. On 5 July 1966, "NVN missile crews launch[ed] an estimated 26-28 SA-2s at USAF aircrews" in what was then the most prolific SAM activity to date. The NVN beat that number four months later, when it launched

94 SA-2s at aircraft on 19 November 1966. The SAM attacks could be lethal, but luckily the pods proved effective. Michel noted, "In 1965, the SAMs shot down one aircraft for every 16 SAMS fired; in 1966 it dropped to about one kill for every thirty-three missiles fired, then decreased to one kill per fifty in 1967 as pods came into general use, and in 1968 it took more than 100 missiles to bring down an Air Force aircraft."

51. PACAF, SEACAAL, iv, V-6. The report noted that the loss of three F-4s to SA-2s "in less than two weeks" prompted the F-4 flight restriction into SA-2 areas.

52. Ibid., V-6. Taking the pods off the F-105s and putting them on the F-4s formed the basis for the famed Operation Bolo mission on 2 January 1967. Led by Col Robin Olds, 8th TFW commander, "14 flights of F-4Cs, 6 flights of F-105 Iron Hand SAM suppressors, and 4 flights of F-104 covering fighters departed from Ubon and Da Nang and converged on Hanoi. . . . The plan was to have them [the F-4Cs] imitate the F-105s and so draw NVAF [North Vietnamese Air Force] MiGs out for a dogfight. Though the force from Da Nang was forced to turn back because of poor weather, the 'bait' from Ubon was challenged by MiGs from Phuc Yen. . . . Three flights from the 8th TFW downed seven MiG-21s 'within 12 minutes of combat.'" Hone, "Southeast Asia," 536.

53. "By mid-1967, [the Air Force] had enough [pods] to equip all strikers and most escorts." Michel, *Clashes*, 121.

54. Ibid., 181 and 186. See Wilcox, *Scream of Eagles*, for a discussion of the Navy's Top Gun training program.

55. Michel, *Clashes*, 120–21 and 163. The failure to address aircrew training concerns would continue to plague the Air Force throughout the war. Cycling pilots through combat after 100 missions put a strain on the pilot inventory and limited the Air Force's ability to collectively garner and apply combat experience to future tactical missions. Michel noted that one Air Force report concluded that the aircrew policy ensured "the Air Force wound up 'fighting seven one-year wars instead of one seven-year war.'" The Navy elected to "not limit the number of combat missions an aircrew could fly over North Vietnam and, since a Navy aircrew's tour of duty on a Pacific Fleet carrier was about three years, it was normal to make two or three cruises to Vietnam during that time." It was "a two-edged sword. While Navy aircrews became very experienced, . . . Navy combat losses over North Vietnam nevertheless were high; soon the pilot supply began to dwindle, forcing the survivors to participate in more combat cruises—which affected their morale."

56. Ibid., 182; and *Project Red Baron II*, 13. The report noted that only one AIM-7E-2 launched before Rolling Thunder concluded.

57. Michel, *Clashes*, 279. The Navy's Sparrow employment statistics matched those of the Air Force.

58. Ibid., 156; and Westrum, *Sidewinder*, 177. The Navy pressed on and developed the AIM-9D. The AIM-9D sported a redesigned gyro optical system, which improved its ability to track a maneuvering target, and a new, cooled lead sulfide target detector cell for more sensitive and discriminate heat-source tracking. The problem with the new detector was that it required high-pressure nitrogen gas to cryogenically cool it to minus 196°Celsius (77 Kelvin). Lacking sufficient room within the missile body to store the nitrogen gas, Navy engineers elected to redesign the missile rail so that it could store a bottle of nitrogen gas and pipe it to the missile seeker. This design allowed the missile to cool for almost four hours.

59. Michel, *Clashes*, 110.

60. Thornborough and Davies, *Phantom Story*, 110.

61. Michel, *Clashes*, 110.

THE GUN RESURRECTED

62. General Olds, oral history interview, 70–74. Olds's opinion of the AIM-4D Falcon never changed. In his posthumously published memoirs, Olds commented, "By the beginning of June, we all hated the new AIM-4 Falcon missiles. I loathed the damned useless things! I wanted my Sidewinders back." Olds, *Fighter Pilot*, 314.

63. Michell, *Clashes*, 111. The USAF *Red Baron II* report concluded that the AIM-4's "performance was degraded due to design limitations, tactical restrictions, and complexity of operation." The AIM-4D was in combat for 10 months before the transition back to the AIM-9B Sidewinder was completed. During that period, "49 firing attempts were made . . . resulting in four MiG-17s and one MiG-21 being downed." Fifty-five percent of the missiles were fired outside of parameters. *Project Red Baron II*, 13.

64. Michel, *Clashes*, 111; and Hargrove, oral history interview, 14. Hargrove, an Air Force pilot who had the benefit of serving with the Navy on a 30-day exchange assignment, described the benefits of the Navy's AIM-9D relative to the Air Force's AIM-9B: "It's [the Navy's AIM-9D] a much better missile. It has better G capability, it has a better look angle, . . . has a better close-in range, so I think the Air Force is definitely missing a big point in not getting the A[IM-]9D."

65. Knaack, *Post–World War II Fighters*, 273.

66. Thornborough and Davies, *Phantom Story*, 108. Thornborough and Davies described the dive-toss method: "Having rolled down, or 'popped up' on to the target heading at the pre-planned altitude, the pilot selected weapons, stations, and fuses . . . and then lined up on the target, wings-level, for the dive-bomb run. In the back seat, the GIB . . . [had only] to lock the radar on to the top of the ground return line, which by then would be moving down the vertex as the pilot entered the dive, 'ready for pickle.' Once accomplished, the radar boresight line (RBL) was lined up with the pilot's gunsight LCOSS 'pipper.' Jiggling into position, usually at an altitude where the necessity for jinking was less problematic, the pilot centered the servoed 'pipper' on the target and pressed and held the firing (bomb release) trigger, thereby telling the WRCS to ingest and hold the radar-generated slant-range information to target, which it used automatically to compute the moment for optimum weapons release (also drawing on computed variables derived from the INS [inertial navigation system] and central air data system). Still keeping the button pressed, . . . the pilot pulled back on the stick up out of the dive and the WRCS, sensing when all the release parameters had been met, sent a signal at the speed of light to the bomb ejector racks (which responded lazily by comparison), to deposit the bombs on target. Bombing patterns could be initiated at least 2,000 feet higher than when employing manual 'down the chute' procedures, keeping them out of small-arms fire." As testimony to the value of the system, Olds, commenting on the F-4 in general, described the new F-4D system: "And that dive toss worked very well. Very, very, well indeed. It improved our bombing accuracy tremendously." General Olds, oral history interview, 78.

67. PACAF, "F-4C Fighter Screen," 3.

68. Ibid., 8.

69. PACAF, SEACAAL, H-14. The report continued: "Making an ID pass places a restriction on the effective use of missiles. In addition, where missiles have been unsuccessful, these attacks might have been followed with an effective gun attack if the F-4 had been equipped with a gun and lead computing sight."

70. Michel, *Clashes*, 105–6; and Blesse, *Check Six*, 121. The Air Force's 1968 *Red Baron I* report, cited by Michel, concurred with the pilots' assessments. Studying 29 F-4 versus MiG engagements, the report concluded that "in 23 of the engagements the F-4s had cannon-firing opportunities, and often the lack of a cannon appeared to have cost a kill. [Furthermore,] the study concluded that in approximately half of the 29 engagements, North Vietnamese fighters

benefitted from the F-4's inability to shoot them at close range, and that even if the only effect of the cannon was to keep the MiGs from getting close, it would help because then the MiG would be in the missile envelope." Blesse noted, "The slower MiG-17s quickly learned of our poor maneuverability and established the procedure of using the tight turn as a defensive haven. We had no gun and couldn't turn with them, so unless we could get a long-range missile shot, they were quite safe. At low altitude the missiles had little success. We needed the gun to be able to take that shot at them and break up their defensive haven."

71. Futrell, *Aces and Aerial Victories*, 118–19; and Michel, *Clashes*, 106. Although significantly outclassed in an air-to-air sense by the MiG-17, by the end of 1966 the F-105 had dispatched five MiG-17s with its 20-mm Vulcan cannon. Additionally, Michel noted that, in the event an F-105 was shot down, the accompanying F-105s often used their cannon "to strafe approaching North Vietnamese to protect the crew until a rescue helicopter arrived."

72. Bugos, *Engineering the F-4 Phantom II*, 158.

73. "What I [Catledge] was proposing was to put guns in the F-4, and the only way to do it since they were already in production was to pod one.... If we could sell the program, someday down the line they would go into production airplanes.... So we spent the money, and we podded the gun. The change in concept came about. We got them into production, and the F-4 came out as the F-4E." Catledge, oral history interview, 32.

74. Burns, oral history interview, 3. The F-X program evolved into the F-15 program. Burns also noted, "This was before the experience of Southeast Asia bore these things out, I might add."

75. Ibid., 17–18 (emphasis added).

76. "Air Force Orders New Jet Fighter," 9.

77. "US Jets Will Reflect Lessons," 3.

78. Mets, *Evolution of Aircraft Guns*, 225–26; and Catledge, oral history interview, 32.

79. Bugos, *Engineering the F-4 Phantom II*, 158.

80. Thornborough and Davies, *Phantom Story*, 114.

81. Ibid. The muzzle problem was not corrected until "an elongated Midas IV shroud" was developed and flight-tested in April 1970.

82. Ibid., 113.

83. Knaack, *Post–World War II Fighters*, 278–80.

84. Ibid., 280.

85. Bugos noted that the introduction on an internal gun to the F-4E "added flexibility in planning and was a powerful ideological statement that Air Force pilots were less missile system managers than gunfighters, capable of dogfighting and strafing ground units like their predecessors in other wars." Bugos, *Engineering the F-4 Phantom II*, 159.

86. Kirk, oral history interview, 7.

87. General Olds, oral history interview, 77.

88. Futrell, *Aces and Aerial Victories*, 118–25 and 157. Aircraft 20 mm gunfire accounted for 40 of the USAF's 137 MiG kills during the Vietnam War. The F-4E contributed one MiG-19 and four MiG-21s to that tally. In contrast, the podded gun system initially put into service at Da Nang for the F-4C and F-4D tallied 10 MiGs, with an 11th MiG shared between an F-105F and an F-4D.

Chapter 6

An Interim Solution

I gnash my teeth in rage to think how much better this wing could have done had we acquired a gun-carrying capability earlier.

—Brig Gen Robin Olds, USAF

In early 1915 a French pilot, aided by his mechanic Jules Hue, affixed a set of steel deflectors to the propeller of his Morane-Saulnier L monoplane and took off in search of German aircraft operating over the western front. Despite saddling the already fragile aircraft with additional weight, the inelegant propeller-mounted steel plates were critical to mission success. Without them, Roland Garros would have shot off his own propeller blades when firing his Hotchkiss machine gun, which he mounted directly in front of the cockpit and squarely behind the spinning prop. The innovation, although certainly unorthodox, worked. In a three-week period, the Frenchman claimed three German airplanes.[1]

More than 50 years later, American pilots of the 366th TFW at Da Nang AB, South Vietnam, slowly meandered around their F-4C Phantom—a machine constructed of advanced metals and capable of speeds in excess of 1,600 miles per hour, which stood in stark contrast to Garros's earlier, fabric-covered machine that maxed out at a blistery 70 miles per hour—and wondered how they would accomplish a similar feat. They also succeeded.

In both instances, a tactical innovation, born of necessity and resourcefulness in the field, made its appearance with little fanfare, but had startling repercussions on the future of air combat. Although the 366th's innovation would by war's end contribute to less than one-thirteenth of the total number of Air Force MiG kills during the Vietnam War, their leap backward to what was thought to be an antiquated form of aircraft armament actually heralded a renewed era in aerial combat that has continued into the twenty-first century.[2]

The Tool at Hand

Spurred by Catledge's efforts at the Pentagon, in 1964 the Air Force began developing an external housing that could hold the GE 20 mm M61 Vulcan cannon, a six-barrel and 6,000-rounds-per-minute Gatling gun then installed on the F-105 Thunderchief fighter-bomber.[3] The resultant SUU-16/A gun pod, powered by a ram-air turbine (RAT) and the aircraft's electrical system,

AN INTERIM SOLUTION

weighed over 1,700 pounds, contained 1,200 rounds of ammunition, and measured 16 feet long.[4]

Air Force Systems Command's Air Proving Ground Center began testing the SUU-16/A on the F-4C in summer 1965. Alternately installing the gun pods on the F-4C's centerline station under the belly of the aircraft and in pairs beneath each wing, the test investigated the effectiveness of the F-4C/SUU-16 combination in a close air support role attacking enemy personnel and vehicles. After the test began, Air Force engineers also decided to study the gun pod's utility in an air-to-air role.[5]

The August 1965 test report concluded that multiple successful air-to-ground firings justified use of the SUU-16/A for close air support missions; the report was less enthusiastic about the gun pod's air-to-air potential. The first three of six air-to-air test missions were deemed unsuccessful when the F-4C did not score a single hit on the target. Aircrews struggled to identify an appropriate aiming reference, and the lack of an accurate air-to-air gunsight was cited as one of the major deficiencies of the system. To help compensate for the poor gunsight, the report recommended "that tracer ammunition be used while employing the F-4C/SUU-16/A combination in an air-to-air situation whenever possible." Despite the limited air-to-air testing and the known deficiencies, the report concluded, "The F-4C/SUU-16/A combination provides a *limited* air-to-air capability."[6]

Based on the demonstrated air-to-ground potential of the SUU-16 system, the Air Force pursued procurement. SUU-16/As began arriving in SEA in April 1967, with initial pods directed to the 366th TFW at Da Nang.[7] Two rationales contributed to the selection of Da Nang. First, because of Da Nang's location in northern South Vietnam, the 366th performed a large number of in-country and near-border missions, including the close-air-support mission for which the pod was tested.[8] Second, the 8th TFW at Ubon, Thailand, was scheduled to receive its first F-4Ds in about a month. In addition to having a lead-computing air-to-air gunsight, the F-4D also had the capability to carry a new gun pod, the SUU-23/A.

The SUU-23 boasted two improvements over its SUU-16 predecessor: the gun was powered not by a RAT but by muzzle gasses, and it had a sleeker design, which theoretically reduced drag and fuel consumption.[9] Despite its better aerodynamics, one former F-4 pilot still lamented, "With the open-ended gun barrels and blast deflector on its front ends, the [SUU-23 gun] pod was indeed cruel to the Phantom II's slipstream and its fuel consumption."[10]

The extra weight and drag associated with the gun pod, and the expected consequent decrease in aircraft maneuverability and increase in fuel consumption, led many pilots to doubt its utility in combat.[11] Aircrews assumed they had to wait for the recently announced F-4E with its internal cannon before they would enjoy a combat-effective gun.

The Gunfighters

"Boots" Blesse knew about employing the gun in air-to-air combat. A two-tour, 123-combat-mission Korean War veteran, Blesse downed 10 North Korean aircraft—nine jet-powered MiG-15s and one propeller-driven LA-9—with his F-86 Sabrejet's six 0.50-inch Colt-Browning M-3 machine guns.[12] Returning to the states in late 1952, Blesse reported to the Air Force's gunnery school at Nellis AFB, Nevada, forerunner to today's USAF Weapons School. While there, Blesse published a popular tactics manual, *No Guts, No Glory*.[13] Also while he was at Nellis, Blesse's aerial gunnery prowess was publically highlighted when he took first place in all six individual events at the USAF Worldwide Gunnery Meet in 1955, an unprecedented accomplishment.[14] After completing National War College in 1966, Blesse volunteered for service in Vietnam, specifically at Da Nang. When the members of the 366th TFW learned that their new deputy commander for operations would be Blesse, they knew that he would play a pivotal role in improving the wing's lackluster tactical performance.[15] Blesse wouldn't have much time.

Shortly after Blesse's arrival at Da Nang in April 1967, Pres. Lyndon Johnson for the first time authorized strikes against both Hanoi's electric power system and North Vietnamese military airfields.[16] North Vietnam responded by ramping up the number of MiG sorties, which prompted the Air Force to dedicate more F-4s to MiGCAP missions to protect the F-105 fighter-bombers.[17] The 366th TFW at Da Nang and Olds's 8th TFW at Ubon were assigned the extra escort missions.[18] Prior to that, the 366th TFW had been executing almost exclusively air-to-ground missions. In fact, when Blesse arrived he bemoaned, "there wasn't anyone in the outfit who had ever fought an enemy aircraft except me."[19] The wing desperately needed a quick refresher on air-to-air tactics, and Blesse went to work providing it, at times even calling upon his 12-year-old *No Guts, No Glory* tactics manual.[20]

Much of the wing's focus was on the F-4's air-to-air armament. As a Korean War air-combat veteran, Blesse had a unique appreciation for the nature of air combat and the "complementary" roles for both missiles and guns in a jet fighter:

> I had felt for years we went the wrong direction in the Air Force when we decided guns no longer were necessary. This was "the missile era," they said. I was told by some pretty high-ranking officers I was wrong, but my experience in Korea seemed to tell me other-

> wise. Missiles don't always work, they had limiting parameters under which they could be fired, they were ballistic (no guidance) for several hundred feet after launch, they didn't arm immediately, and, in general, left a great deal to be desired. In addition, from an operational standpoint, you could be surprised while attacking another aircraft and find yourself in a tight turning battle. High Gs and tight turns are not ideal parameters for firing a missile, and besides, range between aircraft decreases rapidly under those conditions and you could easily find a gun a far more useful weapon. An internal gun also provides a capability at all times for targets of opportunity on the ground. For all these reasons, I found the missile and gun complementary weapons, not weapons that were in competition with each other.[21]

Blesse reasoned that the wing, now tasked with additional MiGCAP missions in NVN and receiving the first of several SUU-16 gun pods, "could take that SUU-16 gun to Hanoi and increase our air-to-air capability."[22] One former 366th pilot recalled that Blesse, pointing to an F-4, once exclaimed, "All I want to do is get a gun on there.... I don't care if we have to ... wire a ... 38-caliber pistol with a string to it, that's what we'll need against those MiGs!"[23] While it did not require such drastic measures, introducing the SUU-16 to F-4 air-to-air combat was nonetheless easier said than done.

Blesse assigned the task of integrating the SUU-16 onto the F-4C for air-to-air employment to the wing's elite weapons section.[24] The first problem the officers encountered was where to hang the gun pod on the aircraft. The F-4 had two pylons attached to the underside of each wing. The outermost wing pylons could carry either a 370-gallon external fuel tank or air-to-ground ordnance (including the SUU-16/23). The innermost wing pylons could carry either two AIM-9 (or on the F-4D, two AIM-4D) missiles or additional air-to-ground ordnance, but not external fuel tanks. The centerline pylon suspended from the belly of the aircraft could carry a larger, 600-gallon external fuel tank or an array of air-to-ground ordnance, including the SUU-16/23. The F-4's four Sparrow missiles were carried underneath the aircraft's fuselage in specially designed, recessed missile stations. During F-4 air-to-air missions early in the war, the preferred configuration included two 370-gallon external fuel tanks, a tank suspended from each outermost wing pylon; four Sidewinder missiles, two attached on either side of the innermost wing pylons; four Sparrow missiles nestled along the belly of the aircraft; and often a 600-gallon fuel tank attached to the centerline of the aircraft. The extra fuel provided by the three external fuel tanks allowed the F-4s to maximize their flight time patrolling the target area.[25] Also, the configuration was symmetric, offering maximum aircraft stability in-flight.

However, the introduction of the external ECM pod on the F-4 in early 1967 required a change to the preferred aircraft configuration. The ECM pod, necessary for aircraft defense against the escalating SAM threat, relied on special wiring that was only available in the outermost wing pylon, a position normally reserved for a 370-gallon external fuel tank. Consequently, the approximately 190-pound ECM pod was mounted on the outermost right wing pylon.[26] Loaded on the opposite pylon was the 370-gallon external fuel tank, which weighed almost 2,400 pounds when full. The 600-gallon fuel tank was carried on the centerline as before, and the Sidewinders and Sparrows likewise maintained their prior positions on the aircraft.

The resultant configuration was far from symmetrical, and it forced the F-4 to fly in a notoriously unstable configuration. Colonel Bolt, the 366th TFW commander at the time, later exclaimed that in that configuration, "Well, the airplane flew sideways! It used up a lot of gas, and it was dangerous."[27] Colonel Olds, the 8th TFW commander, offered a similar appraisal: "When they originally wired the airplane, they put the ECM pod on the right outboard pylon. This put us into a terrible, horrible configuration. . . . You had to carry a 600-gallon centerline tank . . . and your other external tank, your 370-gallon left outboard tank, hanging way out here, in [sic] the outside of the wing, with nothing to balance it on this side. . . . Takeoff was very exciting."[28]

Prior to the arrival of the ECM pods, Olds and others requested that the Air Force address the pending aircraft configuration issue, hoping Air Force engineers would modify the F-4 so that the ECM pod could be hung from an inboard wing pylon. The Air Force's response was disconcerting, "We were told it would take some 12 to 14 hundred man-hours per aircraft to modify our F-4s."[29]

The 366th's weapons section therefore faced a dilemma. On MiGCAP missions, the SUU-16 had to be mounted on the centerline pylon; otherwise, it would be extremely difficult to aim at the MiG target. However, the F-4 could not afford to sacrifice the extra fuel provided by the 600-gallon tank usually mounted on the centerline, especially when the necessary ECM pod precluded the possibility of loading a 370-gallon fuel tank on the right outermost wing pylon. The only solution was to devise a way to move the ECM pod to the inboard pylon in a manner that did not require excessive time or maintenance effort.

Later described by the wing's historian as working under the premise, "You know it can't be done, so now tell us how to do it," a team of pilot tacticians and aircraft maintenance personnel finally developed a solution.[30] Fortunately, it was both inexpensive and relatively easy to implement. Crediting the genius to a particular chief master sergeant, the 366th wing commander later described the proposed fix: "All he did was build a simple harness with two cannon plugs on it and tie it in to the nuclear armament system."[31] After having confirmed the design's potential, wing personnel performed a limited number of pylon and ECM pod modifications so that they could test the new configuration.

Based on these in-house tests, the weapons section concluded that the pod did not appreciably degrade the F-4's performance and maneuverability as once thought.[32] The tests also illustrated that the most effective gun-carrying configuration was to load the flight and element lead aircraft, flying in the #1 and #3 positions, with the SUU-16 gun pod on the centerline, two 370-gallon external fuel tanks on the outermost wing pylons, the ECM pod on the innermost right wing pylon, two Sidewinders on the innermost left wing pylon, and two of the four Sparrow missiles on the aircraft's belly. Although there was still room for four Sparrows, the reduced fuel supply based on substituting the 370-gallon fuel tank for the typical 600-gallon fuel tank and the increased drag associated with the SUU-16 pod led the tacticians to recommend that two of the Sparrow missiles be downloaded to reduce aircraft weight and drag.[33] The wingmen, flying in the number 2 and number 4 positions, retained the previous asymmetric ECM pod configuration and all eight missiles—four Sparrows and four Sidewinders.[34] This allowed the wingmen to carry the larger 600-gallon centerline fuel tank, providing them with more fuel for the mission, which they typically burned trying to maintain formation with the lead aircraft.[35]

Having developed a viable configuration to carry the gun, the 366th's weapons section turned its attention to establishing the procedures to employ the gun in combat. The lack of a lead-computing air-to-air gunsight on the F-4C seriously degraded the gun's effectiveness. Blesse described the wing's expedient solution:

> We decided we could make do with the fixed sight that was installed. With no lead computer, it was useless to put the pipper (aiming dot) on the enemy aircraft because the rounds fired would all end up behind the target. The . . . gun we carried had a very high rate of fire. So high, in fact, that the rounds that came out of this single gun would strike the [target] aircraft only about eight inches apart at 2,000 feet range. We figured, if you put the pipper on the target, then moved it forward about twice as far as you thought necessary before you began to fire, you would over-lead the target. The procedure then was to begin firing as you gradually decreased your amount of lead. This would allow the enemy aircraft to fly through your very concentrated burst. Wherever hits occurred, the rounds stitched through the wing or cockpit area like a sewing machine. Clusters of 20mm rounds striking close together would weaken the wing or whatever it hit, and the violent air and G forces would tear it off the aircraft.[36]

Olds later noted that the procedure entailed "wasting a lot of bullets, but all you need is a few of them to hit and down he goes."[37] Using this imprecise-but-best-available procedure was also thought to take advantage of the other-

AN INTERIM SOLUTION

wise adverse effects on bullet dispersion caused by the gun vibrating on the mounting pylon when it was fired.[38]

With the background research done, Blesse was ready to approach Momyer, Seventh Air Force commander, seeking permission to modify the 366th's entire fleet of F-4Cs. Blesse described the meeting:

> Charts and all, I parked myself in the general's outer office and awaited my turn. Finally the door opened and "Spike" Momyer appeared. With him was Colonel Robin Olds, commander of the 8th Tac[tical] Fighter Wing at Ubon. General Momyer, seeing me waiting and remembering the subject, turned to Robin Olds and invited him to hear my briefing.
>
> So, with my select audience of two, I laid out our ideas, our test results, our method of compensating for the lack of a computing gun sight, and our ideas for air-to-ground use of the gun. It was magnificent, I thought—innovative, thorough, concise. I was quite happy with myself as General Momyer reflectively turned to Colonel Olds and said, "What do you think of that idea, Robin?"
>
> Olds then proceeded to blow me out of the water, hull and all, with the simple statement, "General, I wouldn't touch that with a ten-foot pole!" . . . I was stunned.
>
> General Momyer was more kind. "You and I talked about this a few years ago, Boots, and I didn't think much of the idea then. Maybe things are a little different now, I'm not sure. I think you have a hole in your head but go ahead with your gun project and keep me informed."
>
> It wasn't the whole-hearted support I was shooting for but at least we could go on with it.[39]

Additional configuration testing at the 366th on 3–4 May 1967 focused on evaluating the ECM pod's performance when mounted on the inboard pylon. The subsequent message to Momyer on 5 May 1967 reported:

> 100 percent successful electronic emissions all applicable altitudes and attitudes. Twenty-six man-hours involved [in the aircraft modification]. . . . ALQ-71 [ECM] pod modification makes possible SUU-16 gun installation [on] centerline station for use in Package Six.
>
> *This Wing has lost minimum seven kills in the past ten days because of a lack of kill capability [against targets] below 2,000 feet altitude and inside 2,500 feet range. . . .*
>
> SUU-16 can be carried without degradation of aircraft performance. . . .
>
> Your HQ has 120,000 rounds of 20mm tracer ammo enroute to Da Nang, which we will use on one to eight basis in our ammo load. With a fixed sight, this tracer of utmost importance both for sighter burst and deflection shooting.
>
> *It is interesting to note we are dusting off deflection shooting info published early WW II and Korea for our Mach 2 fighters. . . .*
>
> Request authority to continue modification for entire 366th fleet.[40]

AN INTERIM SOLUTION

Momyer granted the request. Six days later the 366th sent a message to the top aircraft maintenance officer at Seventh Air Force, courtesy-copying Thirteenth Air Force in the Philippines and the two other F-4 wings in SEA (the 8th TFW at Ubon and the 12th TFW at Cam Ranh), outlining the modification procedures and justification in greater detail: "This modification allows the carriage of the SUU-16 gun pod, the only air-to-air weapon that can be employed against very low altitude aircraft. The need for the modification became apparent after a number of pilots reported unsuccessful results after engaging the MiG-17. In all cases, the main reason was the very low altitude the MiG attained after engagement. This station [366th at Da Nang] proposes to add an ECM capability to the right inboard pylon. . . . The aircraft wiring changes are merely a splice made with existing aircraft wiring. The inboard pylon is modified to add one connector. . . . The modification in no way affects the present ECM capability nor any other system on the a[ir]c[ra]ft."[41]

The following day, Blesse and Maj Bob Dilger, a member of the wing's weapons section that had worked on the gun project, took off for a mission "Up North," their F-4Cs toting an ECM pod on the right inboard wing pylon and a SUU-16/A on the centerline—"the first gun-equipped Phantoms into Pack Six."[42] Two days later, the tireless efforts of Blesse and the other members of the 366th TFW, as well as earlier efforts by Catledge and Burns at the Pentagon, finally came to fruition.

After the members of Speedo flight landed at Da Nang following their 14 May 1967 mission, they were mobbed by their compatriots, including Blesse, before being hustled into the intelligence section to debrief the first-ever F-4 air-to-air gun engagements.[43] During the debrief, the flight members praised the SUU-16 "as a very good gun" and "a very good system." Captain Craig from Speedo 3 commented, "The kills with the gun . . . could not have been made with a missile." Maj Hargrove from Speedo 1 reminded his debriefers that he "never had a chance to shoot the SUU-16 air-to-air before this encounter," and added that although he "would like to have had a lead-computing [gun] sight," the use of "tracers [in the future] . . . will help a lot."[44] The message traffic describing the engagement sent across the theater late that night noted, "All members of Speedo [flight] spoke praise for the SUU-16 gunnery system. We think the results speak for themselves."[45]

In a later interview, Hargrove described the combat in more detail: "I opened fire at about 2,000 feet, and he [the North Vietnamese MiG pilot] still—right away—he didn't break, and I guess he probably saw my muzzle flashes with the smoke, and didn't know what that crazy pod was underneath anyway, but he did break at, oh, a thousand feet or so. He broke hard, . . . but it was too late now. I cut him in half with the gun. But had he known, of course, that I had the gun, he would have maneuvered differently. But without the gun—in the fight that we were in—I don't see how I possibly could have gotten a MiG without slowing down and exposing myself considerably more than is smart to do."[46] Hargrove also reportedly chuckled, "I'll bet they [the North Vietnamese] had a tactics meeting at Kep (NVN air base) that night."[47]

Following Speedo flight's successes, news of the 366th and the F-4/SUU-16 weapons combination spread rapidly throughout the Air Force. At 0250 on 14 May 1967, local Hawaii time (17 hours behind Da Nang), the PACAF command center logged the first message about the 366th's engagements, reporting Elgin flight's MiG kill:

> 0250   MiG Shoot Down. 366TFW OPREP-3/011 reports that Elgin Flight, F-4C's, MIG CAP, saw 6 MiG-17s and Elgin Lead shot down one with a Sparrow. AFCP [Headquarters, Air Force Command Post] notified.[48]

Thirteen minutes later, the second message from the 366th arrived:

> 0303   Two MiG-17s Shot Down by F-4Cs. 366 TFW Msg OPREP-3/010 reports that Speedo Flight, while escorting strike flight against Ha Dong Army Barracks, engaged at least 10 MiG-17s and shot down two of them using the SUU-16 gunpods. AFCP notified.[49]

Those initial messages were followed up by more detailed ones approximately four hours later.

> 0715   MiG Shootdown, Elgin Flight. 366 TFW Msg Fastel 448 is detailed report of Elgin Flight engaging MiGs. Comment by pilots indicate [sic] that the SUU-16 would have been more effective against the MiG-17s than any of their missiles.[50]

By 1030 interest in the message traffic, as well as some confusion, extended to Washington:

> 1030   SUU-16 Pods. Col Dunn (AFCP) requested information as to which F-4C MiGCAP aircraft were equipped with SUU-16 pods. Lt Col Hartinger (7AFCC [Seventh Air Force Command Center]) stated that Elgin lead and #3, and Speedo Lead and #3 were equipped with the gun pods. However, Elgin Lead aborted and the spare aircraft was not gun pod equipped. Elgin Lead did shoot down one MiG with a missile

AN INTERIM SOLUTION

and Speedo Lead and #3 each downed a MiG with 20-mm. Passed to AFCP.

1145   <u>Speedo Flight MiG Kills</u>. 366 TFW OPREP-3/Ch1, DTG 14/1800Z May 67, is narrative of the two MiG-17 kills by Speedo flight (4 F-4C MiGCAP against JCS [target] 31). The flight expended 4XAIM-7s and 1XAIM-9, all duds. Both kills were with 20mm cannon.[51]

Two hours later, Seventh Air Force and the PACAF command center were still trying to alleviate confusion surrounding the 366th's exploits:

1345   <u>Configuration for Carrying SUU-16 (20-mm Pod)</u>. Colonel Hartinger (7AFCC) stated two fuel tanks are carried outboard, a QRC-160 pod on the right inboard, two AIM-9s on the left inboard, and the SUU-16 pod carried on the centerline. A minor modification was required to allow the QRC-160 pod to be carried on the right inboard station.[52]

While messages buzzed back and forth between the 366th TFW, Seventh Air Force, Headquarters PACAF, and Headquarters USAF, Blesse received an irate phone call from the 8th TFW commander at Ubon. Responding to Blesse's daily operational summary that quipped, "There will be two pilot meetings tonight. One in Hanoi, the other in the 8th Tac Fighter Wing," Olds shouted into the receiver, "What the hell are you trying to do, you crazy bastard! Don't you realize what kind of a position this puts me in?"[53] Nevertheless, by the end of the month, the 8th TFW had begun modifying its F-4Cs and newly arriving F-4Ds according to the 366th-developed procedures.[54] The 8th downed its first MiG with the 20-mm gun pod on 24 October 1967.[55] The aircraft commander, Maj William Kirk of the 433rd TFS, would later enthusiastically characterize the gun pod as "the finest thing that was ever invented."[56]

As news of the engagement continued to spread, Momyer urged the 366th to send a message to the chief of staff of the Air Force, which they did on 18 May 1967. "Subj[ect]: MiG Engagement Supplement to 366TFW OPREP-3//012 [Speedo flight] . . . . The missiles were fired at minimum ranges and maximum allowable G forces. The missiles were fired at low attitudes and against a cloud background. Upon observing the futility of trying to maneuver for an optimum missile attack, which is virtually impossible against an enemy aircraft that is aware of an attacker's presence, the pilot shot a MiG down using the 20-mm cannon."[57] Two months later, Blesse was in Washington, DC, briefing the Senate Preparedness Investigating Committee and the secretary of defense, touting the gun as "one of our most versatile and effective weapons."[58]

Since arriving at Da Nang, "Boots" Blesse had wanted a nickname for the wing. For example, Olds's boys at the 8th TFW were known as the "Wolfpack." After May 1967, the 366th's prowess with the SUU-16 had earned them one. Their insignia became a "little guy in a black full-length coat wearing tennis shoes and a very large black hat"—the McDonnell Aircraft Company's cartoon Phantom—"carrying a SUU-16 gun pod." Their nickname became "the Gunfighters."[59]

Assessment

The 366th's official wing history from the period recorded that "the desirability of a 20-mm Gatling gun in air-to-air combat was, in large measure, an expression of the limitations of air-to-air missiles."[60] By the end of Rolling Thunder, Blesse's innovation accounted for almost one-third of the wing's air-to-air victories, a significant tally considering the 366th resumed its primary air-to-ground missions after only six weeks of MiGCAP taskings.[61] By the end of the war, the gun on the Air Force's F-4C/D/E aircraft had accounted for 15 and a half of the Air Force's 137 kills.[62] Once deemed an antiquated armament system not worthy of further development in 1957, the gun had proven its value in air combat once again.

The combat results achieved by the external cannon, and a small jab from Blesse in his 14 May 1967 daily operational summary, swayed initial skeptics like Olds.[63] During one interview, Olds characterized the gun pod as "a very, very fine weapon and a very accurate one."[64] In a separate, earlier interview, Olds commented, "Now the old gun—the Vulcan M-61 Gatling gun we've got—is an outstanding development. . . . It's a good close-in weapon. I gnash my teeth in rage to think how much better this wing could have done had we acquired a gun-carrying capability earlier."[65] Other Air Force officers also took note. One report issued after the war concluded, "At low altitude, the air-to-air ordnance which afforded the highest kill probability was the cannon."[66]

Momyer was not so easily convinced, though, as evidenced by his writings after the war. Acknowledging in his book *Air Power in Three Wars* that "the low kill rates for missiles may also be explained in part by the fact that the AIM-7 was designed as an antibomber weapon," Momyer sounded like Gen Emmett O'Donnell of Korean War fame when he next wrote, "The different circumstances of the wars in Korea and the Middle East [referring to the1973 Arab-Israeli War] . . . prevent us from making responsible judgments about

AN INTERIM SOLUTION

the relative quality of pilots or equipment [during Vietnam].... Both political and technological factors tended to depress our kill ratio in Vietnam, with political constraints being probably the most significant factor."[67]

Other documentation reveals Momyer's continued faith in the promise of advanced air-to-air missile technologies. In a 1975 Corona Harvest memorandum, Momyer urged, "There must be a major increase in kill potential of air-to-air missiles employed to what was obtained in Vietnam. More effort is needed in the development of a new radar and dogfight missile that has a capability of kill between seventy and eighty percent."[68] An earlier 1974 memorandum similarly concluded, "The final dogfight phase [of air combat] should be optional."[69] Still, despite emphasizing the primacy of guided missiles in air-to-air combat, Momyer came to recognize the complementary value of an air-to-air gun mounted in, or on, a fighter aircraft, and he urged the Air Force to procure a "new air-to-air gun."[70]

The decision to load an external gun on the F-4C/D and build one into the new F-4E reflected a growing appreciation that, despite the continued promises of the air-to-air missile proponents, air combat could not be reduced to simple missile exchanges at long range. Consequently, aircrews needed better air-to-air training. After surveying the Air Force's air-to-air engagements in Vietnam through 1968, the *Red Baron II* report reached a similar conclusion: "History has shown that the aircrew that is most likely to excel is the one that is the most highly trained. Without adequate training, the capabilities and limitations of the fighting platform are neither recognized nor used effectively."[71] The report recommended that "tactical aircrews ... be provided improved (quantity and quality) ACM [air combat maneuvering] training," which helped spur the Air Force's Red Flag and Aggressor training programs in 1975.[72]

Ultimately, the persistent efforts of determined Air Force officers like Blesse, Burns, and Catledge triumphed over the Air Force's penchant for technological exuberance, embodied in its untenable embrace of poorly performing air-to-air missile technologies and the contexts they informed. In doing so, the gun advocates had to overcome the bureaucraticism and unjustified optimism that had jaded the Air Force's opinions of three interrelated technological systems—the airframe, the armament, and the aircrew training process—that collectively proceeded according to a circular logic trail gone bad:

> Missiles were better suited to shoot down jet aircraft than guns. Jet aircraft were therefore built without guns. Aircrews were therefore trained to shoot down jet aircraft using missiles. Because aircrews were trained to shoot missiles and not guns, the Air Force had to develop better mis-

siles, not guns. Because the Air Force was building better missiles, it needed better aircraft to shoot those missiles, and so on.

Each technological system or process developed according to a technological trajectory, and each reinforced the other. It was not until a few determined individuals began questioning the predicating assumptions—was a Soviet bomber the most likely target? could missiles and guns actually be complementary weapons? and could aircrews be trained to employ both types of armament?—that the Air Force's technological blinkers were finally removed.

The impact is still felt today. Aircrews continue to conduct air-to-air training in the skies north of Nellis AFB during Red Flag exercises, and the newest Air Force fighters, the F-22A Raptor and F-35A Lightning II, are both equipped with internal cannons.[73] Finally, the history of the Air Force's air-to-air armament through Rolling Thunder provides a valuable case study to examine the nature of military innovation.

Notes

1. Morrow, *Great War in the Air*, 91–92.
2. Futrell et al., *Aces and Aerial Victories*, 157. The 366th's innovative solution for employing the podded gun on the F-4C/D eventually accounted for 10 of the USAF's 137 MiG kills during the Vietnam War. (Additionally, an F-4D and F-105F shared a 20-mm gun kill.) In contrast, the F-4E's gun contributed only five MiGs to the tally.
3. Michel, *Clashes*, 102.
4. Mehserle, *Capability of the F-4C Aircraft*, 2.
5. "After the test was started, an additional objective was added, this being a demonstration of the SUU-16/A weapon in the air-to-air role." Ibid., 1.
6. Ibid., 35–36 (emphasis added). In contrast to the paltry six air-to-air missions flown, the test included 67 air-to-ground missions.
7. Davies, *USAF F-4 Phantom II*, 30.
8. Thornborough and Davies, *Phantom Story*, 105.
9. Thornborough, *USAF Phantoms*, 101; Davies, *USAF F-4 Phantom II*, 30; and Thornborough and Davies, *Phantom Story*, 107.
10. Davies, *USAF F-4 Phantom II*, 31.
11. Bakke, interview. Air Force pilots actively reinforced this perception. Retired major Sam Bakke recalled that one Saturday while he was at Nellis AFB, Nevada, for F-4 weapons training, he and his flight commander flew a demonstration flight for "a civilian official of influence" who was flying in the back seat of an accompanying F-4. Bakke's F-4 was loaded with the SUU-16 and two external fuel tanks, the accompanying F-4 with the civilian carried only the two external tanks. Bakke described the flight: "We pulled up side-by-side to demonstrate . . . how the aircraft underperformed when you had extra weight in the centerline area—to imprint on him [the civilian official] that we needed an internal gun."
12. Blesse, *Check Six*, 73–76. Blesse's last kill coincided with his last mission in Korea. While flying back to base after an otherwise uneventful mission, Blesse's flight of F-86s was jumped by a flight of four MiGs. Although the F-86s were low on fuel and, more importantly,

AN INTERIM SOLUTION

the MiGs were out of range, Blesse's young wingman mistakenly turned to defend himself and was unwittingly exposed to the MiGs' attacks. Blesse turned to help his wingman and engaged one of the attacking MiGs, allowing the young pilot to escape. Unfortunately, in the process of shooting down the MiG, Blesse's F-86 ran out of fuel, and he had to eject off the North Korean coast. He was rescued a short time later by an American air-sea rescue plane and was quickly ordered back to the United States, lest the Air Force risk losing one of its leading MiG killers in combat.

13. Ibid., 87; Blesse, *No Guts, No Glory*, ii and iv; and "Maj Gen Frederick C. Blesse Biography." In the foreword to *No Guts, No Glory*, Col George L. Jones, commander of the 359th Combat Crew Training Group (Ftr) at Nellis AFB, Nevada, described the manual as a "clear presentation of a way of flying and a pattern of thought essential to the fighter pilot for survival and victory in air-to-air combat." In the preface, Blesse stated that his goal was to help "produce a pilot who is aggressive *and* well trained." Affirming the manual's popularity, Blesse's official Air Force biography noted that his "book [*No Guts, No Glory*] has been used as a basis of fighter combat operations for the Royal Air Force, Marines, Chinese Nationalist, Korean Air Force, and US Air Force since 1955. As recently as 1973, 3,000 copies were reproduced and sent to tactical units in the field."

14. Blesse, *Check Six*, 91. Blesse's claims are confirmed by his official Air Force biography.

15. Bolt, oral history interview, 187–89. Assuming command one month before Blesse's arrival, the 366th TFW commander, Col Jones Bolt, later described his assessment of the wing's poor morale and lackluster tactical performance: "I was never so disillusioned and low in my life as I was when I got to that base [Da Nang] and took a look around. It was really bad. The morale of the people and the *esprit de corps* were just nonexistent.... Their loss rates were high, both at night and in the daytime. Their loss rate up in Route Pack I was atrocious." Bolt attributed some of the substandard performance to the relative inexperience of the deputy commander for operations prior to Blesse: "The guy that was the DO was a newly promoted colonel, B-57 [Canberra, a twin-engine, light bomber built in 1953] pilot. This doesn't work. The guy was hardly current in the F-4."

16. Hone, "Southeast Asia," 537.

17. "The number of air engagements during recent strikes against JCS numbered targets is indicative of the increasing MiG threat to our forces. Attacks against the remaining jet capable fields... are considered necessary at this time to further harass and disrupt the MiG air defense capability.... [Targets] should be attacked by larger forces in order to saturate the defenses." Message, 300055Z APR 67, CINCPAC to JCS.

18. Michel, *Clashes*, 99.

19. Blesse, *Check Six*, 125; and *Project Red Baron II*, 10. The relative decline in aircrew combat experience cited by Blesse was also noted in the *Red Baron II* report: "The average experience level of aircrew during the early part of the war was relatively high. By the end of the USAF air-to-air activity in March 1968, this experience level had been sharply lowered.... This lessening of experience resulted from the replacement of 'old heads' by recent UPT [undergraduate pilot training] graduates and pilots with ADC [Air Defense Command] and ATC [Air Training Command] backgrounds, or otherwise lacking tactical aircraft experience."

20. Ibid.; and "Historical Data Record," 2. Blesse's recollections are confirmed by the 366th TFW's monthly historical report for the period 1 April 1967 to 1 May 1967: "Commander[']s Conclusion: The bombing of MiG airfields in and around Hanoi has brought the MiGs up in force and confronted this Wing with a new and interesting mission. Intense training, individual squadron briefings on air-to-air fighter tactics by the Director of Operations [Blesse], and

several new ideas to improve the Wing's air-to-air capability all have played their part in changing the personality of the Wing."

21. Blesse, *Check Six*, 120. During one interview, Blesse characterized the limitations of air-to-air missiles: "Show me a missile that is good and I will throw my guns away, but I have not seen any good ones yet. They still require about 1,500 feet, just to arm. I am not interested in something in that range. I don't want a dead area, dead range in there." Blesse, oral history interview, 79.

22. Blesse, *Check Six*, 120.

23. Bakke, interview.

24. Blesse, *Check Six*, 120. Blesse named the following individuals as having "earned their pay and then some experimenting with the gun: Lt Col Fred Haeffner, Majors Jerry Robbinette, Ed Lipsey, Bob Dilger, Sam Bakke, and Captains Bob Novak, Skip Cox, Jim Craig." Of the eight officers recognized by Blesse, four would claim a MiG during the tour; one (Craig) used Blesse's innovation in doing so. Futrell et al., *Aces and Aerial Victories*, 127–31 and 157.

25. During the premission aerial refueling, all of the F-4's fuel tanks, both internal and external, would be filled. En route to the target, the F-4 would burn the fuel in the external tanks first. That way, if the F-4 needed to engage a MiG, the aircrew could jettison the empty (or almost empty) external fuel tanks and still have sufficient internal fuel to fight the MiG before returning home, via a postmission aerial refueling. Less external fuel capacity resulted in the F-4 burning its limited internal fuel supply earlier in the mission, consequently decreasing its available on-station time in hostile airspace.

26. General Olds, oral history interview, 22. Olds cites the 190-pound weight of the ECM pod.

27. Bolt, oral history interview, 195.

28. Olds described the configuration's effect on takeoff: "Now that was a lovely little takeoff configuration, like, maybe, full right rudder as soon as you broke ground." General Olds, oral history interview, 22 and 74–75.

29. Colonel Olds, oral history interview, 43.

30. History, 366th Tactical Fighter Wing, vol. 1, 2.

31. Bolt, oral history interview, 195–97; General Olds, oral history interview, 76; and Bakke, interview. Bolt's recollection of the events is suspect. Seventeen years after his assignment to the 366th, General Bolt claimed almost exclusive credit for the wing's tactical innovations: "I went down to the armament shop and I said, 'Chief, we've got to do something about this configuration on the airplane. Do you think you can rig up a wiring harness where we can put that ECM pod on the left or right inboard station and drop off two of the Sidewinders so we will just have two Sidewinders on the other side and still have the radar-guided missiles but will have the two outboard tanks? Then we don't have to fool with that big old centerline tank.' . . . He said, 'I don't know; let me try. I haven't thought about it.' I said, 'See if you can do it. I'll check back with you later in the day.' I went back down that afternoon. He said, 'I got one made; it will work.' All he did was build a simple harness with two cannon plugs on it and tie it in to the nuclear armament system, which the regulations say you can't touch, so don't ask anybody—just go ahead and do it [*sic*]. If you ask anybody, all they can tell you is no. We put that thing on. I said, 'Okay, take the airplane and configure it. I will fly it tomorrow against our radar sites . . . and we will see if it works like it is supposed to.' So I flew it and it worked great, so we reconfigured all our airplanes. We only had two Sidewinders rather than the four. I said, 'We've got to counter that someway. The only way I know to do it is to put the gun on the centerline.' We got the airplane configured right now [*sic*], and the gun on the centerline will be all right.

AN INTERIM SOLUTION

The gunsight left a lot to be desired in that F-4C. . . . I said, 'I think if we can get behind the MiGs, and we can, we don't really need a gun sight. We can hit him. We can put every tenth round with tracers in there. We can hit him without a gun sight.'" While flattering, the interview's narrative does not agree with Blesse's own narrative in *Check Six*, the 366th Wing's official history from the period in question, or the secondary sources such as Davies and Thornborough. A review of the message traffic also reveals discrepancies and lends more credence to Blesse's narrative. (Even Blesse's narrative, though, contains inaccuracies; see note 42, this chapter, and note 18 in Chapter 1.) This historical interpretation was affirmed during an interview with Bakke, one of the officers in the 366th TFW weapons section in 1967, who stated, "If it was anybody [who deserves credit for putting the SUU-16 on the F-4 for air-to-air], . . . I'd give . . . 100 percent of the credit to 'Boots' Blesse." Olds's after-the-fact recollection of the events is also suspect, as he failed to credit the 366th TFW for devising the wiring solution: "So, it was because of this dadgone [ECM] pod that we were having trouble with the gun and the reason is because you had to hang the gun on the centerline, see. But we had to put the 600-gallon tank on the centerline. It was a mess, so finally we rewired the airplane using the nuke circuitry that's in the bird and were able to put power to the pod on the right inboard pylon, then put the tank back on the right outboard, . . . and then you could hang the gun pod or bombs [on the centerline]." Still, Bolt's self-aggrandizing interpretation of the events possesses some merit because he succinctly and accurately described the technical solution.

32. Bakke, interview. According to Bakke, this was the other major hurdle to gaining pilot acceptance of the pod's combat utility.

33. History, 366th Tactical Fighter Wing, vol. 1, 4; and Dilger, oral history interview, 12. Dilger, one of Blesse's troops working in the 366th tactics section, noted that the new configuration "was capable of out-flying our all-missile configuration." Eventually, the wing resumed loading all four Sparrow missiles on the aircraft, even with the SUU-16 gun pod, as based on aircraft configuration data reported in AFHRA, "1967-22 May; Titus and Zimer." From the outset, the 8th TFW loaded four AIM-7s with the SUU-16/23 gun pod; their December 1967 "Tactical Doctrine" manual listed the "normal fragged configuration of the F-4 performing escort and/or 'sweep' missions" as "4 x AIM-7 Sparrows; 4 x AIM-9 Sidewinders or 3 x AIM-4 Falcons; 1 x QRC-160 or ALQ-71 ECM Pod; 600 Gal Centerline fuel tank or SUU-16/23 Gun Pod; 2 x 370 Gal Outboard fuel tanks." 8th TFW, "Tactical Doctrine," December 1967, 106. The aircraft configuration from the 5 June 1967 engagement verifies this standard 8th TFW configuration, as did former 8th TFW F-4 pilot, Lt Col (ret.) Darrell "D" Simmonds. AFHRA, "1967-5 June; Raspberry and Gullick"; and Simmonds, interview.

34. Although each wingman carried eight missiles, they were rarely able to employ them during combat. The Air Force flew a fluid four/fighting wing formation that assigned primary responsibility for shooting down enemy fighters to the flight or element lead and relegated the wingmen to simply covering the lead aircraft. In contrast, the Navy flew a loose deuce formation that afforded the wingman greater freedom of action and a shared responsibility for offensive missile employment. Michel, *Clashes*, 169–72.

35. Blesse, *Check Six*, 121.

36. Blesse offered his opinion as to how the F-4C came to be manufactured without an adequate gunsight: "After the extremely capable A-1C radar computing gun sight we used 15 years earlier in Korea, it was difficult to understand how we could find ourselves in this situation. Fuzzy thinkers were sure guns no longer were useful in combat, and in some cases even had them removed from the aircraft and destroyed. It was a disease. They pulled guns out of the F-86F and F-104 to name a couple and—what was worse—left them out of new aircraft in the

design stage. Without guns, who needs a gun sight—and that's how our predicament came about." Ibid.

37. Colonel Olds, oral history interview, 42.
38. Bugos, *Engineering the F-4 Phantom II*, 157; and Michel, *Clashes*, 103.
39. Blesse, *Check Six*, 123. Olds's less than enthusiastic response contradicts his previously cited opinion of the gun. Davies noted that in a later personal interview with Olds, "his reservations still held." Davies then quoted Olds at length: "The gun pod wasn't so much a speed penalty as an object of increased drag, and therefore increased fuel consumption. But that wasn't my objection to the gun pod. I refused to carry it for three basic reasons: 1) It took the place of five or six 750-pound bombs; 2) Only my older and more experienced fighter pilots had ever been trained in aerial gunnery, to say nothing of air-to-air fighting. There were perhaps a dozen of them in the 8th TFW; 3) I had no intention of giving any of my young pilots the temptation to go charging off to engage MiG-17s with a gun. They would have been eaten alive. Instead, they fought the MiGs the way I taught them, and I might say they did so with notable success. They learned that there were times to fight and there were times to go home and come back the next day." Davies, *USAF F-4 Phantom II*, 31. Olds's trepidations regarding inadequate pilot training were also evident in his memoir: "I really had to argue with myself about my own desire to carry a gun. I knew I could hit anything I shot at but was damned sure I didn't want to tempt my men to engage a MiG-17 in an old-fashioned dogfight or give them the urge to go down in the mountain passes in Laos to strafe a stupid truck. In either case, I would have lost bunches of them. We needed guns, no doubt about it, but we needed pilots trained to use them even more." Olds, *Fighter Pilot*, 317.
40. Message, C10697, 366 TFW to 7 AF (emphasis added). "Package Six" referred to route package six. See note 2 in chapter 1 for a brief discussion of the Route Pack arrangement in Vietnam.
41. Message, DCO00148, 366 TFW to 7 AF.
42. Blesse, *Check Six*, 123; and Michel, *Clashes*, 104. Blesse stated that the mission took place "around the first week in May." However, other sources, including the previously cited message traffic, suggest that the mission actually occurred on 12 May 1967. Michel came to a similar conclusion, "The podded cannon finally was brought into combat on May 12." The SUU-16 project was well received within the wing. In his squadron monthly history report, 1st Lt John Frazier of the 390th TFS reported, "During May, 390th aircraft were the first in the wing to be modified to carry the ECM pod on an inboard . . . pylon, . . . thus giving the F-4C the capability of carrying two external 370[-gallon] wing tanks as well as a SUU-16 20-mm cannon on the centerline station. This has been a much looked for modification." Frazier, "Historical Summary."
43. AFHRA, "1967-14 May; Hargrove and DeMuth." Carey and Dothard in Speedo 2 did not make it back to Da Nang with the rest of their flight; they ran low on gas, missed their aerial refueling rendezvous, and had to recover at Nakhon Phanom Royal Thai AFB.
44. Ibid.; AFHRA, "1967-14 May; Craig and Talley."
45. Message, 141800Z MAY 67, 366 TFW to NMCC. The message describing Elgin flight's engagement concluded with a similar assessment, "ELGIN Lead, [#]2 & [#]4 all feel that the SUU-16 would have been much more effective against the MiG-17s than any of their missiles." Message, 141410Z May 67, 366 TFW to NMCC.
46. Hargrove, oral history interview, 22–23. A "break" is an aggressive defensive turn away from an attacking fighter.
47. History, 366th Tactical Fighter Wing, vol. 1, 5.

48. PACAF Command Center, Chronological Log, 13-14 May 1967. It was later assessed that Elgin flight faced 10 MiGs that afternoon, not the six originally reported. See note 7 in Chapter 1.

49. Ibid.

50. PACAF Command Center, Chronological Log, 14–15 May 1967. Recall that Elgin 1 was not armed with the SUU-16 gun pod because it was a spare aircraft. See note 4 in chapter 1.

51. Ibid.

52. Ibid., 2.

53. Blesse, *Check Six*, 125. See note 18 in chapter 1 regarding the daily operational summary report's wording. The actual message, not Blesse's recollection, has been cited.

54. As the first in-theater wing to accept the new F-4Ds equipped with the lead-computing gunsight, the 8th TFW did not have to rely on the 366th's primitive aiming techniques.

55. Futrell et al., *Aces and Aerial Victories*, 120–21. Maj William Kirk and 1st Lt Theodore Bongartz flew the F-4D. Their prize was a MiG-21, Kirk's second kill of the war.

56. Kirk, oral history interview, 6. Although he quickly added, "It's too bad it's not internal. It's too bad we have to hang it externally; it's extra drag, extra weight, but we're willing to sacrifice that." The gun pod's air-to-air combat utility was solidified in the minds of the 8th TFW pilots on 6 November 1967 when Capt Darrell "D" Simmonds and 1st Lt George McKinney, Jr., downed two MiG-17s with the pod in less than two minutes. During the engagement, Simmonds, leading Sapphire flight, expended fewer than 500 rounds while destroying the two MiGs. During the second gunshot, Simmonds closed to within 400 feet of the MiG before it exploded. Too close to maneuver away from the disintegrating MiG, Simmonds ended up flying through the fireball. By the time of Simmonds's engagement, there were enough gun pods to equip all the F-4D escort fighters with SUU-23 gun pods. AFHRA, "1967-6 November; Simmonds and McKinney."

57. Message, DCO00157, 366 TFW to CSAF.

58. The SECDEF briefing transcript read, "Our low altitude capability was improved by a field modification here at Da Nang. This modification allowed us the carriage of the 20-mm cannon on cover missions. The gun has exceeded our expectations. We fly all missions to Package VI, including strike missions, with the gun." Responding to one of McNamara's questions, Blesse noted that the gun "is one of our most versatile and effective weapons, air-to-ground and air-to-air, in spite of the lack of a computing sight capability." Message, 191633Z JUL 67, CSAF to PACAF.

59. Blesse, *Check Six*, 126.

60. History, 366th Tactical Fighter Wing, vol. 1, 4.

61. Futrell et al., *Aces and Aerial Victories*, 118–22; Blesse, *Check Six*, 125; and Dilger, oral history interview, 11–13. During that period, the 366th shot down five aircraft with AIM-7s, three aircraft with AIM-9s, and four aircraft with the 20 mm gun and caused one MiG to crash while trying to defend itself against a 366th F-4 attack. Dilger described the 366th's combat results with the gun during a 6 July 1967 interview: "The gun right now has been engaged five different times, and we shot down four MiGs with it. Four out of five is a very good answer, and the man that missed was not—he's never fired a gun before in his life, and his tactics were so gross that you couldn't expect the gun to have done him a good job. But the people that employed the gun properly did very well with it."

62. Futrell et al., *Aces and Aerial Victories*, 157. An F-4D and F-105F shared one kill.

63. Blesse, *Check Six*, 123. Recall that Olds warned Momyer not to allow Blesse and the 366th TFW to continue pursuing employment of the SUU-16 in an air-to-air role. See note 39 this chapter for Olds's later-stated rationale.

64. Olds, like others, would have preferred to have an internal gun. He added, "It's beautiful but still it's an appendage. No fighter should be built without a gun in it. That's basic and then anything else you can add is just Jim Dandy with me." General Olds, oral history interview, 76.

65. Colonel Olds, oral history interview, 42–43.

66. *Project Red Baron II*, 19.

67. Momyer, *Airpower in Three Wars*, 158; and Crane, *American Airpower Strategy in Korea*, 60. Crane cites Air Force major general Emmett O'Donnell's 1951 Congressional testimony, "I think this is a rather bizarre war out there, and I think we can learn an awful lot of bad habits in it." One of the political restrictions which Momyer was referring to was the requirement "for positive visual identification before the pilot could open fire." Momyer more clearly articulated his concern in an earlier Corona Harvest memorandum: "We should, therefore, be cautious about the lessons derived from these limited combats [in Vietnam]. Most certainly, relative performance of aircraft could be judged and restricted conclusions on air-to-air tactics could be deduced, but one should not try to extrapolate these limited experiences in generalizing about the character of an air war in Europe where thousands of fighters would be involved." Momyer to Ellis, "Corona Harvest (Out-Country Air Operations)," 4. Corona Harvest was the Air Force's and AU's comprehensive study of airpower in SEA. Momyer, then retired, was hired in April 1974 as a paid consultant by the vice chief of staff of the Air Force, Gen Richard Ellis, to review the studies and to provide summary memorandums and recommendations for each of the Corona Harvest volumes. Cunningham, "Spike," 123–24. Hone agreed with Momyer's assessment of the political restrictions, although Hone offered a better appraisal of their effect on the air war during Vietnam. Hone emphasized, "The conflict [in the skies over North Vietnam] was between two systems, one of which was hampered by politically motivated constraints. In North Vietnam, the Vietnamese constructed a multifaceted, mutually supporting system of air defense. The burden was on US forces to penetrate it because they were never allowed to totally destroy it. . . . Vietnam was not a conflict of fighter-on-fighter but of offensive systems against defensive systems." Hone, "Southeast Asia," 555.

68. Momyer to Ellis, "Corona Harvest (USAF Operations against North Vietnam)," 4.

69. Momyer continued, "To accomplish this requirement, airborne radar will require extensive improvement in range, resolution, and reliability." Momyer to Ellis, "Corona Harvest (Out-Country Air Operations)," 24.

70. Momyer to Ellis, "Corona Harvest (USAF Personnel Activities)," 4.

71. *Project Red Baron II*, 17. The report also noted that "as the war progressed, the USAF aircrew population with prior tactical experience was diluted over 50 percent; the average in-aircraft time also decreased by a similar proportion. Conversely, the enemy's tactical experience level most probably increased over time. As a result, the USAF loss rate went up, while the NVN's went down; that is, 3.0 MiGs lost per USAF aircraft lost, decreased to 0.85 MiGs per each USAF loss."

72. Ibid., 21; Hannah, *Striving for Air Superiority*; and Skinner, *Red Flag*. The *Red Baron II* report continued, "Aircrews must have extensive initial and continuation ACM training. This training should include instruction on enemy capabilities and limitations. Realistic training can be gained only through thorough study of, and actual engagements with, possessed enemy aircraft or realistic substitutes."

AN INTERIM SOLUTION

73. The F-22A uses a lighter-weight version of GE's M-61 20 mm Vulcan cannon, the same gun built into the F-4E, the F-15, and the F-16. The Air Force version of the F-35 joint strike fighter (JSF) will sport the Air Force's first new fighter-gun design in almost 50 years. The Navy's version of the F-35, however, does not carry an internal gun.

Chapter 7

Military Innovation

C'est l'ancien qui nous empêche de connaître le nouveau.
—Auguste Comte

The human tendency to focus on singular concepts—old or new, intellectual or technological—often obfuscates the broad perspective critical to recognizing evolving strategic contexts. It also impedes timely and innovative adaptation to an emerging situation. While not necessarily more susceptible to this tendency than other institutions, the American military is nevertheless affected more profoundly by it, particularly within the technological realm. Countless volumes have probed the nature of military innovation, seeking a better description of it so that leaders can cultivate a more responsive and flexible organization ready to adapt to the ever-changing conditions of war. Extending the theory of technological dislocations and the Air Force air-to-air armament case study to the larger context of military innovation aids this endeavor.

The Role of Cognitive Consistency

In *Strategy in the Missile Age*, Bernard Brodie chided 1950s' American defense officials' narrow-minded approach to national strategy in the emerging thermonuclear age. Identifying the undue influence of an "intellectual and emotional framework largely molded in the past," Brodie noted that the American military profession was not only unwilling, but also largely unable to comprehend that the proliferation of nuclear weapons rendered many of their hallowed principles of war obsolete and irrelevant.[1] Brodie observed, "We have been forced to revise our thinking about weapons; but unfortunately there is not a comparable urgency about rethinking the basic postulates upon which we have erected our current military structure, which in fact represents in large measure an ongoing commitment to judgments and decisions of the past."[2] Based on his assessment, Brodie called upon August Comte's adage, *"C'est l'ancien qui nous empêche de connaître le nouveau"* (It is the old that prevents us from recognizing the new).[3]

History, however, demonstrates that the reciprocal of Comte's adage can also be true—the *new* can sometimes prevent us from recognizing the *old*. David Edgerton alluded to this phenomenon in his description of "use-based history," noting that the history of technology is often written as though there were no alternatives to a given technology. This dominant perspective ignores the reality that "there is more than one way to skin a cat, to fight a war, to generate energy. Yet, these alternatives are often difficult to imagine, even when they exist."[4] Fascination with technology and a generally uncritical "assumption that the new is clearly superior" often skews judgment of an emerging technology's feasibility and practicality.[5] For example, Edgerton noted that Hitler's obsession with developing the technologically advanced V-2 rocket drained valuable German resources from more practical and potentially more fruitful wartime enterprises.[6] A similar pattern was revealed in the previous case study when the American Air Force's fascination with guided air-to-air missile technology biased its assessment of the combat utility of guns on future fighter aircraft.

Robert Jervis explored these limitations of human cognition within the strategic realm. He noted that an individual's desire to maintain *cognitive consistency* leads to a "strong tendency for people to see what they expect to see and to assimilate incoming information to preexisting images."[7] Whereas this pattern of obstinacy is not new to human history, Jervis was unique in his assertion that this "closed-mindedness and cognitive distortion" takes place at the decision maker's subconscious level.[8] Furthermore, not only does the desire for cognitive consistency restrict individuals to usually studying at most "only one or two salient values" when formulating a strategy, it also entices decision makers to continue pursuing a particular strategy despite evidence that may suggest the policy is ill-conceived and inappropriate.[9] Jervis concluded, "Expectations create predispositions that lead actors to notice certain things and to neglect others, to immediately and often unconsciously draw certain inferences from what is noticed, and to find it difficult to consider alternatives."[10] These inflexible schemas, whether focused on the old or the new, manifest as an inability to effectively innovate.[11]

This tendency is especially pronounced in military organizations. Citing Dean Pruitt, Jervis noted that the more extreme the perceived significance of a schema, the less flexible it becomes. Commitment—"the degree to which [a] way of seeing the world has proved satisfactory and has become internalized"—plays an important role when matters of national security, and consequently choices of life and death, are considered.[12] Moreover, because the real-world opportunities for the military's schemas to be tested are fortunately infrequent, the organization's commitment to its schemas tends to become institutionalized within military culture.[13] Historian Michael Howard described the military's plight: "For the most part, you have to sail on in a fog of peace until the last moment. Then probably, when it is too late, the clouds lift and there is land immediately ahead; breakers, probably, and rocks. . . . Such are the problems presented by 'an age of peace.' "[14]

> Bureaucratic norms and the human need to maintain cognitive consistency, exacerbated by the high stakes associated with national security and the relatively rare data set made available by active warfare, reveal themselves in the dialectical perception of both the technological exuberance and technological skepticism of the American military. Dominant technologies are embraced while alternative technologies, especially revolutionary ones, are shunned. Initially, the Navy preferred battleships to aircraft carriers; the Army preferred cavalry to tanks and aircraft, and single-firing rifles to machine guns; and the Air Force preferred bombers to ICBMs and manned aircraft to unmanned aircraft.[15]

However, the dominant technology need not be the old technology. Indeed, as the preceding case study illustrated, the introduction of a proven-but-assumed-antiquated technology like the air-to-air cannon can also be greeted with technological skepticism. A technological innovation need only diverge from the established technological trajectory to draw the wary eye of the constituency it potentially threatens.

Technological Innovation as Military Innovation

Technological innovation does not always equate with *military* innovation. As Brodie observed, technological innovations in aircraft-delivered nuclear weapons did not induce a corresponding and necessary innovation in American military strategy; military leaders simply incorporated the new *means* into the same *ways* and *ends* equation.[16] Nevertheless, while technology clearly does not dictate strategy, a complex interdependent relationship exists between the two.

Reflecting on this link between technology and military strategy, Colin Gray noted, "Technology, as weaponry or as equipment in support of weaponry, does not determine the outbreak, course, and outcome of conflicts, but it constitutes an important dimension [of strategy]."[17] Howard drew a similar conclusion. Reminiscent of Carl von Clausewitz's trinity of war, Howard believed that strategy "progresses . . . by a sort of triangular dialogue between three elements in a military bureaucracy: operational requirement, technological feasibility, and financial capability."[18] Similarly observing the role of technology and finances within strategy, Brodie asserted, "Strategy in peacetime is expressed largely in choices among weapons systems . . . [and] the military budget is always the major and omnipresent constraint."[19] Jervis likewise acknowledged the strong influence technology can have on military strategy: "The adoption of one weapon . . . often requires changes in other weapons, in tactics, and—in some cases—in strategies and interests."[20] These interpretations all support the assessment that technology and strategy are somehow linked, but the disparity between the individual observations suggests the linkage is amorphous, bound in historical context, and not easily discernible.

For example, one scholar relied heavily on Jervis and organizational theory to support his view of military innovation. Barry Posen observed that "innovations in military doctrine will be rare because they increase operational uncertainty." Posen purported that two powerful catalysts could nevertheless force the military to adapt: military defeat and civilian intervention. Furthermore, he observed that the two catalysts were linked: "Failure and civilian intervention go hand in hand. Soldiers fail; civilians get angry and scared; pressure is put on the military."[21] However, due to their relative unfamiliarity with military doctrine, civilians usually required a military compatriot to provide the necessary specialized knowledge—a "maverick" like Billy Mitchell, Hyman Rickover, or "Bony" Fuller.[22]

While Posen's research was clearly focused on innovation at the doctrinal level, his evidentiary base established a clear link between technological innovation and military innovation. For example, Posen cited the British air defense system of 1940 as "one of the most remarkable and successful military innovations of the pre-atomic machine age."[23] However, whereas this British military innovation was obviously reliant upon the coupling of technological developments in radar and fighter aircraft, the key innovation catalyst according to Posen was the timely intervention of a civilian-military maverick team composed of Henry Tizard, Thomas Inskip, Prime Minister Stanley Baldwin, and Air Chief Marshall Sir Hugh Dowding.[24] The team, cognizant of the changing strategic context of the 1930s when others were not, forced the Royal Air Force to shift its focus from procuring offensive strategic bombers to developing the Chain Home radar system and the corresponding fighter defenses that later proved invaluable during the Battle of Britain.[25]

Posen therefore urged future civilian leaders to actively engage with the military in matters of strategy: "Civilians must carefully audit the doctrines of their military organizations to ensure that they stress the appropriate type of military operations, reconcile political ends with military means, and change with political circumstances and technological developments."[26] Without this civilian intervention, Posen claimed that the military bureaucracy would prefer "predictability, stability, and certainty" over innovation, at potential great cost to national security.[27]

Writing seven years after Posen, Stephen Rosen offered a different assessment of military innovation. While both Posen and Rosen agreed on the importance of developing an appreciation for changes within the strategic environment and overcoming bureaucratic resistance, Rosen vehemently disagreed with the primacy Posen granted to civilian intervention, even labeling Posen's theory a "*deus ex machina*."[28] Rosen viewed the process of military innovation as being far more complex, and consequently, he elected to parse innovation into three more manageable subsets: peacetime, wartime, and technological. Identifying different operative mechanisms within each category, Rosen determined:

> Peacetime innovation has been possible when senior military officers with traditional credentials, reacting not to intelligence about the enemy but to a structural change in the security environment, have acted to create a new promotion pathway for junior officers practicing a new way of war.
>
> Wartime innovation, as opposed to reform, has been most effective when associated with a redefinition of the measures of strategic effectiveness employed by the military organization, and it has generally been limited by the difficulties connected with wartime learning and organizational change, especially with regard to time constraints.
>
> Technological innovation was not closely linked with either intelligence about the enemy, though such intelligence has been extremely useful when available, or with reliable projections of the cost and utility of alternative technologies. Rather, the problems of choosing new technologies seem to have been best handled when treated as a matter of managing uncertainty.[29]

Rosen's catalysts share one common attribute—all require a keen perception of the evolving strategic context. Whether adapting to "a structural change in the security environment," new measures to assess "strategic effectiveness," or technologies pursued to help mitigate uncertainty within the changing strategic context, all of Rosen's mechanisms are hobbled by the frequently obstinate nature of bureaucracies and individuals' search for cognitive consistency.

Other scholars treating military innovation have typically offered variations on the above themes. Owen Coté, Jr., suggested that interservice conflict "can act alone and independently to cause innovative military doctrine."[30] John Nagl focused his research on the military's organizational culture, concluding that an "institutional learning" environment was key to successful innovation, especially during wartime.[31] Barry Watts and Williamson Murray borrowed heavily from Rosen when they concluded, "Without the emergence of bureaucratic acceptance by senior *military* leaders, including adequate funding for new enterprises and viable career paths to attract bright officers, it is difficult, if not impossible, for new ways of fighting to take root within existing military institutions."[32] Allan Millet's study of innovation during the interwar period successfully linked Posen's "civilian intervention," Rosen's "measures of strategic effectiveness," and Coté's "interservice conflict" into a single assessment: "History . . . does demonstrate a relationship between strategic net assessment and changes in military capability. . . . [It] demonstrates the importance of civilian participation in the process of change at two levels, political and technological. Both levels of interaction are important, not the least because they compensate for interservice and intraservice friction. Innovators need allies in the civilian political and technological establishments as well as patrons within their service."[33]

Howard offered his own assessment of military innovation in 1973, more than a decade before Posen published his study. Indeed, Posen's argument seems a reflection of Howard's earlier observation: "One may need a dynamic force of exceptional quality administered from outside the profession to cut through . . . arguments, and with a possible irrational determination, give the order 'You *will* do this.'" Howard also foresaw the potential negative effects technological and bureaucratic complexity would have on innovation. He continued: "It becomes increasingly difficult as warfare becomes more complex, as the bureaucracy becomes more dense, as the problems become harder, for anybody to credibly emerge and impose his will on the debate in this basically irrational manner. Thus, as military science develops, innovation tends to be more difficult rather than less."[34] Howard's observation affirms the critical role knowledgeable and credible individuals play in spurring innovation and, if necessary, disrupting the established technological trajectory.[35] These individuals are well suited to effect technological dislocations.

Technological Dislocations

Critics may contend that the preceding air-to-air case study is too narrowly focused and the innovation too minor to derive worthwhile conclusions regarding the nature of technological innovation, much less military innovation. True, air-to-air gun technology existed on other Air Force aircraft, and rather than threatening the Air Force's pilot constituency, the F-4/SUU-16 technology in effect bolstered the idolization of heroic pilots who generations earlier valiantly dueled over the western front. In addition, the innovation, being relatively inexpensive and requiring little modification to the existing aircraft, did not demand significant capital or resource expenditure. For all these reasons, adding a gun to the Phantom should have been a relatively simple task; even if the bureaucracy was not eager to adopt the innovation, it should have at most been indifferent to it. It was not. The addition of an air-to-air gun on the F-4C was opposed by not only the corporate bureaucracy, but also by many of the practitioners themselves, including combat veterans like Olds. Why?

The paradigm and resultant technological trajectory that shaped this Air Force attitude can be traced back to the first experimental Tiamat guided missile launched in the closing days of World War II. Despite the missile's failure to meet expectations, the Air Force quickly became enamored with the prospect of arming its newest, high-speed jet fighters with advanced, radar- and infrared-guided air-to-air missiles. Initially, the nascent technology had its share of skeptics within the bureaucracy. Facing severe reductions in the postwar defense budget, Air Force officials slashed initial missile funding in favor of the Air Force's higher-priority strategic bomber fleet. While there were some rare missile successes that helped soften bureaucratic resistance, the skepticism that threatened the early missile programs was largely overcome only when the missile proponents linked their technology to the Air Force's dominant strategic assumption and its organizational self-image.

The Air Force of the 1950s marketed itself as *the* technologically minded service. Armed with impressive fleets of high-flying bomber aircraft, the Air Force promised to deliver the newest products of the nation's technological wizardry—its growing nuclear arsenal—on the Soviet Union the moment the president gave the order. However, this vision of future war also required the Air Force to prepare to thwart any Soviet attempts to deliver the same. Within this strategic context and persuaded by the incontrovertible laws of intercept geometry, as well as the ceaseless demand for ever-greater firepower, the Air Force demanded better and faster fighters with longer-range and more destructive armament that could quickly dispatch the Soviet hordes.[36] It demanded guided air-to-air missiles.

As Soviet bomber aircraft capabilities rapidly improved during the 1950s and 1960s, the Air Force responded in kind. American F-86s gave way to F-102s and F-106s, the last of which was capable of sprinting at greater than Mach 2 to intercept Soviet bombers flying nearly 10 miles high. During this period, fighter and air-to-air missile development fell into a rut that channeled future acquisitions in an unchallenged and nearly autonomous fashion. There were improvements in missile design—GAR-1s gave way to GAR-1Ds, then GAR-3s; and GAR-2s eventually transitioned to GAR-4s—but the technological paradigm and the resultant technological trajectory constrained revolutionary, innovative thinking. Incremental technical progress substituted for a conscious evaluation of the evolving strategic context, thereby reinforcing a self-deluding perception that American technological prowess would dominate future conflict. Few Air Force leaders questioned the assumption that the fighters and their missiles would only be required to destroy large, high-flying, nonmaneuvering Soviet bomber aircraft. Even when the assumption proved invalid in the skies over Korea, the demand for cognitive consistency allowed the Air Force to disregard its tactical air-to-air experience in favor of its preferred strategy and its dominant technological trajectory.

Compounding matters, as the missiles spread through the Air Force in the decade prior to Vietnam, technological skepticism gave way to overconfidence and technological exuberance. Lackluster test performance, even against the narrowly focused, bomber-aircraft target set, did not dissuade Air Force leaders from equipping their newest fighter interceptors exclusively with missiles. Guns were seen as archaic, and the methods and techniques for employing them were considered irrelevant in future air combat that would be characterized by long-range missile attacks against unsuspecting enemy aircraft. As such, many senior Air Force leaders deemed continued air combat maneuvering training unsafe and an unnecessary risk to Air Force aircraft. Subjected to a bureaucracy enamored with the promises of missile technology and captivated by its strategic assumptions, pilots' dogfighting skills quickly atrophied.

When the glaring deficiencies in American air combat capability were finally realized in the opening months of Vietnam, the Air Force scrambled to develop technological solutions. It launched numerous studies, including the 1966 Heat Treat Team, but no viable solution readily emerged.[37] The technological paradigm that contributed to many of the deficiencies continued to dominate Air Force thinking; proposed solutions such as the AIM-4D Falcon and the AIM-7E-2 Dogfight Sparrow largely conformed to the already-established technological trajectory. Unfortunately, like their predecessors, the new weapons arrived late and failed to live up to the overhyped expectations. When the Air Force finally broke free from its technological rut and recognized the complementary value of a gun on a fighter, aircrews were instructed to wait patiently for the F-4E.

For Blesse at Da Nang in April 1967, that was unacceptable. Luckily, he benefited from Catledge's earlier advocacy of the SUU-16 podded gun system. Although Catledge desired an F-4 air-to-air gun capability, his decision to instead market his podded gun solution as an air-to-ground weapon successfully avoided the ire of the air-to-air missile mafia that dominated the Air Force's requirements cadre. He believed that continued manufacturing of the gun, even in podded form, would ensure that it could one day be resurrected in an air-to-air role when conditions demanded. Without Catledge's tireless advocacy and ingenious work-around solution, Blesse would have lacked the critical tool necessary to introduce his technological dislocation.

As a heterogeneous engineer, Blesse proved adept at integrating assumed-disparate components into a practical solution.[38] His ad hoc innovation combining the F-4C and the SUU-16 gun pod for air-to-air combat against the North Vietnamese MiGs was in many ways a precursor to today's concept of "recombinative technology."[39] By utilizing off-the-shelf technologies and integrating them in an unforeseen way and with a minimal level of effort, Blesse was able to leverage existing technologies to fill a capabilities void. Shortfalls in the integration, such as the lack of a lead-computing gunsight in the F-4C, were identified, and procedures were developed to mitigate the negative effects. Blesse's cobbled-together F-4/SUU-16 weapons system was not a perfect solution; the F-4E was a better one. However, Blesse's innovation provided a low-cost, effective, and, most importantly, timely solution that the F-4E could not offer.

The story of Blesse and the 366th TFW's mating of the SUU-16 gun pod to the F-4 for air-to-air combat highlights the significant potential of unit-initiated tactical innovation. Granted, Blesse's innovation did not affect the strategic outcome of the Vietnam War, but it did have a dramatic impact on the Air Force's culture, acquisition requirements, and operations well into the twenty-first century. All Air Force fighter aircraft since the Vietnam War have been equipped with both missiles and guns, and today's Air Force fighter pilots routinely practice their dog-fighting skills.

Blesse's innovation also demonstrates the fragility of innovation born at the unit level. Certainly, Blesse's renowned credibility as a tactician and a Korean War ace helped disarm his commanders' skepticism. However, if the Da Nang wing commander, Bolt, or the Seventh Air Force commander, Momyer, had deemed Blesse's project too risky to personnel, equipment, or reputation, they could have simply ordered the project to be abandoned. Blesse would have had little recourse. Surprisingly, had Olds been in command, the program probably would have been terminated.

Therefore, Blesse's technological innovation aptly illustrates the important role that commanders, even those at a relatively low level, play in military and technological innovation. By nature of the military hierarchy, these individuals exert considerable influence on the military's ability to innovate. Their significance is magnified by the fact that the individuals least likely to be gripped by the dominant technological paradigm and thus more open to investigating alternatives typically reside at the lower ranks. But, because bypassing the chain of command is frowned upon, a single supervisor can sound the death knell for an otherwise promising innovation. As Jervis pointed out in his discussion of cognitive consistency, the supervisor's decision need not even be malicious.[40]

The standard military response in these situations has been to wait out the opposition, knowing that eventually all commanders move on or retire. However, waiting can complicate matters as it gives more time for the existing technology to build momentum and the bureaucracy to become even more resistant to change. Catledge's method of disarming the opposition by masking the true intention of the innovation provides one strategy, albeit an ethically questionable one, for innovating in spite of bureaucratic resistance.

The historical case study of the F-4-gun system also affirms the difficulty in identifying a discrete tipping point and its causal factors in a complex technological system befuddled by competing historical interpretations. A strong case can be made that efforts to reintroduce guns to fighter combat reached a tipping point at Da Nang in 1967. However, when dissecting the historical evidence, identifying a *single* causal factor that led to the tipping point is too reductionist and woefully inadequate. While Blesse stands at the forefront, Catledge was also certainly integral to the innovation; without his efforts, guns might not have been ready for the F-4E, and a podded gun would certainly not have been ready for the F-4C/D. Additionally, a variety of other social influences prodigiously aligned themselves at Da Nang in April and May 1967—for example, arrival of the SUU-16 gun pods, President Johnson's decision to attack the more valuable NVN targets, the consequent surge in MiG activity, the decision to assign additional MiGCAP sorties to the 366th TFW, and a receptive Momyer. All contributed to the dislocation in one way or another.

Thus, like Schriever with the American ICBM, Blesse shares credit for his innovation with others. But, also akin to Schriever's role in ICBM development, it was Blesse's unique credibility and his heterogeneous engineering skills that allowed him to associate these varied influences into a practical solution. In doing so, Blesse successfully introduced a socially constructed dislocation, disrupting the deterministic technological trajectory that, for more than two decades, had been constraining Air Force air-to-air armament design.

The preceding case study did not validate the individual innovation catalysts as described by Posen, Rosen, or Coté. Although some might consider Blesse a military maverick based on his unwavering zeal for guns, Blesse's innovation did not require his pairing with a civilian official to garner bureaucratic acceptance as Posen suggested was necessary. Rosen's model of innovation also fails to adequately explain the 366th TFW's innovation. Granted, the Air Force recognized a substandard level of effectiveness in its missiles, but the institution's solution was to wait for the F-4E, not to load the SUU-16 onto the existing F-4C/Ds for use in air combat. Coté's model of innovation likewise falls short. Although the history of guided-missile development is colored by varying degrees of interservice rivalry between the Air Force and Navy, especially with regard to the Air Force's AIM-4D Falcon and the Navy's AIM-9D Sidewinder, there is little evidence to suggest that interservice rivalry encouraged the Air Force to develop the F-4E or spurred the 366th TFW to develop the F-4C/SUU-16 procedures.

It is possible that Posen's, Rosen's, and Coté's models of innovation apply only to innovation in doctrine. If true, then a significant theoretical gap exists in describing the influential mechanisms that spur innovation at the tactical and technical level. The lack of a suitable model at this level does not diminish its importance. Often, tactical innovations can have operational repercussions. It is also feasible that innovation at the tactical level could bubble up to the strategic level, although regrettably Blesse's innovation did not affect the strategic outcome of the Vietnam War.

The model of technological dislocations and the notions of competing technological skepticism and technological exuberance within a military organization help fill this theoretical void. While the proposed model lacks specific technological forecasting ability, it offers a method of conceptualizing and describing innovation at all levels, including the tactical. It also provides a vocabulary that describes the intermingling of society's influence on technology and vice versa, stimuli that continue throughout the life of a technological system. Furthermore, by identifying those key contingencies in history where a dominant technological trajectory is dislocated, the theory of technological dislocations focuses research to better inform scholars and practitioners of the relative merits of specific innovation strategies.

From this vantage point, the different innovation mechanisms described by Posen, Rosen, Coté, and others can be more accurately assessed. Absence of any of these specific catalysts does not diminish their potential analytic utility in another historical example. Their absence merely reaffirms the observation that the history of technology and the assessment of society's influence on it and vice versa are complex and open to varied interpretation.

This particular case study illustrated the value of keen marketing in outmaneuvering bureaucratic skepticism and the benefits of adopting a strategy of innovative systems integration vice outright systems acquisition, particularly when time is critical. Success or failure of this type of technical, tactical innovation hangs on the decisions of individual commanders. Thus, the review of Air Force air-to-air missile development, post–World War II through Rolling Thunder, leads to the conclusion that absent credible, innovative individuals and courageous commanders willing to act on their subordinates' recommendations, the military will regrettably tend to plod along according to a technological trajectory, reinforced by a bureaucracy skeptical of technologies that threaten it and overconfident in existing technologies that reinforce it. This constitutes an important lesson for the future.

Lesson for the Future

The Air Force, by continuing to market itself as a technology-minded service, is particularly susceptible to the allure of technological exuberance and the potential trap of an unchallenged technological trajectory. One current example of this trend is the Air Force's continued enthusiasm for stealth technology.

Initially secreted in a *black* program, the radical F-117 stealth fighter was spared much of the bureaucratic skepticism that often stymies emerging revolutionary technologies.[41] After proving its worth during Desert Storm, stealth technology quickly became the dominant theme guiding future Air Force aircraft design.[42] In October 1991, Gen Merrill McPeak, the Air Force chief of staff, proclaimed that "it will be very difficult for the Air Force to buy ever again another combat aircraft that doesn't include low-observable qualities."[43]

Unfortunately, stealth technology is expensive, and the Air Force's nascent stealth programs of the 1990s, such as the B-2 bomber and the F-22 fighter, languished because of it.[44] In particular, acquisition problems, cost overruns, and claims that "the F-22 represents technological overkill" that is "irrelevant to the wars of today" plagued the $65 billion F-22 Raptor program.[45] Amidst the criticism over the two-decade-long Raptor program, the Air Force pared its requests from 740 aircraft to 381, and then to 243. It reluctantly settled on only 183.[46]

The Air Force's next stealth fighter, the F-35 Lightning II, is now experiencing similar cost overruns and production delays that doomed the earlier F-22. Touted as "the future centerpiece of the US military's approach to waging war in the skies," the massive F-35 program developed "a troubling performance record," according to then-secretary of defense Robert Gates.[47] Despite facing a per-aircraft cost rocketing upwards of $100 million and a production delay extending beyond two years, defense officials remain committed to the program.[48] In February 2010 Gates announced that there were "no insurmountable problems, technological or otherwise, with the F-35.... We are in a position to move forward with this program in a realistic way."[49]

The Air Force has chained its future to F-35 success. In their support of the decision to halt F-22 production, former Air Force secretary Michael Donley and former chief of staff Gen Norton Schwartz jointly endorsed the F-35 and affirmed its exigency to the Air Force's future, proclaiming, "Much rides on the F-35's success, and it is critical to keep the Joint Strike Fighter on schedule and on cost."[50] Unfortunately, failure to do just that now burdens the service with what one scholar termed "the single greatest threat to the future Air Force's strategic viability," one that "risks bleeding the Air Force white over the next twenty years."[51]

While the problems associated with F-35 development are disconcerting, the Air Force's apparent refusal to reexamine the stealth aircraft's strategic utility is more alarming. Few deny the importance of maintaining a sizable fleet of stealth fighters (F-22A) and stealth fighter-bombers (F-35A) to deter potential conflict with a near-peer competitor (and if deterrence fails, to be victorious in combat). However, the simple, repeated chorus that all Air Force fighters require stealth technology does not suggest that a careful strategic assessment has been performed. An all-stealth fighter fleet would certainly simplify contingency planning. Likewise, it would be far simpler to maintain a fighter fleet that consisted of only two types of fighter aircraft. But what is the opportunity cost to other capabilities and requirements? Furthermore, what happens if a potential adversary develops a counter to American stealth technology? Even as stealth technology was introduced to the world in dramatic fashion during Desert Storm, Airmen and scholars noted that the United States would not enjoy this product of technological mastery forever.[52]

The Air Force appears reluctant to address these mounting fiscal constraints and shifting strategic contexts. Granted, the Air Force must revitalize its aging fleet. However, in its strategy to do so, the Air Force seems trapped in a technological trajectory that has yet to be sufficiently stressed and, if necessary, dislocated. Just as an Air Force armed with 740 F-22s became absurd as the strategic environment evolved during the 1990s, an Air Force equipped with more than 1,700 F-35s defies logic today. Yet the Air Force continues to demand a full inventory of stealthy F-35s at the expense of procuring, or even considering procuring, lower-cost alternatives such as the latest F-15 Silent Eagles or F-16 Block 60s that could complement a smaller, more cost-effective inventory of advanced stealth-fighter aircraft. Echoing these concerns, one independent study concluded, "The F-35 represents a classic 'middle-weight' capability—excessively sophisticated and expensive for persistent strike operations in the benign air environment of the developing world and most irregular warfare operations, yet not capable enough to contribute effectively to a stressing campaign against a nation employing modern anti-access/area-denial defenses."[53]

The Air Force's current, single-minded focus on a vision of future air combat and its dogged pursuit of the tools deemed necessary for that air war's conduct seem eerily reminiscent of Air Force attitudes toward air-to-air guided missiles in the 1950s and 1960s. Air Force officials must guard against the seduction of a promising but unchallenged and contextually bankrupt technological trajectory, lest we one day find the world's premiere air force ill-equipped to face the nation's future adversaries. The assumption that new technology is always better than old technology is not always valid. Boots Blesse and the 366th TFW "Gunfighters" proved it.

Notes

1. "One of the barriers . . . is the general conviction, implicit throughout the whole working structure and training program of the military system, that strategy poses no great problems which cannot be handled by the application of some well-known rules or 'principles' and that compared with the complexity of tactical problems and the skills needed to deal with them, the whole field of strategy is relatively unimportant. . . . The professional officer, stimulated always by the immediate needs of the service to which he devotes his life, becomes naturally absorbed with advancing its technical efficiency and smooth operation. . . . It is therefore hard for the professional soldier to avoid being preoccupied with means rather than ends." Brodie, *Strategy in the Missile Age*, 391 and 11–17.

2. Ibid., 408.

3. Ibid., 391. Brodie could have also cited Machiavelli, as Rosen did: "There is nothing more difficult to carry out, nor more doubtful of success, nor more dangerous to handle, than to initiate a new order of things. For the reformer has enemies in all those who profit by the old order, and only lukewarm defenders in all those who would profit by the new order. . . [because of] the incredulity of mankind, who do not truly believe in anything new until they have had actual experience of it." Rosen, *Winning the Next War*, 1.

4. "A central feature of use-based history, and a new history of invention, is that alternatives exist for nearly all technologies: there are multiple military technologies, means of generating electricity, powering a motor car, storing and manipulating information, cutting metal or roofing a building. Too often histories are written as if no alternative could or did exist." Edgerton, *Shock of the Old*, xiii and 7.

5. Ibid., 8.

6. Ibid., 17. Edgerton noted that the resources Germany allocated toward development of its anemic V-2 rocket forces could have produced 24,000 fighter aircraft. While an impressive statistic, Edgerton's argument does not consider the fact that Germany did not have a pilot force capable of manning that many aircraft. A more telling statistic is that for every one enemy civilian killed in the V-2 rocket attacks, German officials sacrificed two laborers developing the V-2 and building its underground production facilities.

7. Jervis, *Perception and Misperception*, 117.

8. Ibid.

9. Jervis noted that often "inconsistent premises are used to support a conclusion." Additionally, Jervis asserted that in their search for cognitive consistency, "decision makers are purchasing psychological harmony at the price of neglecting conflicts among their own values and are establishing their priorities by default." Ibid., 137–40.

10. Ibid., 145.

11. Ibid., 201; and Jullien, *A Treatise on Efficacy*. According to Jervis, "If commitment to an image inhibits the development of a new one, those who are most involved in carrying out policies guided by the old image will be the least able to innovate." Jullien attributed this inflexibility to a Western way of thinking.

12. "The flexibility of an image seems to be an inverse function of the extremity of its level. The higher the level of trust or distrust, the lower its flexibility." Jervis, *Perception and Misperception*, 195–96.

13. Schein, "Defining Organizational Culture," 373–74. Schein defined an organization's culture as "a pattern of shared basic assumptions" that guide individual perceptions, thoughts, and behaviors within the organization.

14. "The greater the distance from the last war, the greater become the chances of error in this extrapolation. Occasionally, there is a break in the clouds: a small-scale conflict occurs somewhere and gives you a 'fix' by showing whether certain weapons and techniques are effective or not; but it is always a doubtful fix." Howard, "Military Science," 4. Howard provided an example later in the lecture: "After 1918 we [the British] did little better. We had a navy which absurdly underrated the effectiveness of air power. We had an air force which equally absurdly overrated it." Howard's analogy is an adaptation of Carl von Clausewitz's popular "fog of war" adage: "The general unreliability of all information presents a special problem in war: all action takes place, so to speak, in a kind of twilight, which like fog or moonlight, often tends to make things seem grotesque and larger than the[y] really are." Clausewitz, *On War*, 140.

15. See earlier discussion in Chap. 2, 20–21, regarding the Air Force's response to ICBMs and unmanned aircraft and the Army's response to machine guns. On the Navy's response to aircraft carriers, see Reynolds, *Fast Carriers*, and Hone, Friedman, and Mandeles, *American and British Aircraft Carrier Development*. On the Army's response to armored tanks, see Johnson, *Fast Tanks*.

16. "It is therefore hard for the professional soldier to avoid being preoccupied with means rather than ends." Brodie, *Strategy in the Missile Age*, 17. For an example of military innovation conducted largely independent of a corresponding technological innovation, see Thomas G. Mahnken's discussion of the Army's AirLand Battle doctrine. Mahnken, *Technology and the American Way*, 127–31.

17. Gray, *Modern Strategy*, 37. Technology constitutes one of Gray's 17 dimensions of strategy.

18. Howard, "Military Science," 5; and Clausewitz, *On War*, 89. Clausewitz's trinity of war is "composed of primordial violence, hatred, and enmity, which are to be regarded as a blind natural force; of the play of chance and probability within which the creative spirit is free to roam; and of its element of subordination, as an instrument of policy, which makes it subject to reason alone."

19. Brodie, *Strategy in the Missile Age*, 361. Brodie's chapter was aptly titled, "Strategy Wears a Dollar Sign."

20. Jervis, *System Effects*, 22.

21. Posen, *Sources of Military Doctrine*, 54–55 and 57.

22. Ibid., 57; Howard, "Military Science," 5; and Jervis, *Perception and Misperception*, 199. Howard alluded to the importance of "military mavericks": "Therefore the problem of encouraging and rewarding original thinkers—men like Bony Fuller who have insights of near genius into the nature of their profession and the problem of war but who do not combine these insights with other professionally desirable qualities—presents genuine problems of a kind which

MILITARY INNOVATION

laymen tend to underrate." Jervis likewise asserted, "Within the military, those who propose major innovations are often outside the mainstream of the profession."

23. Posen, *Sources of Military Doctrine*, 175.

24. Ibid., 171–73; and Beyerchen, "From Radio to Radar," 265–99. Tizard chaired the Committee for Scientific Study of Air Defense; Inskip was the minister of coordination for defense; Dowding was the head of the Royal Air Forces' Fighter Command.

25. Overy, *Battle of Britain*.

26. Posen, *Sources of Military Doctrine*, 241.

27. Ibid., 46.

28. "Failure in war has not been necessary or sufficient for peacetime innovation.... Civilian intervention is an appealing *deus ex machina* that might explain innovation in peacetime military bureaucracies. But observations of the difficulties civilian leaders, up to and including the president of the United States, have had in bending the military to their desires should again lead us to be cautious." Rosen, *Winning the Next War*, 9–10.

29. Ibid., 251.

30. Coté, "Politics of Innovative Military Doctrine," 13.

31. Nagl, *Learning to Eat Soup*.

32. Watts and Murray, "Military Innovation in Peacetime," 409 (emphasis in original).

33. Millett, "Patterns of Military Innovation," 336 and 359.

34. Howard, "Military Science," 6 (emphasis in original).

35. Rosen and Posen echo Howard's requirement for credible innovators. In Posen's model, the military maverick provides the necessary credibility; in Rosen's theory, senior officers choose to extend their credibility to junior officers. Posen, *Sources of Military Doctrine*; and Rosen, *Winning the Next War*.

36. In this vein, Air Force missile armament development reached a pinnacle with the GAR-11, the nuclear-armed version of the Falcon guided air-to-air missile.

37. PACAF, "F-4C Fighter Screen," 10. The 1966 Heat Treat team's findings were discussed in chapter 5. The team of Air Force and industry specialists concluded that even "assuming proper maintenance of both aircraft and missiles, the probability of kill with the Sparrow can be expected to be low." As Michel highlighted, the Navy's efforts to improve pilot training and experience were regrettably not aggressively pursued within the Air Force, another illustration of the Air Force's inability to break free from the constraints of its technological paradigm. Michel, *Clashes*, 181–86.

38. To review, John Law suggested that "'heterogeneous engineers' seek to associate entities that range from people, through skills, to artifacts and natural phenomena. This is successful if the consequent heterogeneous networks are able to maintain some degree of stability in the face of the attempts of other entities or systems to disassociate them into their component parts." Law, "Technology and Heterogeneous Engineering," 129.

39. Hasik, *Arms and Innovation*.

40. Jervis, *Perception and Misperception*, 117.

41. Schmitt, "Stealth Technology," A5. An October 1991 *New York Times* article noted that the F-117 program suffered serious setbacks, including the crashes of two early prototypes and the first production aircraft. In the article, Air Force general Joseph Ralston was quoted as saying, "The way things are conducted today, a successful program like the F-117 would have been cancelled."

42. "Fifty F-117s made up only 2.5 percent of the US combat planes deployed in the [Desert Storm] operation, but they attacked 31 percent of the targets on the first day of the air war." Kaplan, "General Credits Air Force."

43. Schmitt, "Stealth Technology," A5.

44. Ibid. B-2 bomber production was halted after only 15 aircraft, at a cost of $865 million apiece. Maj Gen Stephen B. Croker ascribed stealth's high cost not to "the physics of stealth, but the problems of producibility." He noted that the B-2 had to be constructed within tolerances of 1/10,000th of an inch.

45. Wayne, "Air Force Jet Wins Battle," C1; and Dilanian and Brook, "Raptor in Dogfight," B1. A July 2006 Government Accountability Office (GAO) report stated, "The F-22 acquisition history is a case study in increased cost and schedule inefficiencies."

46. Donley and Schwartz, "Moving beyond the F-22," A15; and Bender, "President Wins on Defense," A1. The Air Force secretary and chief of staff penned the Air Force's response to the mounting pressure threatening the F-22 program. After the Senate voted 58 to 40 to halt F-22 production in July 2009, Pres. Barack Obama praised the decision: "Every dollar of waste in our defense budget is a dollar we can't spend to support our troops, or prepare for future threats, or protect the American people. . . . Our budget is a zero-sum game, and if more money goes to F-22s, it is our troops and citizens who lose."

47. Whitlock, "Gates to Major General," A4; and Hedgpeth, "GAO Analyst Says Cost Overruns," A13. With the Department of Defense planning to acquire nearly 2,400 aircraft, the F-35 is the largest and most expensive acquisition program in US history. Its "troubling performance record" led Gates in February 2010 to fire the Marine general charged with managing the F-35 program.

48. Hedgpeth, "Price Tag for F-35 Jets," A4; Hedgpeth, "GAO Analyst Says Cost Overruns," A13; and Rolfsen, "Despite Problems."

49. Hedgpeth, "Price Tag for F-35 Jets," A4. However, others have noted, "The secretary of defense reluctantly supports this program because he has no alternative. . . . The [Joint Strike Fighter] is like a sweater. . . . You pull any thread, like pushing back on full-rate production, and things can fall apart very quickly." International participation in the program—nine US allies have staked a future on the F-35—further complicates domestic decision-making.

50. Donley and Schwartz, "Moving beyond the F-22," A15.

51. Ehrhard, *Air Force Strategy*, 88.

52. "Soviets Can Detect Stealth Bomber"; and Schmitt, "Stealth Technology," A5. Returning from an eight-day official visit to the Soviet Union in October 1991, Gen Merrill McPeak, Air Force chief of staff, responded to reporters' questions regarding the long-term viability of the then-embroiled B-2 stealth bomber. McPeak answered, "By the way, I expect that certain parts of their [Soviet] air defense setup would be able to detect the B-2 today, so we don't have to wait ten years." He quickly qualified his remarks, "No one has ever said that the B-2 is invisible or immortal. What we've argued is that it is a very hard target to shoot down, and I expect that'll still be true ten years from now." A few days later, a senior military specialist at the Library of Congress noted, "The ability to keep stealth technology sacrosanct over a protracted period of time will be nil."

53. Ehrhard, *Air Force Strategy*, 87.

Chapter 8

Conclusion

*While decision-makers do not learn most from reading about history
... they may learn best from these sources.*

—Robert Jervis

History reveals a Janus-faced, nearly schizophrenic military attitude toward technological innovation. On the one hand, there is an image of a military wedded to technology, aptly evidenced during the cybernetic and chaoplexic revolutions in military affairs of the 1960s and 1980s. On the other hand, there is a competing and equally vivid image of a military institution frustratingly slow to adapt to technological change.[1] Stories of obstinate bureaucratic resistance stymieing promising new technologies such as the British steamship in the 1800s, the American airplane in the 1900s, or the Air Force's unmanned aircraft entering the 2000s are but a few examples of the latter image.[2] Careful historical analysis, however, divulges a pattern in which revolutionary technologies that threaten bureaucratic constituencies are often shunned in favor of evolutionary technological improvements that bolster the organizational culture. Because of its prominent techno-savvy self-image, this trend is especially pronounced in the Air Force.

The Wright brothers' aircraft was originally greeted with significant bureaucratic skepticism. Less than 60 years later, the institution's exuberance for its manned, strategic bomber fleets jaded its assessment of promising alternative technologies such as the ICBM.[3] In a similar pattern, but occurring over a much shorter period, the Air Force transitioned from questioning the combat capabilities of its new air-to-air guided missiles to relying exclusively upon them in air combat.

This pattern of alternating skepticism and exuberance can have a deleterious effect on strategic decision making. Entering the self-proclaimed "Air Age" of the 1950s, Air Force leaders were entranced by visions of gleaming B-36 bombers soaring high across the sky, armed with the atomic weapons that American scientific wizardry had bequeathed to the nation.[4] However, this fascination with technologically advanced bombers largely bankrupted the nascent service's capability to perform more limited, tactical action. When the Korean War revealed this failure in strategic planning, Air Force leaders simply dismissed the experience as an anomaly and continued to pursue the gadgetry that reinforced their interpretation of the strategic environment.[5]

CONCLUSION

The Air Force followed a similar pattern during Vietnam. Despite the failure of its air-to-air guided missiles in combat against the small North Vietnamese MiG fighters, the Air Force remained enthralled with the missiles' technological potential. Rather than investigating alternative technologies such as the assumed-anachronistic air-to-air cannon, the Air Force bureaucracy focused its efforts on developing a new generation of more complex missiles, such as the AIM-4D Falcon, that were unfortunately just as ineffective. In both the Korean and Vietnam Wars, the Air Force's exuberant embrace of the dominant technology and wary assessment of potential alternatives clouded its strategic vision.

A parallel to this historical phenomenon within the social science realm informs the current discussion. Social constructivists suggest that society shapes technology; technological determinists contend that technology shapes society. Thomas Hughes attempted to enjoin the two interpretations into a comprehensive theory of technological momentum. Unfortunately, his effort failed to address the contextual nuances and historical contingencies that often intervene in technological development. While suggesting that technologies can be both shaped by society and shaping of society, Hughes unfortunately drew an artificial and time-dependent distinction between the two that is unrepresentative of reality.[6]

Incorporating Giovanni Dosi's descriptions of technological paradigms and technological trajectories, the theory of technological dislocations attempts to close the conceptual gap between Hughes's theory and reality.[7] Rather than suggesting that a discrete tipping point divides social influences from technologically deterministic influences, or skepticism from exuberance, the theory of technological dislocations facilitates a more holistic historical appreciation. Technological systems are born of social influences, but the technology quickly begins to exert a deterministic influence on society in the form of a technological paradigm. Within that technological paradigm, a trajectory develops that guides further technological progress. However, that same technological paradigm and the corresponding trajectory can constrain revolutionary, innovative thinking as the bureaucracy becomes bound by its dominant technology. Compounding matters, the incremental, nearly autonomous evolutionary technical development that takes place according to the technological trajectory is often misconstrued as innovative, responsive adaptation. Using Michael Howard's analogy, when the "fog of peace" finally lifts, the disparity is revealed.[8] Even then, exuberance for the dominant technology can continue to exert a profound influence on an organization's decision makers.

A technological dislocation is therefore required to jar the bureaucracy from its technological rut. The catalysts that converge to affect the dislocation and the mechanisms by which it alters the dominant technological trajectory are contextually dependent. Posen, Rosen, and Coté all offered slightly different assessments of military and technological innovation, focusing on civilian influence, strategic assessment, and interservice rivalry, respectively.[9] However, the evidence from the preceding study of Air Force air-to-air armament did not support any of these individual interpretations. Rather, the case study suggested its own influential mechanisms; namely, the importance of keen marketing, innovative systems integration, and credible, innovative individuals and courageous commanders willing to act on their subordinates' recommendations.

While the technological dislocation model does not grant decision makers the power to pre-identify critical technologies, it does offer them a tool to analyze past technological development and extract appropriate lessons for future application. One of the advantages of the theory of technological dislocations is that it accommodates a variety of influential mechanisms in the description of how technological innovation occurs. In fact, the particular method of interposing a dislocation into a technological trajectory is not especially important; the strategies suggested earlier by Posen, Rosen, and Coté retain their relevance.

The more significant value of the technological dislocations model lies in its ability to facilitate decision makers' understanding of the obstinate nature of bureaucratic institutions, despite superficial appearances to the contrary. Bureaucracies will exuberantly innovate, but without a technological dislocation to jar them from their preferred technological trajectory, the incremental technical progress they cultivate only yields an illusion of thoughtful strategic reflection and adaptation.

A careful review of history provides the decision maker with a unique appreciation for the role of technological dislocations in organizations. It also forms a bank of lessons that, appreciating their contextual nuances, can be drawn upon when required. Citing Jervis, "While decision-makers do not learn most from reading about history, . . . they may learn best from these sources."[10]

Technological progress is not a substitute for strategic analysis. Unfortunately, the allure of the *new* often clouds accurate assessment of a technology's feasibility and practicality. The Air Force has succumbed to technological exuberance in the past, and the pattern continues today with the F-35. To counter these ill effects, Airmen and civilians alike must challenge the Air Force's strategic assumptions guiding its technological acquisitions. If necessary, they must be ready to introduce a technological dislocation. Air Force leaders in turn must be open to such criticism and potential disruption. Recognizing and removing the technological blinkers that obscure strategic vision provide a vital first step in conducting a meaningful strategic dialogue.

Unlike Goethe's Faust, who was at the last moment spared eternal demise, the Air Force's future should not rely solely on the angels of providence.[11] When tempted by a technological Mephistopheles, the Air Force should instead embrace well-reasoned foresight and open strategic dialogue. Choose well, Air Force.

Notes

1. Bousquet, *Scientific Way of Warfare*, 33–34. Bousquet defined cybernetic warfare as a "self-proclaimed 'science of communications and control'" that "promised to manage chaos and disruption through self-regulating mechanisms of information feedback" in war. General Westmoreland's battlefield of the future (see note 21, chapter 2) was reflective of the cybernetic way of warfare and was largely realized in the Igloo White program during Vietnam. Conversely, Bousquet linked the "principles of chaos and complexity (referred to together as chaoplexity)" and the notions of "non-linearity, self-organization, and emergence" into the term *chaoplexic* warfare. He suggested that, based on their advocacy of technologies that would enable the self-synchronization of forces without additional command and control mechanisms, netcentric warfare proponents subscribed to the chaoplexic vision of warfare. Both the military cyberneticists and the chaoplexists heralded a revolution in military affairs. See Lonsdale, *Nature of War in the Information Age*.

2. McNeil, *Pursuit of Power*; Morrow, *Great War in the Air*; and Singer, *Wired for War*. McNeil discussed the British resistance to the steamship; Morrow outlined the Army's resistance to the Wright Flyer; and Singer provided a rich narrative of the Air Force's resistance to unmanned drone aircraft.

3. Sheehan, *Fiery Peace*.

4. Barlow, *Revolt of the Admirals*, 46. Air Force general Tooey Spaatz announced the arrival of the "Air Age" in October 1945.

5. Crane, *American Airpower Strategy*, 60. Recall Air Force major general Emmett O'Donnell's testimony to Congress in 1951: "I think this is a rather bizarre war out there [in Korea], and I think we can learn an awful lot of bad habits in it."

6. "A technological system can be both a cause and an effect; it can shape or be shaped by society." Delineating the difference, Hughes continued, "The social constructivists have a key to understanding the behavior of young systems; technical determinists come into their own with the mature ones." Hughes, "Technological Momentum," 112.

7. Dosi, "Technological Paradigms," 152–53. Dosi defined a technological trajectory as the "direction of advance within a technological paradigm." He also noted that "technological paradigms have a powerful exclusion effect: the efforts and the technological imagination of engineers and of the organizations they are in are focused in rather precise directions while they are, so to speak, 'blind' with respect to other technological possibilities."

8. Howard, "Military Science," 4.

9. Posen, *Sources of Military Doctrine*; Rosen, *Winning the Next War*; and Coté, "Politics of Innovative Military Doctrine."

10. Jervis, *Perception and Misperception*, 246.

11. Goethe, *Faust*.

Abbreviations

AAA	antiaircraft artillery
AAF	Army Air Force
ACM	air combat maneuvering
ACT	air combat tactics
ADC	Air Defense Command
AFHRA	Air Force Historical Research Agency
AFRI	Air Force Research Institute
AIM	air intercept missile
ARDC	Air Research and Development Command
ATS	Air Tactical School
AU	Air University
CIA	Central Intelligence Agency
CORDS	Coherent-on-Receive Doppler System
DARPA	Defense Advanced Research Projects Agency
ECM	electronic countermeasure
FFAR	folding-fin aerial rocket
FY	fiscal year
GAO	Government Accountability Office
GAR	guided air rocket
GCI	ground-controlled intercept
GE	General Electric
GIB	guy in back
ICBM	intercontinental ballistic missile
INS	inertial navigation system
ISR	intelligence, surveillance, and reconnaissance
JCS	Joint Chiefs of Staff
kt	kiloton
LCOSS	Lead Computing Optical Sight Set
MiGCAP	MiG combat air patrol
MIRV	multiple independently targetable reentry vehicle
mm	millimeter
NASA	National Aeronautics and Space Administration
NVN	North Vietnam
OST	Office of Science and Technology
PACAF	Pacific Air Forces

ABBREVIATIONS

R&D	research and development
RAT	ram-air turbine
RBL	radar boresight line
recce	reconnaissance
ROE	rule of engagement
RPA	remotely piloted aircraft
RTU	replacement training unit
SAASS	School of Advanced Air and Space Studies
SAC	Strategic Air Command
SAM	surface-to-air missile
SDI	Strategic Defense Initiative
SEA	Southeast Asia
SEACAAL	Southeast Asia Counter-Air Alternative
SVN	South Vietnam
TAC	Tactical Air Command
TFS	Tactical Fighter Squadron
TFW	tactical fighter wing
UAV	unmanned aerial vehicle
UCLA	University of California, Los Angeles
WRCS	Weapons Release Computer System

Bibliography

Books, Journals, and Newspapers

"Air Force Orders New Jet Fighter." *New York Times*, 19 October 1966.

Ambrose, Stephen E. *Eisenhower: Soldier and President*. New York: Simon & Schuster, 1990.

Ammer, Christine, and Dean S. Ammer. *Dictionary of Business and Economics*. Revised and expanded ed. New York: The Free Press, 1984.

Anderegg, C. R. *Sierra Hotel: Flying Air Force Fighters in the Decade after Vietnam*. Washington, DC: Air Force History and Museums Program, 2001.

Barlow, Jeffrey G. *Revolt of the Admirals: The Fight for Naval Aviation, 1945–1950*. Washington, DC: Government Reprints Press, 2001.

Bender, Bryan. "President Wins on Defense Spending: With Prodding, Senate Halts Order for F-22 Jets." *Boston Globe*, 22 July 2009.

Beyerchen, Alan. "From Radio to Radar: Interwar Military Adaptation to Technological Change in Germany, the United Kingdom, and the United States." In *Military Innovation in the Interwar Period*, edited by Williamson Murray and Allan R. Millett, 265–99. New York: Cambridge University Press, 1996.

Bijker, Weiebe E., Thomas P. Hughes, and Trevor Pinch, eds. "Common Themes in Sociological and Historical Studies of Technology: Introduction." In *The Social Construction of Technological Systems: New Directions in the Sociology and History of Technology*, 7–15. Cambridge, MA: MIT Press, 1989.

Blakemore, J. S. *Solid State Physics*. 2nd ed. Cambridge: Press Syndicate of the University of Cambridge, 1960.

Blesse, Maj Frederick C., USAF. *No Guts, No Glory*. Nellis AFB, NV: 3596th Combat Crew Training Squadron. In *USAF Fighter Weapons School*, no.1, 1955, R358.4 A29833n, Muir S. Fairchild Research Information Center (MSFRIC), Maxwell AFB, AL.

Blesse, Maj Gen Frederick C., USAF, retired. *Check Six: A Fighter Pilot Looks Back*. Mesa, AZ: Champlin Fighter Museum Press, 1987.

Bousquet, Antoine. *The Scientific Way of Warfare: Order and Chaos on the Battlefields of Modernity*. New York: Columbia University Press, 2009.

Boyne, Walter J. *Phantom in Combat*. Washington, DC: Smithsonian Institution Press, 1985.

Brauer, Jurgen, and Hubert Van Tuyll. *Castles, Battles, and Bombs: How Economics Explains Military History*. Chicago: University of Chicago Press, 2008.

Brinkley, Douglas. *Wheels for the World: Henry Ford, His Company, and a Century of Progress, 1903–2003*. New York: Viking, 2003.

Brodie, Bernard. *Strategy in the Missile Age*. New RAND ed. Santa Monica, CA: RAND, 2007.

Browne, Malcolm W. "Fusion in a Jar: Announcement by Two Chemists Ignites Uproar." *New York Times*, 28 March 1989.

———. "Physicists Debunk Claim of a New Kind of Fusion." *New York Times*, 3 May 1989.

Bugos, Glenn E. *Engineering the F-4 Phantom II: Parts into Systems*. Annapolis, MD: Naval Institute Press, 1996.

Campbell, Stephen A. *The Science and Engineering of Microelectronic Fabrication*. New York: Oxford University Press, 1996.

Cherny, Andrei. *The Candy Bombers: The Untold Story of the Berlin Airlift and America's Finest Hour*. New York: G. P. Putnam's Sons, 2008.

Clausewitz, Carl von. *On War*. Edited and translated by Michael Howard and Peter Paret. Princeton, NJ: Princeton University Press, 1976.

Constant, Edward W., II. *The Origins of the Turbojet Revolution*. Baltimore, MD: Johns Hopkins University Press, 1980.

Crane, Conrad C. *American Airpower Strategy in Korea, 1950–1953*. Lawrence: University Press of Kansas, 2000.

Davies, Peter. *USAF F-4 Phantom II MiG Killers 1965–68*. Oxford: Osprey Publishing, 2004.

Diamond, Jared. *Guns, Germs, and Steel: The Fates of Human Societies*. New York: W. W. Norton and Company, 1997.

Dilanian, Ken, and Tom Vanden Brook. "Raptor in Dogfight for Its Future: Powerful Critics Take Aim at Costly F-22 Fighter Jet." *USA Today*, 26 February 2009.

Donley, Michael, and Norton Schwartz. "Moving beyond the F-22." *Washington Post*, 13 April 2009.

Dosi, Giovanni. "Technological Paradigms and Technological Trajectories." *Research Policy* 11, no. 3 (June 1982): 147–62.

Dreazen, Yochi J. "Gates Ousts Top Leaders of Air Force after Gaffes." *Wall Street Journal*, 6 June 2008.

Edgerton, David. *The Shock of the Old: Technology and Global History since 1900*. New York: Oxford University Press, 2007.

Ehrhard, Thomas P. *An Air Force Strategy for the Long Haul*. Washington, DC: Center for Strategic and Budgetary Assessments, 2009.

Ellis, John. *The Social History of the Machine Gun*. New York: Pantheon Books, 1975.

Enthoven, Alain C., and K. Wayne Smith. *How Much Is Enough? Shaping the Defense Program, 1961–1969.* New York: Harper & Row Publishers, 1971.

Fino, Steven A. "Breaking the Trance: The Perils of Technological Exuberance in the U.S. Air Force Entering Vietnam." *The Journal of Military History* 77, no. 2 (April 2013): 625–55.

Futrell, R. Frank, William H. Greenhalgh, Carl Grubb, Gerald E. Hasselwander, Robert F. Jakob, and Charles A. Ravenstein. *Aces and Aerial Victories: The United States Air Force in Southeast Asia 1965–1973.* Edited by James N. Eastman, Jr., Walter Hanak, and Lawrence J. Paszek. Washington, DC: Office of Air Force History, 1976.

Gladwell, Malcolm. *The Tipping Point: How Little Things Can Make a Big Difference.* New York: Back Bay, 2002.

Goethe, Johann Wolfgang von. *Faust: A Tragedy.* Edited by Cyrus Hamlin. Translated by Walter W. Arndt. 2nd ed. New York: W. W. Norton & Company, 2000.

Gray, Colin. *Modern Strategy.* New York: Oxford University Press, 1999.

Hannah, Craig C. *Striving for Air Superiority: The Tactical Air Command in Vietnam.* College Station: Texas A&M University Press, 2002.

Hasik, James. *Arms and Innovation: Entrepreneurship and Alliances in the Twenty-First-Century Defense Industry.* Chicago: University of Chicago Press, 2008.

Hedgpeth, Dana. "GAO Analyst Says Cost Overruns, Delays Plague F-35 Program." *Washington Post*, 12 March 2010.

———. "Price Tag for F-35 Jets Expected to Rise; Air Force Secretary Says Weapons Program Is Also Likely to Face Delays." *Washington Post*, 3 March 2010.

Holmes, Erik. "Why USAF's Top Two Were Forced Out." *Defense News*, 9 June 2008.

Hone, Thomas C. "Korea." In *Case Studies in the Achievement of Air Superiority*, edited by Benjamin Franklin Cooling, 453–504. Washington, DC: Center for Air Force History, 1994.

———. "Southeast Asia." In *Case Studies in the Achievement of Air Superiority*, edited by Benjamin Franklin Cooling, 505–62. Washington, DC: Center for Air Force History, 1994.

Hone, Thomas C., Norman Friedman, and Mark D. Mandeles. *American and British Aircraft Carrier Development, 1919–1941.* Annapolis: Naval Institute Press, 2003.

Howard, Michael. "Military Science in an Age of Peace." *RUSI Journal for Defence Studies* 119, no. 1 (March 1974): 3–11.

BIBLIOGRAPHY

Hughes, Thomas P. *American Genesis: A Century of Invention and Technological Enthusiasm 1870–1970*. New York: Viking, 1989.

———. "The Evolution of Large Technological Systems." In *The Social Construction of Technological Systems: New Directions in the Sociology and History of Technology*, edited by Weiebe E. Bijker, Thomas P. Hughes, and Trevor Pinch, 51–82. Cambridge, MA: MIT Press, 1989.

———. *Human Built World: How to Think about Technology and Culture*. Chicago: University of Chicago Press, 2004.

———. "Technological Momentum." In *Does Technology Drive History? The Dilemma of Technological Determinism*, edited by Leo Marx and Merritt Roe Smith, 101–13. Cambridge, MA: MIT Press, 1994.

———. "Technological Momentum in History: Hydrogenation in Germany 1898–1933." *Past and Present*, no. 44 (August 1969): 106–32.

Jervis, Robert. *Perception and Misperception in International Politics*. Princeton, NJ: Princeton University Press, 1976.

———. *System Effects: Complexity in Political and Social Life*. Princeton, NJ: Princeton University Press, 1997.

Johnson, David E. *Fast Tanks and Heavy Bombers: Innovation in the US Army, 1917–1945*. Ithaca: Cornell University Press, 2003.

Jullien, François. *A Treatise on Efficacy between Western and Chinese Thinking*. Translated by Janet Lloyd. Honolulu: University of Hawaii Press, 2004.

Kaplan, Fred. "General Credits Air Force with Iraqi Army's Defeat." *Boston Globe*, 16 March 1991.

Knaack, Marcelle Size. *Post–World War II Fighters, 1945–1973*. Washington, DC: Office of Air Force History, 1975.

Kuhn, Thomas S. *The Structure of Scientific Revolutions*. 3rd ed. Chicago: University of Chicago Press, 1962.

Kurzweil, Ray. *The Age of Spiritual Machines: When Computers Exceed Human Intelligence*. New York: Penguin Books, 1999.

Law, John. "Technology and Heterogeneous Engineering: The Case of the Portuguese Expansion." In *The Social Construction of Technological Systems: New Directions in the Sociology and History of Technology*, edited by Weiebe E. Bijker, Thomas P. Hughes, and Trevor Pinch, 111–34. Cambridge, MA: MIT Press, 1989.

Lonsdale, David J. *The Nature of War in the Information Age: Clausewitzian Future*. New York: Frank Cass, 2004.

Luttwak, Edward N. *Strategy: The Logic of War and Peace*. Revised and expanded ed. Cambridge, MA: Belknap Press of Harvard University Press, 2003.

MacKenzie, Donald. *Inventing Accuracy: A Historical Sociology of Nuclear Missile Guidance.* Cambridge, MA: MIT Press, 1990.

Mahnken, Thomas G. *Technology and the American Way of War since 1945.* New York: Columbia University Press, 2008.

Martel, William C. *Victory in War: Foundations of Modern Military Policy.* New York: Cambridge University Press, 2007.

McDougall, Walter A. *The Heavens and the Earth: A Political History of the Space Age.* Baltimore, MD: Johns Hopkins Press, 1985.

McNeil, William H. *The Pursuit of Power: Technology, Armed Force, and Society since A.D. 1000.* Chicago: University of Chicago Press, 1982.

McPherson, James. *Battle Cry of Freedom: The Civil War Era.* New York: Oxford University Press, 1988.

Medhurst, Martin J. *Dwight D. Eisenhower: Strategic Communicator.* Westport, CT: Greenwood Press, 1993.

Michel, Marshall L., III. *Clashes: Air Combat over North Vietnam 1965–1972.* Annapolis, MD: Naval Institute Press, 1997.

Millett, Allan R. "Patterns of Military Innovation in the Interwar Period." In *Military Innovation in the Interwar Period*, edited by Williamson Murray and Allan R. Millett, 329–68. New York: Cambridge University Press, 1996.

Momyer, Gen William W., USAF, retired. *Airpower in Three Wars.* Washington, DC: US Government Printing Office, 1978.

Morrow, John H., Jr. *The Great War in the Air: Military Aviation from 1909 to 1921.* Tuscaloosa: University of Alabama Press, 1993.

Murray, Williamson, and Allan R. Millet, eds. *Military Innovation in the Interwar Period.* New York: Cambridge University Press, 1996.

Nagl, John A. *Learning to Eat Soup with a Knife: Counterinsurgency Lessons from Malaya and Vietnam.* Chicago: University of Chicago Press, 2002.

Neufeld, Michael J. *Von Braun: Dreamer of Space, Engineer of War.* New York: Vintage Books, 2007.

"New Missile Ready." *New York Times*, 24 September 1958.

Olds, Robin, Christina Olds, and Ed Rasimus. *Fighter Pilot: The Memoirs of Legendary Ace Robin Olds.* New York: St. Martin's Press, 2010.

Overy, Richard. *Battle of Britain: The Myth and the Reality.* New York: W. W. Norton & Company, 2000.

"Pilots Describe Downings of MiGs." *New York Times*, 12 July 1965.

Posen, Barry R. *The Sources of Military Doctrine: France, Britain, and Germany between the World Wars.* Ithaca, NY: Cornell University Press, 1984.

Randolph, Stephen P. *Powerful and Brutal Weapons: Nixon, Kissinger, and the Easter Offensive.* Cambridge, MA: Harvard University Press, 2007.

BIBLIOGRAPHY

Reynolds, Clark G. *The Fast Carriers: The Forging of an Air Navy.* Annapolis: Naval Institute Press, 1968.

Rhodes, Richard. *Arsenals of Folly: The Making of the Nuclear Arms Race.* New York: Alfred A. Knopf, 2007.

———. *Dark Sun.* New York: Touchstone Books, 1996.

———. *The Making of the Atomic Bomb.* New York: Simon & Schuster, 1986.

Ritchie, Lt Col Steve, USAF. "Foreword." In *Phantom in Combat*, by Walter J. Boyne, 6. Washington, DC: Smithsonian Institution Press, 1985.

Rolfsen, Bruce. "Despite Problems, AF Plans to Stick with F-35." *Air Force Times*, 19 April 2010. Accessed 12 May 2010. http://www.airforcetimes.com/news/2010/04/airforce_f35s_041310w/.

Rosen, Stephen Peter. *Winning the Next War: Innovation and the Modern Military.* Ithaca, NY: Cornell University Press, 1991.

Schaffel, Kenneth. *The Emerging Shield: The Air Force and the Evolution of Continental Air Defense, 1945–1960.* Washington, DC: Office of Air Force History, 1990.

Schein, Edgar. "Defining Organizational Culture." In *Classics of Organization Theory*, edited by Jay M. Shafritz and J. Steven Ott, 490–502. 5th ed. Belmont, CA: Wadsworth Publishing Company, 2001.

Schmemann, Serge. "Summit in Moscow: Bush and Yeltsin Sign Pact Making Deep Missile Cuts; It Would Reduce Atom Arsenals about 75%." *New York Times*, 4 January 1993.

Schmitt, Eric. "Stealth Technology: Elusive in More Ways than One." *New York Times*, 13 October 1991.

Shanker, Thom. "Gates Says New Arms Must Play Role Now." *New York Times*, 14 May 2008.

Sheehan, Neil. *A Fiery Peace in a Cold War: Bernard Schriever and the Ultimate Weapons.* New York: Random House, 2009.

Singer, P. W. *Wired for War: The Robotic Revolution and Conflict in the 21st Century.* New York: Penguin Press, 2009.

Skinner, Michael. *Red Flag: Air Combat for the '80s.* Novato, CA: Presidio Press, 1984.

Smith, Merritt Roe. "Introduction." In *Military Enterprise and Technological Change: Perspectives on the American Experience*, edited by Merritt Roe Smith, 1–37. Cambridge, MA: MIT Press, 1985.

———. "Technological Determinism in American Culture." In *Does Technology Drive History? The Dilemma of Technological Determinism*, edited by Merritt Roe Smith and Leo Marx, 1–35. Cambridge, MA: MIT Press, 1994.

Smith, Merritt Roe, and Leo Marx. "Introduction." In *Does Technology Drive History? The Dilemma of Technological Determinism*, edited by Merritt Roe Smith and Leo Marx, ix–xv. Cambridge, MA: MIT Press, 1994.

"Soccer Ball Molecules." *New York Times*, 21 March 1989.

"Soviets Can Detect Stealth Bomber, US Air Force Chief Says." *Boston Globe*, 10 October 1991.

Spires, David N. *Beyond Horizons: A Half Century of Air Force Space Leadership*. Revised ed. Maxwell AFB, AL: Air University Press, 1998.

Stimson, George W. *Introduction to Airborne Radar*. 2nd ed. Mendham, NJ: SciTech Publishing, Inc, 1998.

Swaminathan, V., and A. T. Macrander. *Materials Aspects of GaAs and InP Based Structures*. Englewood Cliffs, NJ: Prentice Hall, 1991.

Taleb, Nassim Nicholas. *The Black Swan: The Impact of the Highly Improbable*. New York: Random House, 2007.

Thoreau, Henry David. *Walden*. Philadelphia: Courage Books, 1990.

Thornborough, Anthony M. *USAF Phantoms*. New York: Arms & Armour Press, 1988.

Thornborough, Anthony M., and Peter E. Davies. *The Phantom Story*. London: Arms and Armour, 1994.

"US Flier Says 2 of His Missiles Struck MiG-21 over Vietnam." *New York Times*, 28 April 1966.

"US Jets Will Reflect Lessons Learned in Vietnam." *New York Times*, 24 November 1966.

Waldrop, M. Mitchell. *Complexity: The Emerging Science at the Edge of Order and Chaos*. New York: Simon & Schuster Paperbacks, 1992.

Walker, Karen. "Air Force Firings: The Right Decision." *Defense News*, 9 June 2008.

Watts, Barry, and Williamson Murray. "Military Innovation in Peacetime." In *Military Innovation in the Interwar Period*, edited by Williamson Murray and Allan R. Millett, 369–415. New York: Cambridge University Press, 1996.

Wayne, Leslie. "Air Force Jet Wins Battle in Congress." *New York Times*, 28 September 2006.

West, Anthony R. *Solid State Chemistry and Its Applications*. New York: John Wiley & Sons, 1984.

Westrum, Ron. *Sidewinder: Creative Missile Development at China Lake*. Annapolis, MD: Naval Institute Press, 1999.

Whitlock, Craig. "Gates to Major General: You're Fired; Defense Secretary Not Happy with F-35 Program's Performance." *Washington Post*, 2 February 2010.

BIBLIOGRAPHY

Wilcox, Robert K. *Scream of Eagles: The Creation of Top Gun—And the US Air Victory in Vietnam.* New York: John Wiley & Sons, Inc., 1990.

Wildenberg, Thomas. "A Visionary ahead of His Time: Howard Hughes and the US Air Force—Part III: The Falcon Missile and Airborne Fire Control." *Air Power History* 55, no. 2 (Summer 2008): 4–13.

York, Herbert F. *The Advisors: Oppenheimer, Teller, and the Superbomb.* San Francisco, CA: W. H. Freeman and Company, 1976.

Unpublished and Archival Materials

Unpublished Manuscripts

8th Tactical Fighter Wing (TFW). "Tactical Doctrine." 1 March 1967. In History, 8th Tactical Fighter Wing, vol. 2, January–June 1967. K-WG-8-HI, Air Force Historical Research Agency (AFHRA), Maxwell AFB, AL.

———. "Tactical Doctrine." December 1967. In History, 8th Tactical Fighter Wing, vol. 2, October–December 1967. K-WG-8-HI, AFHRA.

Air Tactical School (ATS). "Air-to-Air Guided Missiles." Lecture manuscript. Tyndall AFB, FL: AU, ATS, 16 April 1948. K239.716721-49, AFHRA.

Boyd, Capt John R., USAF. "Aerial Attack Study." 11 August 1964. M–U 43947-5, Muir S. Fairchild Research Information Center, Maxwell AFB, AL.

Coté, Owen Reid, Jr. "The Politics of Innovative Military Doctrine: The US Navy and Fleet Ballistic Missiles." PhD diss., Massachusetts Institute of Technology, February 1996.

Cunningham, Maj Case A., USAF. "Spike: A Biography of an Airpower Mind." Master's thesis, School of Advanced Air and Space Studies, AU, June 2007.

McMullen, Richard P. "History of Air Defense Weapons, 1946–1962" (U). Air Defense Command Historical Study Number 14. Historical Division, Office of Information, HQ Air Defense Command. K410.041-14, AFHRA. (Formerly restricted data) Information extracted is unclassified.

Mets, David R. "The Evolution of Aircraft Guns (1912–1945)." Eglin AFB, FL: Air Force Systems Command, Office of History, Armament Division, 1987. K243.04-57, AFHRA.

Pacific Air Forces (PACAF). "F-4C Fighter Screen and Escort." *PACAF Tactics and Techniques Bulletin*, no. 44 (14 July 1966). K717.549-1, AFHRA.

US Air Force. "Maj Gen Frederick C. Blesse Biography." Accessed 20 April 2010. http://www.af.mil/information/bios/bio.asp?bioID=4712.

AFHRA Aerial Victory Credit Folders

"1965-10 July; Holcombe and Anderson." K238.375-8, AFHRA.
"1966-23 April; Cameron and Evans." K238.375-10, AFHRA.
"1966-29 April; Dowell and Gossard." K238.375-13, AFHRA.
"1966-30 April; Golberg and Hardgrave." K238.375-15, AFHRA.
"1966-16 September; Jameson and Rose." K238.375-21, AFHRA.
"1967-14 May; Bakke and Lambert." K238.375-59, AFHRA.
"1967-14 May; Craig and Talley." K238.375-58, AFHRA
"1967-14 May; Hargrove and DeMuth." K238.375-57, AFHRA
"1967-22 May; Titus and Zimer." K238.375-65, AFHRA.
"1967-5 June; Raspberry and Gullick." K238.375-69, AFHRA.
"1967-6 November 1967; Simmonds and McKinney." K238.375-78, AFHRA.

Interviews

Bakke, Maj Sam, USAF, retired. Interview by the author, 24 April 2010.

Blesse, Maj Gen F. C., USAF. Oral history interview, by Lt Col Gordon F. Nelson, 14 February 1977. K239.0512-1077, AFHRA.

Bolt, Maj Gen Jones E., USAF. Oral history interview, by Col Frederic E. McCoy III, 6–7 December 1984. K239.0512-1624, AFHRA.

Burns, Maj Gen John J., USAF. Oral history interview, by Jack Neufeld, 22 March 1973. K239.0512-961, AFHRA.

Catledge, Maj Gen Richard C., USAF. Oral history interview, by 1st Lt Wayne D. Perry, 30 September 1987 and 9 December 1987. K239.0512-1768, AFHRA.

Dilger, Maj Robert G., USAF. Oral history interview, 6 July 1967. K239.0512-202, AFHRA.

Hargrove, Maj James A., USAF. Oral history interview, by Lt Col Robert Eckert and Maj Harry Shallcross, 19 September 1967. K239.0512-020, AFHRA.

Hildreth, Maj Gen James R., USAF. Oral history interview, by James C. Hasdorff, 27–28 October 1987. K239.0512-1772, AFHRA.

Kirk, Maj William L., USAF. Oral history interview. K239.0512-206, AFHRA.

Olds, Col Robin, USAF. Oral history interview, 12 July 1967. K239.0512-160, AFHRA.

Olds, Brig Gen Robin, USAF. Oral history interview, by Major Geffen and Major Folkman, 29 September 1969. K239.0512-051, AFHRA.

Peck, Col Gail, USAF, retired. To the author. E-mail, 12 April 2010.

Simmonds, Lt Col Darrell, USAF, retired. Interview by the author, 19 May 2010.

BIBLIOGRAPHY

Memorandums

Bakke, Maj Samuel O., USAF, 366 TFW/DOTW. To 366 TFW Enemy Aircraft Claims Board. Memorandum. In AHRA, "1967-14 May; Bakke and Lambert." K238.375-59, AFHRA.

Carey, Capt William, USAF. To 366 TFW (DCO). Attachment to Maj Hargrove, 480th Tactical Fighter Squadron. Statement. In AFHRA, "1967-14 May; Hargrove and DeMuth." K238.375-57, AFHRA.

Craig, Capt James T., USAF. To 366 TFW Enemy Aircraft Claims Board. Memorandum. In AFHRA, "1967-14 May; Craig and Talley." K238.375-58, AFHRA.

Hargrove, Maj James A., USAF, 480th Tactical Fighter Squadron. To 366 TFW (DCO). Memorandum. In AFHRA, "1967-14 May; Hargrove and DeMuth." K238.375-57, AFHRA.

Momyer, Gen William W., USAF, retired. To General Ellis. "Corona Harvest (Out-Country Air Operations, Southeast Asia, 1 January 1965–31 March 1968)." Memorandum, 23 July 1974. K239.031-98, AFHRA.

———. To General Ellis. "Corona Harvest (USAF Operations against North Vietnam, 1 July 1971–30 June 1972)." Memorandum, 1 April 1975. K239.031-99, AFHRA.

———. To General Ellis. "Corona Harvest (USAF Personnel Activities in Support of Operations in Southeast Asia, 1 January 1965–31 March 1968)." Memorandum, 14 August 1974. K239.031-96, AFHRA.

Messages

7 AF. Wrap-Up Reports, 1–15 May 1967. K740.3422, AFHRA.

Message. 141320Z MAY 67. 388 TFW. To NMCC, et al. 388 TFW OPREP-3/014, 14 May 1967. In PACAF Command Center, Chronological Log, 13–14 May 1967. K717.3051-1, AFHRA.

Message. 141410Z MAY 67. 366 TFW. To NMCC, et al. 366 TFW OPREP-3/011, 14 May 1967. In PACAF DO Read File, 13–15 May 1967. K717.312, AFHRA.

Message. 141430Z MAY 67. 366 TFW. To 7 AF CC. Subject: "Daily Operations Wrap up Summary," 14 May 1967. In 7 AF, Wrap Up Reports, 1–15 May 1967. K740.3422, AFHRA.

Message. 141800Z MAY 67. 366 TFW. To NMCC, et al. 366 TFW OPREP-3/010, 14 May 1967. In PACAF DO Read File, 13–15 May 1967. K717.312, AFHRA.

Message. 180515Z MAY 67. PACAF CC. To 7 AF and 13 AF. In PACAF DO Read File, 17–18 May 1967. K717.312, AFHRA.

Message. 191633Z JUL 67. CSAF. To PACAF, et al. "Presentation to Sec Def by Col FC Blesse." K717.1622, AFHRA.
Message. 300055Z APR 67. CINCPAC. To JCS. Subject: "MiG Threat," 30 April 1967. In PACAF DO Read File, 29 April–1 May 1967. K717.312, AFHRA. (Secret) Information extracted is unclassified.
Message. C10697. 366 TFW. To 7 AF. Subject: "F-4C Air to Air Modification Program," 5 May 1967. In History, 366th Tactical Fighter Wing, vol. 1, 1 April 1967–30 September 1967. K-WG-366-HI, AFHRA.
Message. DCO00148. 366 TFW. To 7 AF, et al. Subject: "AFR 57-4 Mod I Proposal," 11 May 1967. In History, 366th Tactical Fighter Wing, vol. 1, 1 April 1967–30 September 1967. K-WG-366-HI, AFHRA.
Message. DCO00157. 366 TFW. To CSAF, et al. Subject: "MiG Engagement Supplement to 366 TFW OPREP-3//012," 18 May 1967. In History, 366th Tactical Fighter Wing, 1 April 1967–30 September 1967 1. K-WG-366-HI, AFHRA.
PACAF Command Center. Chronological Log. 13–14 May 1967. K717.3051-1, AFHRA.
———. Chronological Log. 14–15 May 1967. K717.3051-1, AFHRA.
PACAF. DO Read File. 29 April–1 May 1967. K717.312, AFHRA.
———. DO Read File. 13–15 May 1967. K717.312, AFHRA.
———. DO Read File. 17–18 May 1967. K717.312, AFHRA.

USAF Reports

Air Research and Development Command (ARDC). *Evaluation Report on GAR-1 Weapon System*. Holloman AFB, NM: Holloman Air Development Center, March 1956. Microfilm 31792, Frame 344, K280.1056, AFHRA.
Hiller, R. E. Briefing. *SEACAAL (Southeast Asia Counter-Air Alternatives) Briefing for Presentation at HQ USAF, Volume I–Text*. Hickam AFB, HI: Assistant for Operations Research, Headquarters PACAF, 10 February 1967. K717.310-2, AFHRA.
Holloman Air Development Center. *Test Report on GAR-1*, no. 25. Holloman AFB, NM: Holloman Air Development Center, 29 Oct 1956. Microfilm 31792, Frame 253. K280.1056, AFHRA.
HQ PACAF. *SEACAAL: Southeast Asia Counter-Air Alternatives*. Hickam AFB, HI: HQ PACAF, 31 December 1966. K717.310-1, AFHRA.
HQ Tactical Air Command (TAC). *F-4C Limited Evaluation (DRAFT)*. TAC Test 63-50. Langley AFB, VA: TAC, December 1964. In History, 12th Tactical Fighter Wing, vol. 3, 1 July–31 December 1964. K-WG-12-HI, AFHRA.

BIBLIOGRAPHY

Mehserle, Capt H. J., USAF. *Capability of the F-4C Aircraft/SUU-16/A Gun Pod in the Close Support Role.* Eglin AFB, FL: Air Force Systems Command, Air Proving Ground Center, August 1965. K240.04-99, AFHRA.

US Air Force. *Project Forecast: Limited War Report* (U), vol. 1. Washington, DC: US Air Force, January 1964. Microfilm 40868, AFHRA. (Secret) Information extracted is unclassified.

———. *Project Red Baron II: Air to Air Encounters in Southeast Asia*, vol. 1: Overview of Report. Nellis AFB, NV: USAF Tactical Fighter Weapons Center, January 1973. M-U 42339-15a, MSFRIC.

USAF Unit Histories

8th TFW, January–June 1967, vol. 2. K-WG-8-HI, AFHRA.
8th TFW, October–December 1967, vol. 2. K-WG-8-HI, AFHRA.
12th TFW, 1 July–31 December 1964, vols. 1 and 3. K-WG-12-HI, AFHRA.
15th TFW, 1 January–30 June 1965, vol. 1. K-WG-15-HI, AFHRA.
366th TFW, 1 April 1967–30 September 1967, vols. 1 and 8. K-WG-366-HI, AFHRA.
Frazier, 1st Lt John, USAF. "Historical Summary—390th Tactical Fighter Squadron, 1 May–31 May 67." In History, 366th Tactical Fighter Wing, vol. 8, 1 April 1967–30 September 1967. K-WG-366-HI, AFHRA.
"Historical Data Record, from 1 Apr 67 to 1 May 67." In History, 366th Tactical Fighter Wing, vol. 8, 1 April 1967–30 September 1967. K-WG-366-HI, AFHRA.